Understanding Nanomaterials

Understanding Nanomaterials

Edited by **Rich Falcon**

Willford Press

New York

Published by Willford Press,
118-35 Queens Blvd., Suite 400,
Forest Hills, NY 11375, USA
www.willfordpress.com

Understanding Nanomaterials
Edited by Rich Falcon

International Standard Book Number: 978-1-68285-167-8 (Hardback)

Printed in China.

Contents

Preface

Every book is a source of knowledge and this one is no exception. The idea that led to the conceptualization of this book was the fact that the world is advancing rapidly; which makes it crucial to document the progress in every field. I am aware that a lot of data is already available, yet, there is a lot more to learn. Hence, I accepted the responsibility of editing this book and contributing my knowledge to the community.

Nanomaterials are widely used across various fields because of their unique structures and properties. The significant topics in the field of nanomaterials like nanoparticles, modeling nanostructures, nanocomposites, nanocoatings, emerging nanotechnologies in different fields, etc. have been thoroughly elucidated in this book. For all those who are interested in nanomaterials and nanotechnology, the researches and examples included in this book will serve as an excellent guide to develop a comprehensive understanding of the field. It brings forth novel topics for detailed analysis and discussion that students and researchers engaged in the field can take up further.

While editing this book, I had multiple visions for it. Then I finally narrowed down to make every chapter a sole standing text explaining a particular topic, so that they can be used independently. However, the umbrella subject sinews them into a common theme. This makes the book a unique platform of knowledge.

I would like to give the major credit of this book to the experts from every corner of the world, who took the time to share their expertise with us. Also, I owe the completion of this book to the never-ending support of my family, who supported me throughout the project.

Editor

1

A Novel Arch-Shape Nanogenerator Based on Piezoelectric and Triboelectric Mechanism for Mechanical Energy Harvesting

Chenyang Xue [1,2,†,*], Junyang Li [1,2,3,†], Qiang Zhang [1,2], Zhibo Zhang [1,2], Zhenyin Hai [1,2], Libo Gao [1,2], Ruiting Feng [1,2], Jun Tang [1,2], Jun Liu [1,2], Wendong Zhang [1,2] and Dong Sun [1,2,3]

[1] Key Laboratory of Instrumentation Science and Dynamic Measurement of Ministry of Education, North University of China, Taiyuan 030051, China; E-Mails: junyangli3-c@my.cityu.edu.hk (J.L.); leeyfrank@gmail.com (Q.Z.); markbzhang@gmail.com (Z.Z.); Haizhenyin@gmail.com (Z.H.); nucgaolibo@gmail.com (L.G.); fengruiting401@gmail.com (R.F.); tangjun16@nuc.edu.cn (J.T.); liuj@nuc.edu.cn (J.L.); wdzhang@sxedu.gov.cn (W.Z.); medsun@cityu.edu.hk (D.S.)

[2] Science and Technology on Electronic Test and Measurement Laboratory, North University of China, Taiyuan 030051, China

[3] Department of Mechanical and Biomedical Engineering, City University of Hong Kong, Kowloon 999077, Hong Kong

† These authors contributed equally to this work.

* Author to whom correspondence should be addressed; E-Mail: xuechenyang@nuc.edu.cn

Academic Editor: Thomas Nann

Abstract: A simple and cost-effective approach was developed to fabricate piezoelectric and triboelectric nanogenerator (P-TENG) with high electrical output. Additionally, pyramid micro structures fabricated atop a polydimethylsiloxane (PDMS) surface were employed to enhance the device performance. Furthermore, piezoelectric barium titanate (BT) nanoparticles and multiwalled carbon nanotube (MWCNT) were mixed in the PDMS film during the forming process. Meanwhile, the composition of the film was optimized to achieve output performance, and favorable toughness was achieved after thermal curing. An arch-shape ITO/PET electrode was attached to the upper side of the polarized composite film and an aluminum film was placed under it as the bottom electrode. With periodic external force at 20 Hz, electrical output of this P-TENG, reached a peak voltage of 22 V and current of 9 μA with a peak current density of 1.13 $\mu A/cm^2$, which was six times that of the triboelectric generator without BT and MWCNT nanoparticles. The nanogenerator can be directly used to lighten 28 commercial light-emitting diodes (LEDs) without any energy storage unit or rectification circuit under human footfalls.

Keywords: arch-shape; piezoelectric; triboelectric; mechanical energy

---

1. Introduction

As an attractive substitution for conventional power supply, harvesting energy from the ambient environment is an effective way to achieve self-powered systems and further fulfill large-scale energy needs [1–3]. A nanogenerator converting vibrational and mechanical energy to electric energy based on the piezoelectric effect has been proved to be an effective approach by Wang and Song [1] and Lee's groups [4]. Wang and co-workers have used piezoelectric nanowire arrays to develop nanogenerators, which showed the feasibility to power commercial light-emitting diode (LED) [5], liquid crystal displays [6], and wireless data transmission [7]. $BaTiO_3$ (BT) nanoparticles and carbon nanotubes have been utilized to obtain a piezoelectric nanocomposite generator [8]. Subsequently, technology has thus been further introduced to employ a system that can convert frictional energy sources into electrical energy [9–13].

Recently, different configurations that accommodate the needs of harvesting energy that has been deprived from diverse mechanical motions, such as human walking, liquid-wave moving, and wind blowing, were introduced by changing the structure of the nanogenerators. Although piezoelectric cantilevers and structured PDMS film have been combined to enhance the structure performance [14], there have been few reports about the technology of combining piezoelectric and triboelectric mechanism to optimize the material property and thereby obtain high voltage output.

Here, we developed a hybrid nanogenerator based on piezoelectric and triboelectric principle by employing surface-nanostructured polydimethylsiloxane (PDMS) film, barium titanate (BT) and multiwalled carbon nanotube (MWCNT) nanoparticles. The BT and MWCNT were integrated to PDMS film to increase the output voltage. Doctor-blade coating technology was used to fabricate the device and the output performances were investigated.

2. Results and Discussion

2.1. Frequency Effect

Performance of the arch-shape triboelectric nanogenerator (P-TENG) with surface-structured PDMS under the compressive force at different frequencies is shown in Figure 1. As the external force frequency increased from 15 to 20 Hz, the open-circuit voltage increased gradually from 10 to 22 V. Open-circuit voltage was maximized at a frequency of about 20 Hz. However, the open-circuit voltage declines gradually to 5 V at 25 Hz because of the under-releasing of the P-TENG. This is mainly because, when the cycle of the compressive force is too short or the frequency is too high, the nanogenerator cannot recover to the original position before the next force impact [13].

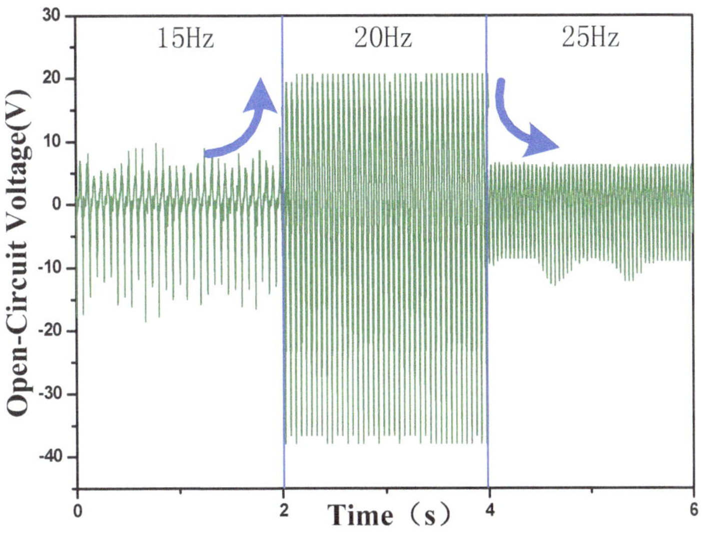

Figure 1. Output performance of the arch-shape piezoelectric and triboelectric nanogenerator (P-TENG) with nanostructured polydimethylsiloxane (PDMS) film under external forces at different frequencies.

2.2. Electrical Output Performance

At 20 Hz, open-circuit voltages of the arch-shaped P-TENG with different *N* (The volume ratio of BT to MWCNT) and *V* (the volume percentage of BT and MWCNT mixtures/PDMS matrix) are shown in Figure 2a. More specifically, the output voltage was very low at about 1 V at 1%. Then the voltage increased gradually from *V* = 2% to 9%, peaking at 22 V with *N* = 2:3 and *V* = 9%. After this point, the output voltage declined slowly to about 20 V until *V* = 11%. Hence, *N* = 2:3 and *V* = 9% are the optimal doping ratio, which is further verified in the Figure 2b. From sample A of Figure 2b, it can be seen that the output trend is consistent with that in Figure 2a, showing a constant improvement. Sample B of Figure 2b shows that open-circuit voltage of the generator based on the triboelectric effect was slightly changed at the beginning *(V* = 1% to *V* = 4%) and then it gradually increased after *V* = 5% to the end. This may be because, as the value of *V* increased, some particles inside the composite film could pass through the surface of film, resulting in an increase in the triboelectric effect. Sample C of Figure 2b shows that there was a gradual increase from *V* = 1% to *V* = 8% and then the output voltage witnessed a sharp increase at *V* = 9%, finally it declined slightly. When *V* = 10% or 11%, since the density of particles in the composite film was very high, the piezoelectric particles contacted, thus further weakening the overall output performance. Hence, the P-TENG shown in sample C of Figure 2b displayed superior performance.

The output performances of the arch-shape nanogenerator P-TENG, with certain doping conditions (*N* = 2:3, *V* = 9%), are shown in Figure 2c. In brief, the open-circuit voltage and short-circuit current were enhanced after adding piezoelectric nanoparticles. It has been demonstrated that triboelectric effect can enhance the output performance most effectively [13]. Here, a P-TENG were introduced, which not only increased the effective roughness (compared with sample B of Figure 2c and sample C of Figure 2c but also enhanced the nanogenerator performance by introducing piezoelectric effect. The advantages of the P-TENG had been displayed by comparing the output performances of five samples (A–E) in Figure 2c. More specifically, the PDMS film with micro pyramid arrays, unpolarized BT nanoparticles and CNT in sample C of Figure 2c has enhanced the open-circuit voltage and short-circuit current by 66.7% and 166.7%. Moreover, the novel P-TENG shown in sample E of Figure 2c exhibited more attractive performance, whose voltage and the current were further enhanced by 367% and 500% as compared with in sample B of Figure 2c. Meanwhile, the open-circuit voltage and the short-circuit current were increased by 244% and 240%, as compared with that of the PDMS with polarized BT nanoparticles and CNT in sample D of Figure 2c.

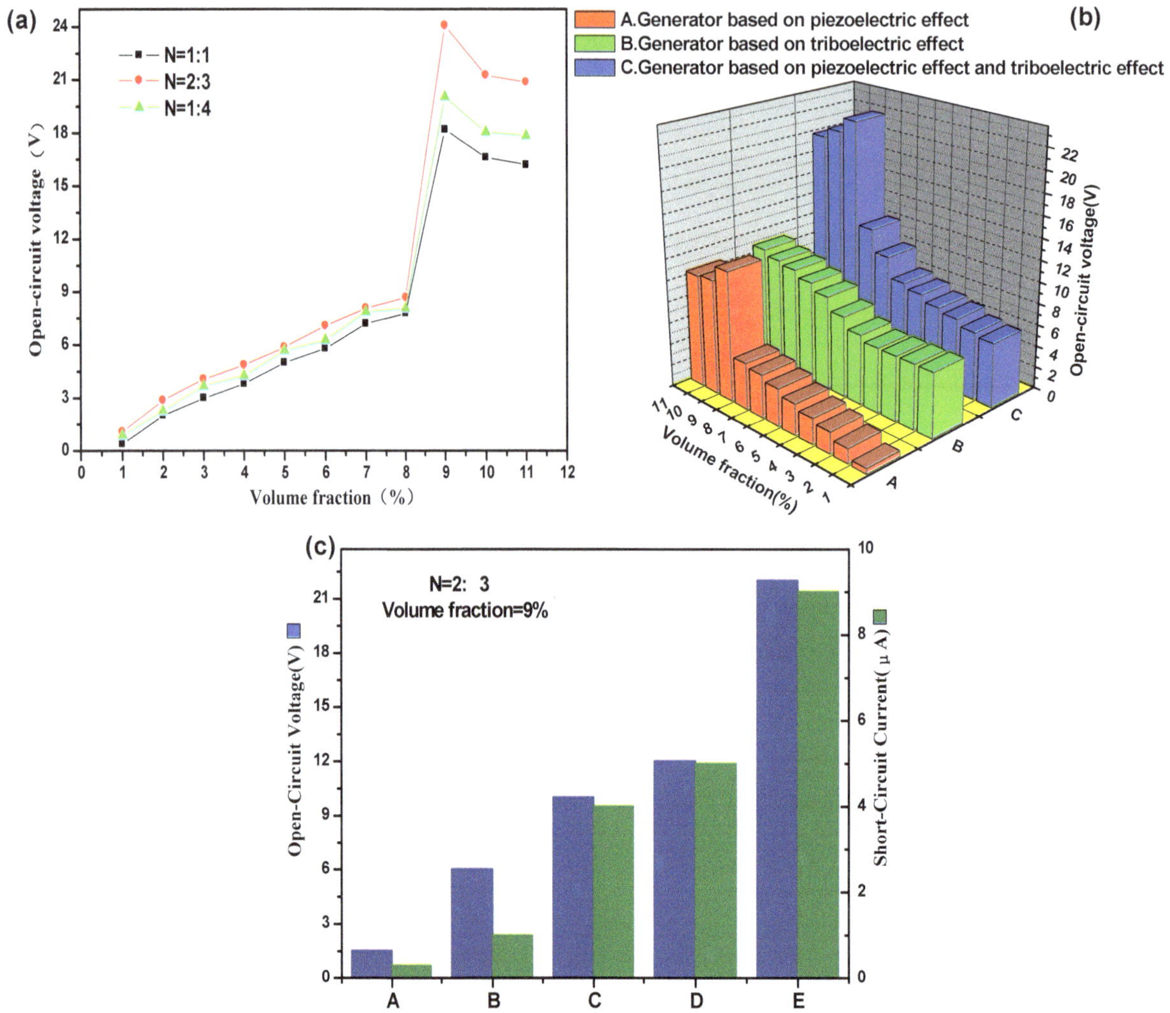

Figure 2. (**a**) Open-circuit voltage of the arch-shaped P-TENG with different *N* at 20 Hz. (**b**) Open-circuit voltage of the P-TENG with different compositions based on *N* = 2:3, including: A. generator based on piezoelectric effect; B. generator based on triboelectric effect; C. Generator based on both the piezoelectric and triboelectric effect; (**c**) Open-circuit voltage and short-circuit current of the arch-shape nanogenerator with different compositions, when the doping radio was fixed (*N* = 2:3, *V* = 9%), including: sample A: flat PDMS film; sample B: PDMS film with only micro pyramid arrays; sample C: PDMS film with micro pyramid arrays, unpolarized barium titanate (BT) nanoparticles and carbon nanotube (CNT); sample D: PDMS with only polarized BT nanoparticles and CNT; sample E: PDMS film with micro pyramid arrays, polarized BT nanoparticles and CNT (P-TENG).

Actually, in the mixed system, granular polarized BT particles may form aggregates and chains. The MWCNTs with great aspect ratio can form a long-conducting channel while fibrous MWCNTs could introduce a bridging effect between the BT-chain or aggregates. Similarly, a BT particles chain or aggregates can be connected to adjacent MWCNTs. Through synergy, a relatively complete network structure in a specific doping ratio was formed and thereby the conductivity of the composite material performance was enhanced [15]. However, the composite output voltage declined slightly after $N = 2:3$, $V = 9\%$. This may be because the density of particles in the composite film was increased, resulting in the piezoelectric particles contacting each other directly and further weakening the overall output performance of the film.

2.3. Working Principle

The working principle of this arch-shape nanogenerator can be depicted as followed in Figure 3, combining piezoelectric and triboelectric mechanism. The operation of P-TENG is realized by applying a cycled compressive force onto the whole area of the device, so that the bending plates will be periodically pressed to contact closely with each other. In order to simplify the description, the upper surface of ITO/BET film is labeled as layer A and the surface of polarizing film is marked as layer B while the lower surface of the Al foil is labeled as layer C.

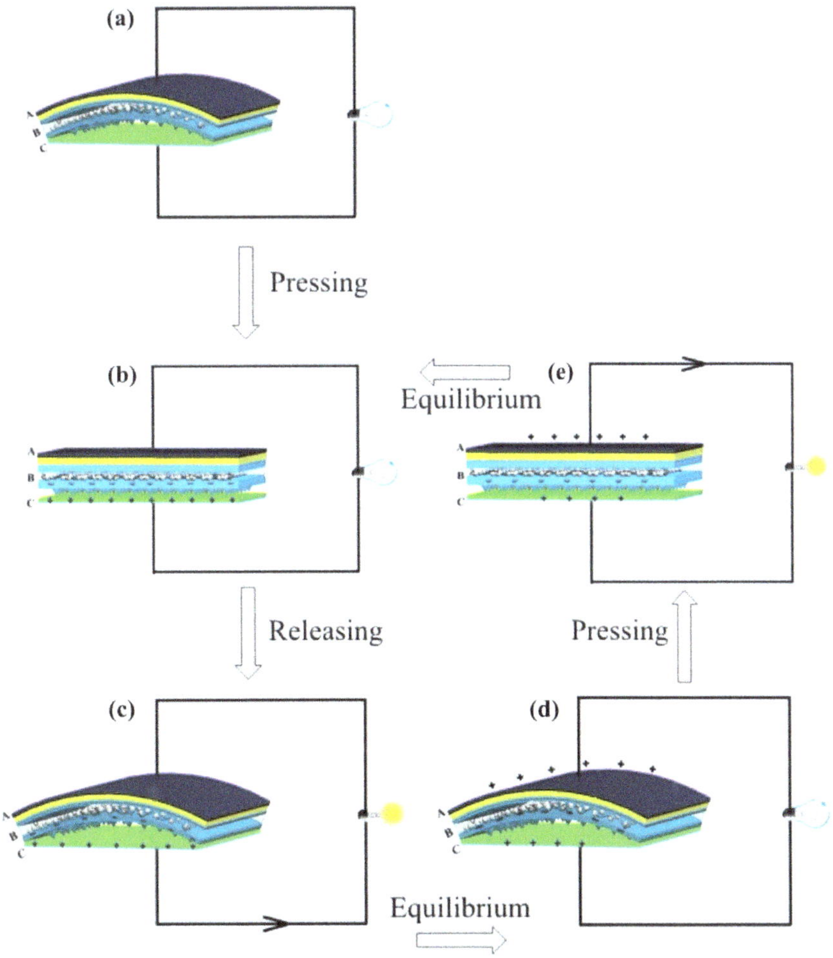

Figure 3. Working principle of this arch-shape nanogenerator. (**a**–**e**) Schematic diagram shows the working principle of the arch-shape P-TENG.

The distribution of triboelectric charge (σ_T) and piezoelectric charge (σ_P) of A, B, C three layers in one pressing cycle was shown in Table 1. At the origin state, the polarized piezoelectric film is in the bending state, so layer A induces negative piezoelectric charge $-\sigma_P$, but there is no friction effect and no charge transferred as well, thus, no electric potential. In the pressing process, the charge $-\sigma_P$ still exists in layer A and electrons will be injected from Al to the PDMS surface because of friction effect, leaving positive triboelectric charges $+\sigma_T$ and piezoelectric charge $+\sigma_P$ on the layer C while the negative triboelectric charges will be preserved on the layer B surface due to the character of the insulator [16]. Such a charge transfer process will continue in the initial few hundreds of cycles until the accumulated charges reach saturation and it is depicted in the part (b) of Table 1. When the pressing force disappears, the P-TENG will return to its original arch shape immediately so that an air gap will form again between the layer B and layer C, resulting in a much higher potential on the layer C than layer B [10]. As it is illustrated in Figure 3c, this potential difference will drive the stream of positive charges from layer C to layer A through the external load until the potential difference is fully counteracted by the transferred charges ($\Delta\sigma$). This will give rise to the layer A with a surface charge density of $+\Delta\sigma - \sigma_P$ and the layer C is left with $+\sigma_T - \Delta\sigma$ as shown in Figure 3d. Soon after, when the P-TENG is pressed again to let the layers B and C contact, these redistributed charges will conversely establish a positive potential on layer A in Figure 3e, which will drive the $\Delta\sigma$ to flow back to the layer C. In this way, a period is achieved and the device will return to the equilibrium state in Figure 3b. This is a whole cycle of electricity generation based on piezoelectric and triboelectric mechanism. Therefore, we can see from Table 1 that when a friction effect exists on its own, the charge transfer of this structure is $\Delta\sigma$ while the charge transfer is σ_P for piezoelectric effect only through combining piezoelectric effect and triboelectric effect. We can easily establish that the charge transfer of arch-shape structure is $\sigma_P + \Delta\sigma$.

Table 1. The distribution of triboelectric charge (σ_T) and piezoelectric charge (σ_P) of A, B, C three layers in one pressing cycle. $\Delta\sigma$: transferred charges.

Layer	(a)		(b)		(c)		(d)		(e)	
	piezo	**tribo**	**piezo**	**tribo**	**piezo**	**tribo**	**piezo**	**tribo**	**piezo**	**tribo**
A	$-\sigma_P$	0	$-\sigma_P$	0	$-\sigma_P$	0	$-\sigma_P$	$+\Delta_\sigma$	$-\sigma_p$	$+\Delta_\sigma$
B	0	0	0	$-\sigma_T$	0	$-\sigma_T$	0	$-\sigma_T$	0	$-\sigma_T$
C	0	0	$+\sigma_P$	$+\sigma_T$	$+\sigma_P$	$+\sigma_T$	0	$+\sigma_T - \Delta\sigma$	0	$+\sigma_T - \Delta\sigma$

2.4. Application

The output performance of this P-TENG has a highest peak voltage of 22 V and current of 9 μA (The output was measured using an oscilloscope with a 10 MΩ probe, and the output current was measured by a low-value series resistance of 9 kΩ). This output is large enough to power many micro/nanosystems, even some commercial electronic devices. Figure 4 shows that the P-TENG was capable of lighting up as many as 28 commercial LED bulbs simultaneously, by converting commonly ambient mechanical energy such as human footfalls.

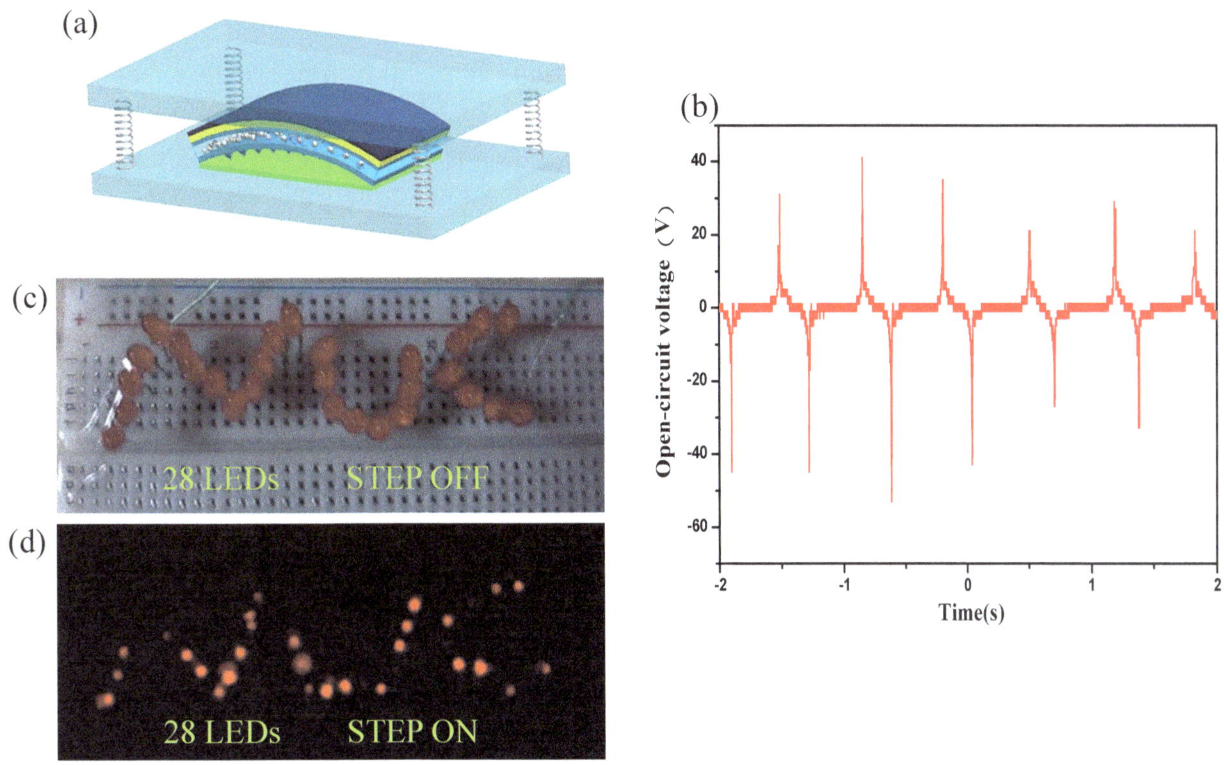

Figure 4. Applications of the arch-shape P-TENG. (**a**) Diagram of the P-TENG between two plexiglasses with springs; (**b**) The output open circuit voltage is about 30 V under footstep. (**c**,**d**) When footstep falls on the P-TENG, 28 paralleled commercial LEDs were lightened without using any energy storage device or rectification circuit. Notes: All LEDs are connected in serial.

3. Experimental

3.1. Reagents and Apparatus

$BaTiO_3$ nanoparticles (BT, average particle diameter of 30 nm) and multiwalled carbon nanotubes (MWCNT, average particle diameter of 20 nm) were provided by DK nano technology Co. Ltd. (Beijing, China) and PDMS (Sylgard 184, Dow Corning) was purchased from Zhenkonghang adhesive technology Co. Ltd. (Shenzhen, China). Pure alcohol was supplied by Sinopharm Chemical Reagent Co. Ltd. (Shanghai, China). ITO/PET was purchased from Kaiwei Electronic Components Co. Ltd. (Zhuhai, China). The thickness of the composite film was controlled by using a doctor-blade coating machine (AC-150, Hefei, China) and then cured in a vacuum oven (ZK-1S type, Changzhou, China). With a JEOL JEM-1200EX transmission electron microscope (Tokyo, Japan) and S4700 scanning electron microscope (Tokyo, Japan), TEM and scanning electron microscope (SEM) measurements were performed. Open-circuit voltage and short-circuit current of the arch-shape P-TENG were measured with a digital oscilloscope (Tektronix DPO 2024 type, Beaverto, OR, USA). A vibration system, including a waveform generator (RIGOL DG1022, Beijing, China), a power amplifier (SINOCERA YE5871A, Shanghai, China) and a vibration platform (SINOCERA, Shanghai, China), was assembled to apply a cycled compressive force with controllable frequency.

3.2. Preparation of the Composites

Figure 5 shows the schematic view of structural design and fabrication process flowchart of the P-TENG. In this work, we fabricated pyramids on the polymer surfaces to increase the triboelectric power output. To make patterned PDMS films, Si wafer molds were first fabricated by using traditional photolithography method, following by dry or wet etching process to achieve pyramids. $BaTiO_3$ NPs were mixed with MWCNTs in specific volume proportions including $N = 1{:}1$, $N = 2{:}3$ and $N = 1{:}4$. The mixtures were then stirred for approximately 5 h in ethanol with a magnetic stirrer. Liquid PDMS elastomer and cross-linker were mixed and degassed to achieve quantitative PDMS matrix (Figure 5a). After the subsequent drying and granulation, the well-mixed nanomaterials with three different proportions were poured into PDMS matrix, with the volume ratios of mixtures/PDMS matrix at $V = 1\%$ to $V = 11\%$. Then, 33 P-BM composite films (PDMS mixtured with $BaTiO_3$ NPs and MWCNTs) were obtained. The composite films were polarized at 3.6 kV/mm, and 90 °C for 60 min.

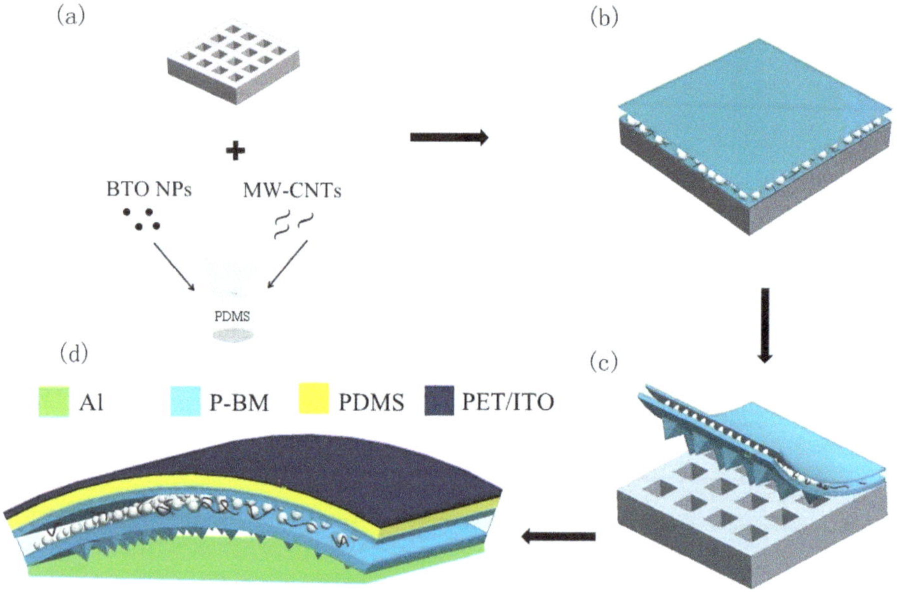

Figure 5. (a–d) Schematic view of the structural design and fabrication process flowchart of the P-TENG device (the total size of the P-BM composite films is 2 cm × 4 cm × 0.3 mm).

Then, the surface of the molds was treated with trimethylchlorosilane to prevent the PDMS film from sticking to the molds, while a doctor-blade coating machine was utilized to enable the P-BM samples to be on the surface of the mold (Figure 5b). After thermal curing, a uniform PDMS layer doped with piezoelectric materials was peeled off, with the inverse of the original pattern features on the surface of the mold, with a thickness of 300 μm (Figure 5c). Then the piezoelectric thin film was placed closely in two aluminum foil pieces and put in a polarization set-up to be polarized. Subsequently, the unstructured side of the composite film was fixed on the insulation surface of a clean baked ITO/PET membrane (uniformly baked by soldering irons for several seconds to form the optimal planar separation distance between the top and bottom surface [16]) by a thin PDMS bonding layer, and then the aluminum foil was used as a bottom electrode to form an arch-shape device (Figure 5d). A detailed fabrication protocol and typical structure of the nanogenerator are schematically shown in Figure 4. Figure 6 shows SEM images of the arch-shape P-TENG. From

Figure 6e, it can be seen that there are some microparticles on surface of the P-BM films besides the pyramid micro structures, which may improve the surface friction effect further. The entire preparation process of the device is simple and low-cost, making it possible for large-scale production and practical applications.

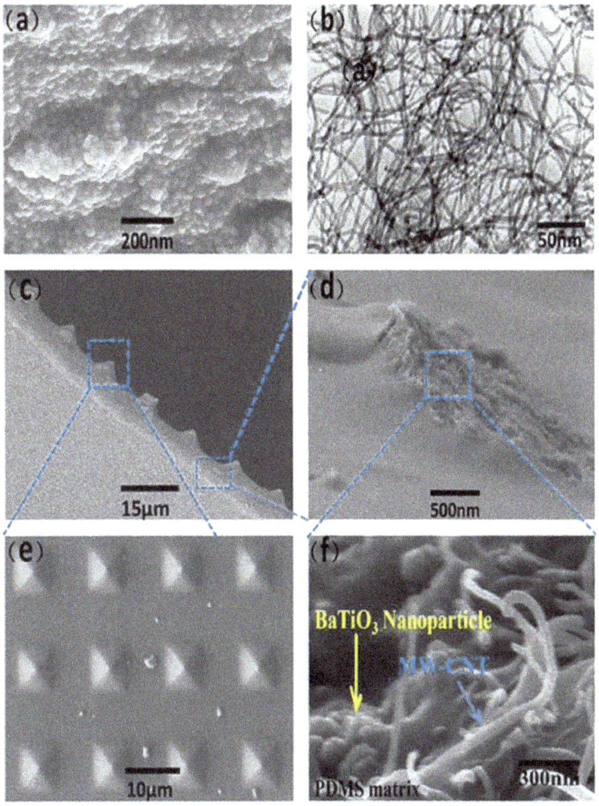

Figure 6. Structure characterization of the P-BM composite films. (**a**) SEM image of the $BaTiO_3$ NPs; (**b**) TEM image of multiwalled carbon nanotube (MWCNTs) with a diameter of 20 nm and a length of 20 μm; (**c**) SEM image of the pyramid PDMS thin film with $BaTiO_3$ NPs and MWCNTs; (**d**) Enlarge photograph of the PDMS thin film tore locally by tweezers; (**e**) High-magnification SEM image of the patterned PDMS surfaces with pyramids features; (**f**) The high-magnification SEM image of mixing effect about $BaTiO_3$ NPs and MWCNTs.

4. Conclusions

A P-TENG based on the piezoelectric-triboelectric process was demonstrated and it can be used as an energy harvester for human footfalls. This design can significantly improve the output efficiency of the nanogenerator by combining the piezoelectric and triboelectric effect. For a typical P-TENG with polarized BT and MWCNT nanoparticles, the electrical output achieved a peak voltage of 22 V and current of 9 μA with a peak current density of 1.13 μA/cm^2, which is six times as high as that of the generator without BT and MWCNT nanoparticles. Owing to the simple and low-cost production process, the P-TENG exhibits great advantages in industrial production and practical applications as well as manufacturability and capability of integration with other processing technologies. Finally, with such a power output, we envision future energy harvesting by this P-TENG not only from human footfall but also from rolling wheels, wind power and ocean waves.

Acknowledgments

This research was supported by merit-funded science and technology project for returned oversea scholars from Ministry of Human and Social Security of Shanxi province, Natural Science Foundation (No. 2011011013-2) and Youth Foundation (No. 2011021020-2) of Science and Technology Agency of Shanxi province, China. National Natural Science Foundation of Key Support Project (91123036), State 863 Hi-Tech Project (2013AA041109). Outstanding Youth Fund of National Natural Science Fund Project (51225504), Basic Research of Scientific Instruments Project (61127008).

Author Contributions

Chenyang Xue, Junyang Li, Qiang Zhang, Jun Tang, Jun Liu, Wendong Zhang and Dong Sun designed the experiments; Junyang Li, Zhenyin Hai and Libo Gao performed the experiments; and Junyang Li, Zhibo Zhang and Ruiting Feng analyzed the data and wrote the paper. All authors discussed the results and commented on the manuscript.

Conflicts of Interest

The authors declare no conflict of interest.

References

1. Wang, Z.L.; Song, J. Piezoelectric nanogenerators based on zinc oxide nanowire arrays. *Science* **2006**, *312*, 242–246.
2. Beeby, S.P.; Tudor, M.J.; White, N.M. Energy harvesting vibration sources for microsystems applications. *Meas. Sci. Technol.* **2006**, *17*, R175–R195.
3. Wang, Z.L. Self-powered nanosensors and nanosystems. *Adv. Mater.* **2012**, *24*, 280–285.
4. Park, K.I.; Jeong, C.K.; Ryu, J.; Hwang, G.T.; Lee, K.J. Flexible and large-area nanocomposite generators based on lead zirconate titanate particles and carbon nanotubes. *Adv. Energy Mater.* **2013**, *3*, 1539–1544.
5. Chen, C.Y.; Zhu, G.; Hu, Y.; Yu, J.W.; Song, J.; Cheng, K.Y.; Peng, L.H.; Chou, L.J.; Wang, Z.L. Gallium nitride nanowire based nanogenerators and light-emitting diodes. *ACS Nano* **2012**, *6*, 5687–5692.
6. Hu, Y.; Zhang, Y.; Xu, C.; Zhu, G.; Wang, Z.L. High-output nanogenerator by rational unipolar assembly of conical nanowires and its application for driving a small liquid crystal display. *Nano Lett.* **2010**, *10*, 5025–5031.
7. Hu, Y.; Zhang, Y.; Xu, C.; Lin, L.; Snyder, R.L.; Wang, Z.L. Self-powered system with wireless data transmission. *Nano Lett.* **2011**, *11*, 2572–2577.
8. Park, K.I.; Lee, M.; Liu, Y.; Moon, S.; Hwang, G.T.; Zhu, G.; Kim, J.E.; Kim, S.O.; Kim, D.K.; Wang, Z.L.; *et al*. Flexible nanocomposite generator made of $BaTiO_3$ nanoparticles and graphitic carbons. *Adv. Mater.* **2012**, *24*, 2999–3004.
9. Zhu, G.; Chen, J.; Liu, Y.; Bai, P.; Zhou, Y.S.; Jing, Q.; Pan, C.F.; Wang, Z.L. Linear-grating triboelectric generator based on sliding electrification. *Nano Lett.* **2013**, *13*, 2282–2289.

10. Zhu, G.; Pan, C.; Guo, W.; Chen, C.Y.; Zhou, Y.; Yu, R.; Wang, Z.L. Triboelectric-generator-driven pulse electrodeposition for micropatterning. *Nano Lett.* **2012**, *12*, 4960–4965.
11. Bai, P.; Zhu, G.; Lin, Z.H.; Jing, Q.; Chen, J.; Zhang, G.; Ma, J.S.; Wang, Z.L. Integrated multilayered triboelectric nanogenerator for harvesting biomechanical energy from human motions. *ACS Nano* **2013**, *7*, 3713–3719.
12. Yang, Y.; Lin, L.; Zhang, Y.; Jing, Q.; Hou, T.C.; Wang, Z.L. Self-powered magnetic sensor based on a triboelectric nanogenerator. *ACS Nano* **2012**, *6*, 10378–10383.
13. Zhang, X.S.; Han, M.D.; Wang, R.X.; Zhu, F.Y.; Li, Z.H.; Wang, W.; Zhang, H.X. Frequency-multiplication high-output triboelectric nanogenerator for sustainably powering biomedical microsystems. *Nano Lett.* **2013**, *13*, 1168–1172.
14. Han, M.D.; Zhang, X.S.; Liu, W.; Sun, X.M.; Peng, X.H.; Zhang, H.X. Low-frequency wide-band hybrid energy harvester based on piezoelectric and triboelectric mechanism. *Sci. China Technol. Sci.* **2013**, *56*, 1835–1841.
15. Pötschke, P.; Kretzschmar, B.; Janke, A. Use of carbon nanotube filled polycarbonate in blends with montmorillonite filled polypropylene. *Compos. Sci. Technol.* **2007**, *67*, 855–860.
16. Wang, S.; Lin, L.; Wang, Z.L. Nanoscale triboelectric-effect-enabled energy conversion for sustainably powering portable electronics. *Nano Lett.* **2012**, *12*, 6339–6346.

2

DNA under Force: Mechanics, Electrostatics, and Hydration

Jingqiang Li [1], **Sithara S. Wijeratne** [1], **Xiangyun Qiu** [2,*] **and Ching-Hwa Kiang** [1,3,*]

[1] Department of Physics and Astronomy, Rice University, Houston, TX 77005, USA; E-Mails: jingqiang.li@rice.edu (J.L.); sitara@rice.edu (S.S.W.)

[2] Department of Physics, George Washington University, Washington, DC 20052, USA

[3] Department of Bioengineering, Rice University, Houston, TX 77005, USA

* Authors to whom correspondence should be addressed; E-Mails: xqiu@gwu.edu (X.Q.); chkiang@rice.edu (C.-H.K.)

Academic Editor: Stephen Ralph

Abstract: Quantifying the basic intra- and inter-molecular forces of DNA has helped us to better understand and further predict the behavior of DNA. Single molecule technique elucidates the mechanics of DNA under applied external forces, sometimes under extreme forces. On the other hand, ensemble studies of DNA molecular force allow us to extend our understanding of DNA molecules under other forces such as electrostatic and hydration forces. Using a variety of techniques, we can have a comprehensive understanding of DNA molecular forces, which is crucial in unraveling the complex DNA functions in living cells as well as in designing a system that utilizes the unique properties of DNA in nanotechnology.

Keywords: DNA; mechanics; electrostatics; hydration

---

1. Introduction

With the advances in nanomanipulation techniques, researchers are able to directly manipulate and measure the molecular force of DNA at the single molecule level, which expands our understanding of DNA mechanics [1–3]. With DNA being a long and thin chain of nucleotides, single molecule techniques are uniquely advantageous in pulling on this linear chain to allow us to quantify the forces and fit the data to theoretical models in order to understand its behavior. On the other hand, molecular forces involved in compacting DNA into high densities have been traditionally studied via ensemble approaches, e.g., pushing DNA strands together by simply concentrating DNA solution [4] or adding crowding osmolytes [5]. Particularly, the ease of condensing DNA via multivalent cations [6], as well as the challenge of its mechanistic elucidation, has spurred broad interests in DNA-inspired electrostatics [7,8] and the roles of hydration in molecular interactions [9]. Single molecule and ensemble studies, acting synergistically, have yielded a comprehensive understanding of the physics of DNA molecular forces that is unlikely to be reproduced for other biomolecules.

2. Mechanical Forces in DNA

The conformation of DNA and its resulting mechanical properties are crucial in a variety of biological processes, such as replication, transcription, gene regulation, and genome compaction. The intra- and inter-molecular forces of DNA play a significant role in the operation of cellular machinery, including how it wraps around histones, packs into phage heads, and interacts with proteins. The potential of using DNA in nanotechnology has been explored due to its favorable characteristics, such as its capability of programmable self-assembly and resulting stiffness due to base-pairing [10–12]. For example, DNA can be used as a building block for complex nanostructures as well as performing computation [11,13–17]. Over the last two decades, with well-developed piconewton instrumentation, researchers were able to directly measure the molecular force of DNA at the single molecule level, which incredibly expands our understanding of DNA mechanics.

Both double-stranded DNA (dsDNA) and single-stranded DNA (ssDNA) may be viewed as a polymer. Single molecule manipulation has been used extensively to study the force response of a biopolymer [1,18–22] (Figure 1). Examples of several widely used techniques, which can measure piconewton level forces, are atomic force microscopy (AFM) [2], optical tweezers [1,23], magnetic tweezers [24], glass microneedles [25], and biomembrane force probes [26]. These techniques involve stretching a single molecule while monitoring its force response. They have been used to characterize the mechanical properties and the forces associated with the conformational changes of ssDNA and dsDNA molecules. Melting transitions were observed by repeatedly stretching and relaxing double-stranded λ-DNA molecules [27], while the unwinding forces have been determined by torque studies [28–32].

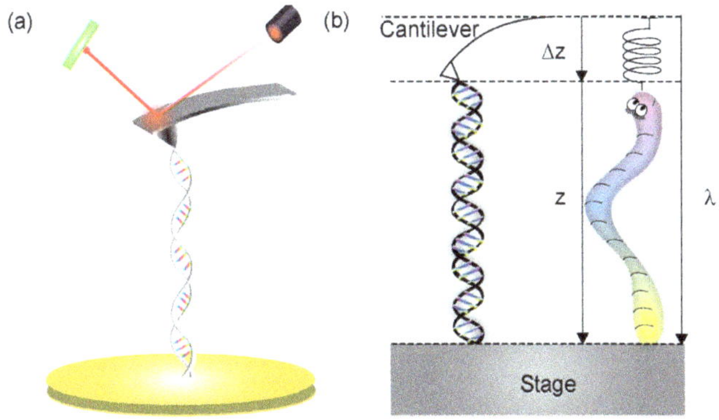

Figure 1. Illustration of a single molecule manipulation experiment using atomic force microscopy (AFM). (**a**) One end of the DNA molecule is attached to a substrate while the other end is pulled by the AFM cantilever tip; (**b**) The cantilever spring obeys Hooke's law and the elasticity of DNA follows the wormlike chain (WLC) model. The stage position, λ, is related to the molecular end-to-end distance, z, by $z = \lambda - \Delta z$. Adapted from [33].

Polymer physics models were successfully used to describe the mechanical behavior of stretched DNA obtained from single molecule force measurement [34]. The elasticity of DNA can be described by using one-dimensional polymer physics models. wormlike chain (WLC) and freely joined chain (FJC) models [35] are usually used to interpret the force-extension curves of DNA stretching. The FJC model assumes a polymer chain consisting of n segments of characteristic length l_k (Kuhn length), connected via freely-rotating joints with contour length $l_c = nl_k$:

$$z(F) = l_c\left[\coth\left(\frac{Fl_k}{k_BT}\right) - \frac{k_BT}{Fl_k}\right] \tag{1}$$

where F is the force; k_B is the Boltzmann constant; and T is the temperature. Stretching ssDNA and dsDNA can be described by extensible FJC (eFJC) model [23]:

$$z(F) = l_c\left[\coth\left(\frac{Fl_k}{k_BT}\right) - \frac{k_BT}{Fl_k}\right]\left(1 + \frac{F}{k_{seg}l_k}\right) \tag{2}$$

which assumes the additional extension by modeling each segment as an elastic spring with segment elasticity k_{seg}. On the other hand, the WLC model treats a polymer molecule as a homogenous elastic rod, or a wormlike chain, characterized by its contour length, l_c, and persistence length, l_p, which characterizes the bending stiffness of the WLC:

$$F(z) = \frac{k_BT}{l_p}\left[\frac{1}{4\left(1 - {}^{z}/_{l_c}\right)^2} - \frac{1}{4} + \frac{z}{l_c}\right] \tag{3}$$

For lengths longer than l_p, it is assumed that the correlation between tangents to the polymer is lost.

The mechanical stretching of dsDNA is best described by the extensible WLC model (eWLC) [36,37]:

$$z(F) = l_c\left[1 - \frac{1}{\sqrt{4l_pF/k_BT}} + \frac{F}{K_{ds}}\right] \tag{4}$$

where K_{ds} is the elastic stretch modulus for dsDNA. In single molecule experiments, for the part of the curve that is below the plateau, the following values are obtained: l_p = 50, and K_{ds} = 1200 pN, which are consistent with theory [18,38].

2.1. Mechanics of Double-Stranded DNA

Single molecule force experiments have revealed in detail the mechanical properties of dsDNA. These studies have provided significant insights into the intra-molecular force interactions of DNA. The studies on short dsDNA focus on the force-induced melting and the sequence-dependent effects [39–42]. Those on long dsDNA address its overall properties, particularly elasticity [1,2,18,38,43–47].

2.1.1. Short dsDNA

Short dsDNA refers to DNA with a length of fewer than 100 base pairs (bp). To characterize the melting of a short dsDNA, single molecule techniques measure rupture force as a function of pulling velocity (Figure 2a) [48,49]. Kinetic melting information is obtained by fitting the data to the Bell's model. On the other hand, the rupture force of short dsDNA increases with the number of base pairs.

For 12–20 base pairs, the rupture force scales linearly with the DNA length [42]. The rupture force was found to reach a limit of 61 pN, when the base pair number increases over 30 (Figure 2b) [39].

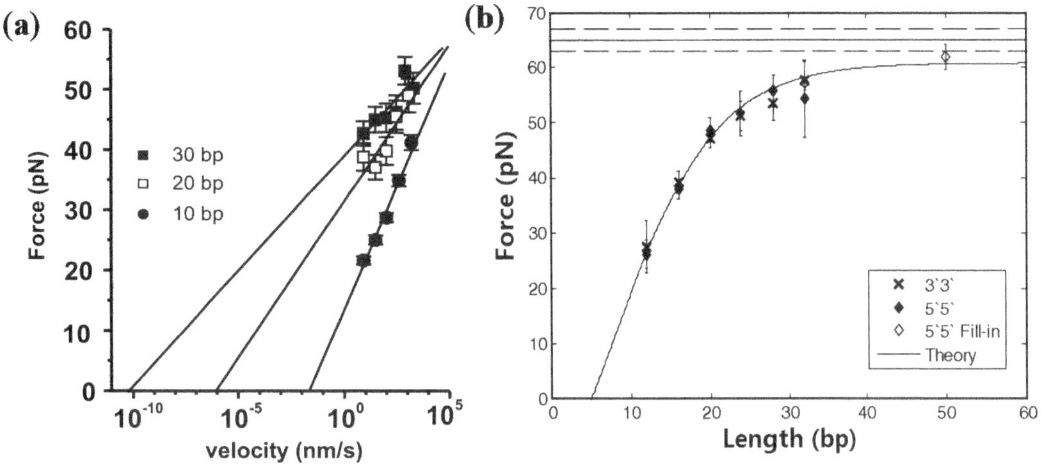

Figure 2. (**a**) Velocity dependence of the rupture force. Data showing the most probable rupture force as a function of stretching velocity for the short dsDNA of 30 bp, 20 bp, and 10 bp. The rupture force scales linearly with the logarithm of velocity. Adapted from [48]. Copyright (1999) National Academy of Sciences, USA; (**b**) Dependence of the rupture force on the length of dsDNA. The central horizontal line is the critical force of the overstretching transition measured in the λ-phage DNA. The upper and lower lines correspond to 10% and 90% of overstretching transition, respectively. Adapted from [39]. Reprinted with permission. Copyright (2005) American Physical Society.

2.1.2. Long dsDNA

A typical force-extension curve of λ-phage DNA with a length distribution of 117–8454 bp obtained from a single molecule force experiment is shown in Figure 3. It exhibits three states of the double helix during the stretching. In the first regime, the force is extremely small, and the DNA is in its B-form. Fitting the force-extension curve to the WLC model gives a persistence length of 53 nm (Figure 3a) [1] and the extensible WLC model gives a persistence length of 50 nm and a stretch modulus of 1200 pN (Figure 3b) [27]. Continuing stretching, DNA undergoes a B-S transition to its overstretching state indicated by the constant force regime at 65 pN. The length stretches up to 1.7 times its B-form length and the DNA conformation changes to the S-form [1,50–55]. After the overstretching transition, DNA transitions into single strands or possibly melts at forces of 150 pN. The extensible FJC is used to fit the high force regime, and the resulting persistence length and stretch modulus are consistent with that of ssDNA.

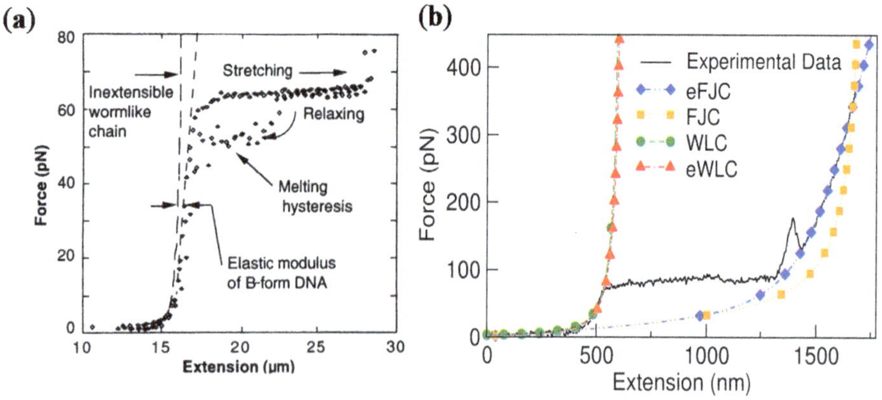

Figure 3. Experimental force-extension data for the stretching of λ-phage DNA. (**a**) The data are obtained from optical tweezers and fitted with the WLC model. Adapted from [1]. Reprinted with permission from AAAS; (**b**) Data are obtained from AFM single molecule experiments and fitted with different one-dimensional polymer models. Adapted from [33]. Reprinted with permission from Springer Science and Business Media.

2.2. Mechanics of Single-Stranded DNA

Single-stranded DNA does not have the base pairing interaction found in dsDNA. However, some ssDNA shows base stacking interaction, which can significantly affect its elasticity and conformation. Among the four bases of DNA, base stacking is strongest among adenine (A) bases and weakest among thymine (T) bases [56,57]. Figure 4 is a typical force-extension curve of poly(dA) and poly(dT), showing the two plateaus in the force-extension curve of poly(dA). The first plateau begins at 23 pN and overstretches about 75% of its original stacked state length, which agrees well with the prediction of a theoretical model [58–60]. Poly(dA) force curves show multiple plateaus and multiple pathways when stretched, whereas poly(dT) force curves can be fitted with a simple FJC model. This suggests the existence of complex intra-molecular interactions in poly(dA). The existence of A-tract in DNA may be related to its unique mechanical properties [61].

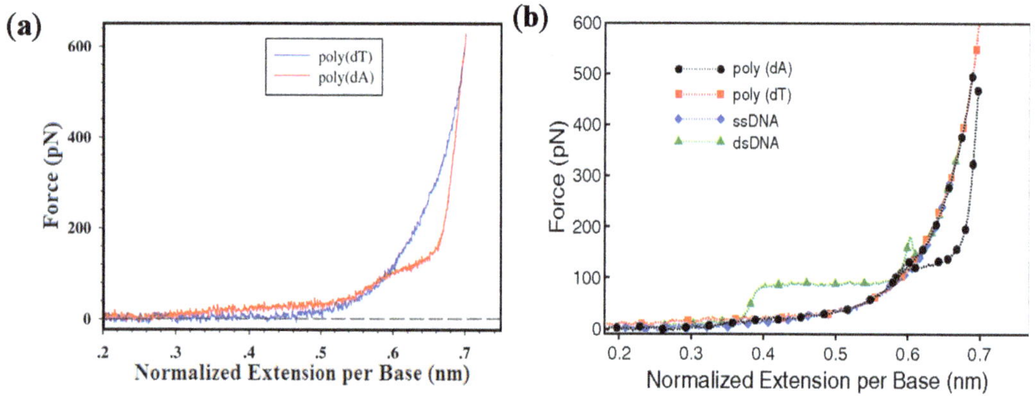

Figure 4. Force-extension curves for ssDNA. (**a**) Comparison of pulling curves between poly(dT) and poly(dA). Reprinted with permission from [60]. Copyright (2007) by the American Physical Society; (**b**) Force-extension curves for different forms of dsDNA and ssDNA. Reprinted with permission from [3]. Copyright (2010) by the American Physical Society.

3. Electrostatic and Hydration Forces in DNA

Consider two parallel dsDNA helices approaching each other from a distance in solution, as illustrated in Figure 5; the force between them as a function of inter-axial distance defines the analytical form and physical origin of DNA–DNA interactions. At inter-axial distances greater than 3 nm (*i.e.*, ≥1 nm surface separation), DNA–DNA interactions are dominated by long-range electrostatic forces due to the highly charged nature of DNA. The dependence on the relative azimuthal angle of the two helices is minimal, and each DNA is free to rotate along its cylindrical axis. Within the final nanometer of surface separation, overlapping of the respective hydration shell of opposing DNA helices leads to (de)hydration forces. In addition, the dielectric discontinuity between DNA and its surroundings results in fluctuation-induced van der Waals interactions, which, however, are comparatively rather weak and not discussed here [62]. As the DNA helices (e.g., less than 0.4 nm surface separation) approach each other, DNA–DNA interactions show significant dependence on the relative azimuthal angle as the "ruggedness" of DNA surfaces becomes relevant [4,63,64]. The strong force at such short distances strongly disrupts the hydration shells and is able to induce conformational changes of DNA [65]. Therefore, a complete and quantitative description of DNA–DNA force–distance relationships necessitates physical understanding of all pertinent components of the system: ions, solvent (water), and helical DNA. While it is largely true that researchers have reached consensus on the fundamental physical interactions at play (e.g., electrostatics, hydration, charged surfaces), quantitative accounts of these interactions are yet to be realized.

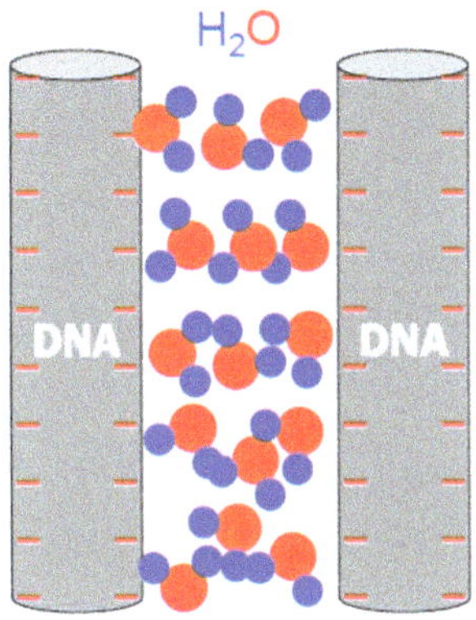

Figure 5. A cartoon illustration of DNA–DNA interactions in a side-by-side configuration.

With the goal of quantitative understanding of DNA–DNA interactions (and nucleic acid interactions in general), extensive studies, both experimental and theoretical, have been carried out in the past decades. These efforts have resulted in a large body of experimental observations and insightful theoretical models that have substantially improved our knowledge of the behavior of highly charged biomolecules in ionic solutions. Due to the limited scope of this review, readers are referred to many excellent review articles on nucleic acid interactions or polyelectrolyte behaviors for in-depth discussions (e.g., a few recent review articles, as in [7,66–74]). In what follows below, rather than being exhaustive, we aim to provide a general overview of relevant physical interactions at play and to promote productive thinking and discussion of the contentious issues in our current understanding.

Before getting started, it is useful to define several important quantities that convert familiar experimental conditions to relevant physical parameters:

(1) Bjerrum length $\lambda_B = \frac{e^2}{4\pi\varepsilon\varepsilon_0 k_B T}$, where e is the electron charge, ε is the dielectric constant of the solvent (e.g., 78.3 for water at 25 °C), and $k_B T$ is the thermal energy. Mathematically, when two ions of elemental charge are separated by Bjerrum length, the electric potential energy is the same as the thermal energy, giving a measure of the length scale of ion–ion correlations which is about 0.7 nm in water at 25 °C. It is worth noting that for ions of higher valence, the effective Bjerrum length is reduced by a factor of the product of the valences of the ion pairs.

(2) Ionic strength $I = \frac{1}{2}\sum_i z_i^2 n_i$, where the sum runs over all ionic species; z_i is the valence of ion; i, and n_i is the concentration of ion i. To put the role of ionic valence in perspective, 100 mM NaCl and $MgCl_2$ solutions will give an ionic strength of 100 mM and 300 mM, respectively. Without resorting to any physical formalism, ionic strength can be considered as an empirical way to characterize the effect of ion screening.

(3) Screening Debye length $\lambda_D = \sqrt{\frac{\varepsilon\varepsilon_0 k_B T}{\sum_i z_i^2 e^2 n_i}} = \sqrt{\frac{1}{4\pi\lambda_B \sum_i z_i^2 n_i}} = \sqrt{\frac{1}{8\pi\lambda_B I}}$. Debye length thus considers the roles of both thermal motion and ion screening. Mathematically, it is the decay length of the electrostatic field in ionic solutions that takes on an exponential form rather than a power law in dielectric medium as a result of ionic screening. For example, λ_D is ~1 nm at 100 mM monovalent salt. However, it should be noted that Debye length is strictly speaking only applicable to weak electrostatic fields, as it is derived from a linearized version of the Poisson-Boltzmann (PB) mean-field treatment of electrostatic interactions.

(4) Charge densities of charged biomolecules, linear charge density η, and surface charge density σ. In the case of dsDNA of 1 nm radius and 2 e per 0.34 nm, η is 5.88 e/nm and σ is 0.94 e/nm^2.

3.1. The Role of Ions

Without ions, electrostatic repulsion between DNA helices would dominate DNA–DNA interactions. An instructive example is given by the energetics of DNA packaging in bacterial viruses. Taking the case of λ phage with 48,502 bp dsDNA and 30 nm radius, a simple model calculation based on a uniformly charged sphere gives 5×10^8 pN·nm electric potential energy if stored in water alone, which is thousands of times greater than the measured work during DNA packaging [75,76]. It is not surprising that these bacterial viruses will burst open due to the strong internal DNA forces at low salt. The presence of ions in physiological conditions can therefore reduce the electrostatic penalties by several orders of magnitude! In addition to the role of screening DNA–DNA repulsive forces, multivalent ions of valence ≥3 are able to induce attractive forces between DNAs at sub-millimolar concentrations, giving rise to the phenomenon of DNA condensation [77]. Such delicate dependence on ionic valence has put ions at the center of studies of DNA-inspired electrostatics. We would like to note that this review focuses on the ions that are generally considered to non-specifically interact with DNA, e.g., Na^+, K^+, Mg^{2+}; conversely, a wide variety of metal ions can associate with and condense DNA through specific bindings that often distort or completely disrupt the double helical structure [78].

3.1.1. Screening of DNA–DNA repulsion

The seminal DLVO theory (named after Derjaguin, Landau, Verwey, and Overbeek [79]), originally developed to describe the behavior of charged colloids, provides a convenient starting point to quantitate the electrostatic repulsion between DNA. The potential energy between charged molecules is given as the Debye-Hückel (DH) potential $U(r) = \begin{cases} \infty, r < D \\ \frac{Z^2}{\varepsilon(1+\kappa D/2)^2}\frac{e^{-(r-D)}}{r}, r \geq D \end{cases}$, where D is the diameter of the molecule, κ is the inverse Debye length; and Z is the molecular charge. Figure 6 illustrates the shapes of DH potentials for two charged spheres in pure water and 10 mM monovalent salt. While the analytical form above is only true for two charged spheres, adapting it to cylindrical charge, non-trivial analytically [80], can be carried out numerically or by treating DNA as spherical beads on a chain. However, the more deeply rooted problem is that the DLVO theory is based on a linearized PB equation. In the PB approach, describing the thermodynamic properties of ions in an electric field, the role of ion-mixing entropy is captured by the Boltzmann factor; it is a mean field approach as the ions are treated as continuous charge densities or spatial distribution probabilities $\rho(r)$ as below:

$$\nabla^2\varphi(r) = -\frac{4\pi}{\varepsilon}\sum_i \rho_i(r) = -\frac{4\pi}{\varepsilon}\sum_i \rho_i(\infty)e^{-\beta Z_i \varphi(r)}, \tag{5}$$

where $\varphi(r)$ is the electric potential and β is $1/k_BT$. With appropriate boundary conditions, such as the dimension and charge density of DNA and the ion concentrations at infinity $\rho_i(r)$, the non-linear PB (NLPB) equation can be solved and give a complete description of the electric field and ion distribution in the space. The NLPB equation is linearized in the DLVO theory due to the relatively weakly charged nature of the colloids in consideration and the convenience of an analytical solution by taking $e^{-\beta z_i \varphi(r)} \sim 1 - \beta z_i \varphi(r)$. However, this is no longer justified for highly charged DNA where the $\beta z_i \varphi(r)$ is not small but even larger than 1 in its vicinity. A linear PB (LPB) equation consequently underestimates the extent of ionic screening, to which many remedial solutions have been proposed.

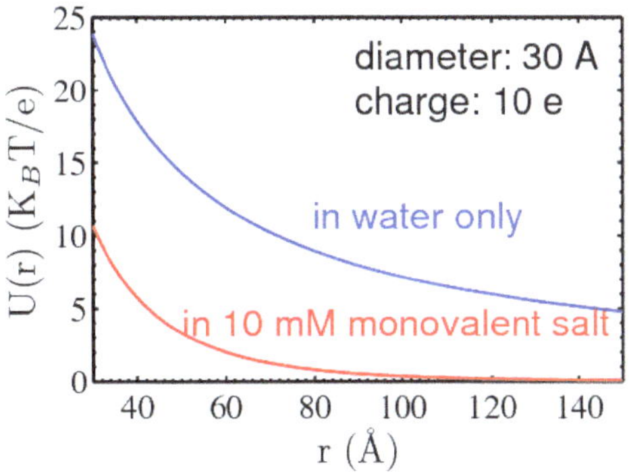

Figure 6. Visualization of Debye–Hückel potentials between two charged spheres of 3 nm diameter and 10 e bare charge. The effect of salt screening is shown for the case of 10 mM monovalent salt.

Counterion condensation, or Manning condensation, represents a significant conceptual advance towards the understanding of electrostatics of highly charged polyelectrolytes such as DNA [81,82]. The criterion for the occurrence of counterion condensation is based on a dimensionless Coulomb coupling strength or the Manning ratio, $\Gamma = \lambda_B \eta$: If Γ is greater than 1, *i.e.*, more than one charge is present in a span of Bjerrum length, counterions can condense around the polyelectrolyte to neutralize part of its charge. Then, the LPB approach can be used to describe the electrostatics of the partially neutralized polyelectrolyte. The extent of neutralization depends on the counterion valence z, given as $1 - \frac{1}{z\Gamma}$. For dsDNA, Γ is 4.1, giving the DNA charge neutralization by mono-, di-, and tri-valent counterions of 76%, 88%, and 92%, respectively. The counterion condensation model is similar in origin to the termed Stern layer in describing the zeta potential of charged colloids. In light of its simplicity and underlying assumptions, the counterion condensation should be viewed only as a first-order approximation of the non-linear screening of counterions. The rich physics of ion atmospheres near polyelectrolyte surfaces has promoted a series of theoretical and experimental studies to achieve quantitative understanding of the non-linearity and its intricate dependence on the ion size and valence and molecular surfaces.

The charge renormalization ansatz took the idea of partial charge neutralization further and sought a thermodynamically more rigorous method to determine the effective charge [83–86]. Instead of considering a single polyelectrolyte surrounded by ions, it analyzes a cell model of polyelectrolyte dispersions, for example, spherical cells for anisotropic solutions and cylindrical cells for nematically ordered polyelectrolytes. In order to obtain the effective charge to be used in the DH potential, the charge renormalization formalism proposed to match the electrical potential at the cell boundary between LPB (with a so-called renormalized charge) and NLPB (with the bare charge) solutions. Such renormalized charge suited to LPB approaches is expected to yield the same osmotic pressure as the full NLPB treatment. Contrary to the counterion condensation, which predicts the same charge neutralized regardless of polyelectrolyte or ion concentration, the charge renormalization formalism predicts the dependence of renormalized charges on both polyelectrolyte and ion concentrations.

Early experimental guidance and validation for many of these theoretical advances came from the fields of colloidal interactions and general polyelectrolyte behaviors. Here the dispersion of DNA helices provided an ideal physically and biochemically well-defined system for quantitative measurements and modeling of polyelectrolyte interactions. As there is no practical method to hold two DNA helices in space and move them towards each other to accurately measure their forces (note that, while such geometry can be realized experimentally, the forces between two DNA helices at long distances will be too weak for reliable determination), DNA–DNA interactions have been probed by a variety of thermodynamic measurements such as salt-dependent persistence length of dsDNA [87,88] and the osmotic pressure of DNA dispersions [89]. A more recent series of studies employed small angle X-ray scattering (SAXS) to measure the structure factor of dispersions of oligomeric DNAs of uniform sequence and length [90,91]. As demonstrated in Figure 7, the structure factor encodes the spatial correlation of dispersed DNAs, which in turn is determined by the DNA–DNA force-distance relationship, which can be obtained via model fitting. These studies verified that the DH functional form is able to quantitatively reproduce the measured structure factors with one fitting parameter, the effective charge z_{eff}. As shown in Figure 8, comparisons with the theoretical renormalized charges showed fairly good agreement given the rather crude nature of the theoretical model, treating dsDNA as uniform cylinders [91]. But significant deviations were observed for even divalent salts [92], which will be discussed later. Overall, the electrostatics of DNA dispersions in monovalent salts has been largely understood.

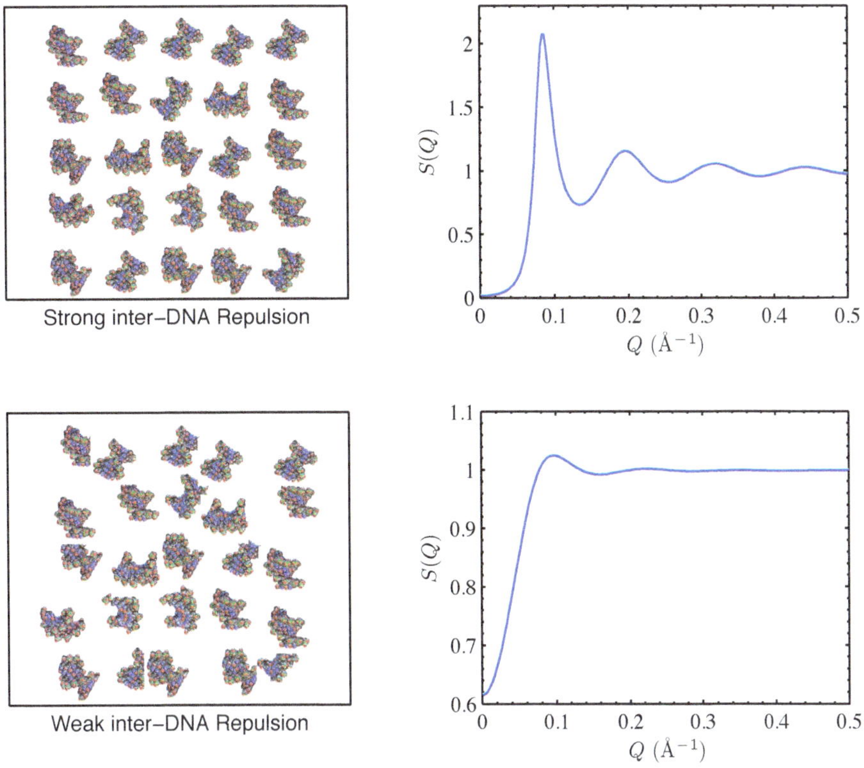

Figure 7. Application of small angle X-ray scattering (SAXS) to measure the structure factor *S*(*Q*) of semi-dilute dispersions of oligomeric DNAs. The top panels show the case of strong DNA–DNA repulsion, giving rise to spatial ordering of DNA strands and a structure factor with pronounced correlation peaks. The bottom panels show the case of weak DNA–DNA interaction, giving rise to random dispersions and a much suppressed structure factor.

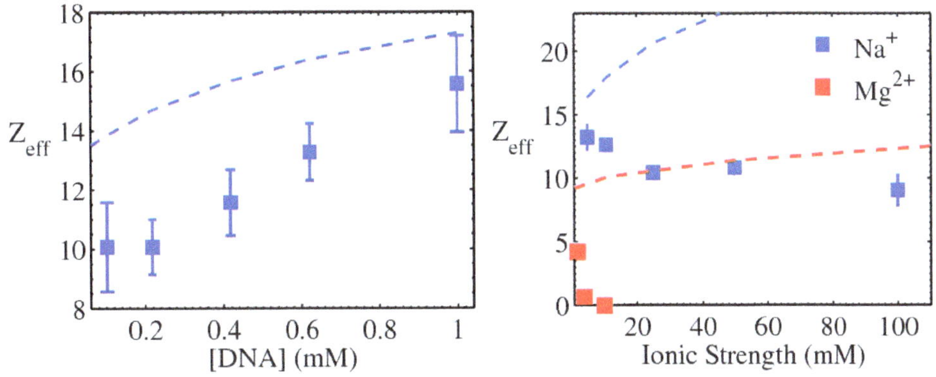

Figure 8. Effective charges (z_{eff}) determined by SAXS measurements in conjunction with physical modeling. The oligomeric DNA has a bare charge of 48 e. Symbols are experimental values and lines are the renormalized charges. Details of experimental conditions and modeling procedures are described in [91]. Adapted from [91]. Copyright (2006) by the American Physical Society.

3.1.2. Mediation of DNA–DNA attraction

Even for divalent ions such as Mg^{2+}, our understanding is far from adequate. In the studies of oligomeric DNA dispersions discussed above, large discrepancies exist between theory and experiment in low salt ($[Mg^{2+}] < 10$ mM) when DNA–DNA forces are repulsive (see Figure 8), and, furthermore, DNA–DNA attraction was observed at $[Mg^{2+}] > 10$ mM. While later studies showed that such attraction may be due to end-to-end stacking of the oligomeric dsDNA [93], counterion-mediated attraction between dsDNA helices is undoubtedly established for multivalent cations with valence ≥3, often referred to as "like-charge attraction" or DNA condensation [6,94,95].

The glaring disparity between DNA condensation and DNA–DNA repulsion invariably predicted by mean-field theories has stimulated vast scientific interest and intense studies in the past decades. The most commonly studied DNA-condensing counterions are trivalent Cobalt Hexammine (CoHex), trivalent Spermidine, and tetravalent Spermine [65,96–98]. It is known that the condensed DNA helices are packed side by side (*i.e.*, in parallel) in hexagonal arrays, as illustrated in Figure 9, and the as-condensed arrays have inter-axial distances between 2.7 and 3 nm, leaving 0.7–1 nm interstitial space between dsDNA surfaces [99–101]. Multivalent cations (and solvent) thus reside in the interstitial space and mediate DNA–DNA attraction. Adding mono- or divalent cations reduces the attraction and eventually redissolves condensed DNA, suggesting an electrostatic-driven attraction. However, the physical origin of attraction is still under debate, as well as the true behavior of ions in such a tight space of extremely high electrostatic fields. To put the composition of the interstitial space in perspective, the effective concentrations of mono-, di-, and trivalent cations, if the only counterion species to neutralize DNA charge, are in the orders of 2000, 1000, and 700 mM, respectively.

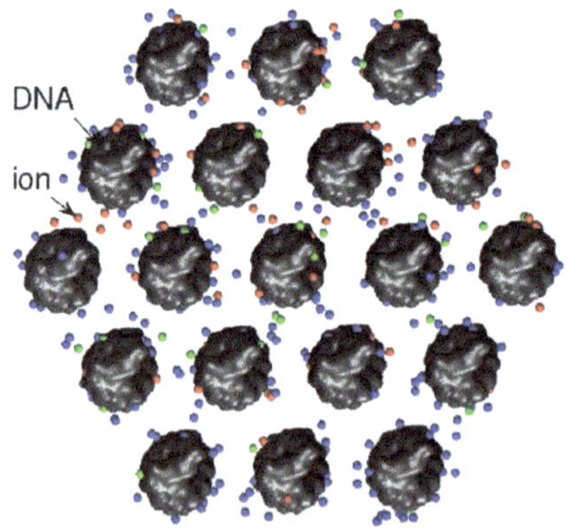

Figure 9. Condensed dsDNA helices packaged in hexagonal arrays, as viewed from the axis. Reproduced from [102]. Copyright (2013) with permission from Elsevier.

It is generally accepted that the mean field description of continuous and static ionic clouds is no longer valid for DNA condensation. One of the early theoretical models involves density fluctuations of these "territorially bound" interstitial counterions, leading to a van der Waals-like attraction [103–105]. However, there has been no report of cation-charged density waves in condensed dsDNA arrays. It is worth noting that charge density waves were observed with divalent-cation-condensed F-actin [106], which is, however, complicated by the periodic patches of charges on the F-actin surface. On the other hand, the discrete nature of ions has been considered in a number of competing theoretical models.

When DNA is still considered as a uniformly charged cylinder, the "Wigner lattice" model argues for the formation of crystal-like ordering of counterions due to strong ion–ion correlations [107–110], in a way similar to the ordering of electron gases confined on a positively charged surface. The lattice cohesive energy, as a result of ordering that lowers ion–ion correlation energy, creates correlation holes that lead to attraction between two such lattices. This ion–ion correlation-based model further predicts over-compensation of DNA charge by counterions, *i.e.*, overcharging. DNA overcharging has successfully explained the redissolvation of condensed DNA helices at higher counterion concentrations [111], though there exist alternative explanations involving counterion–coion pairing [101]. Consideration of the discrete nature of charged DNA groups further led to several distinct physical mechanisms for DNA–DNA attraction. The strong Coulomb coupling theory proposes close confinement of counterions and demonstrated net attraction via additive electrostatic interactions of DNA and ions [112,113]. Without taking the strong coupling limit, it has been shown that the undulations of the electrostatic fields due to the molecular nature can result in attraction through optimization of approaching surfaces [114,115]. Moreover, the transient localization of ions near charged groups can result in charge inversion and/or ion bridges. The tightly bound ion model considers such possibility and showed significant DNA–DNA attraction through exhaustive sampling and subsequent NLPB calculations [116,117]. Instead of charge localization near charge groups, the "DNA-ion zipper" model takes into account the DNA major grooves that can accommodate counterions, forming helical spirals of alternating charges [118]. A zipper can thus be constructed between opposing DNA surfaces. The necessitated geometric commensuration has inspired the conjecture and modeling of sequence-dependent DNA–DNA forces, predicting the electrostatics-driven recognition between homologous sequences [119].

3.2. The Role of Solvent

While its essential role is widely acknowledged, water is more often than not taken for granted as the universal matrix biomolecules live in. Being a polar solvent, water simply acts as a dielectric medium to attenuate electric fields, and it may seem only necessary to introduce minor corrections for likely non-ideal dielectric breakdowns in strong fields. However, in the tight interstitial spaces of thickness $\leq$ 1 nm, its molecular origin can no longer be neglected. Through a series of pioneering osmotic stress measurements and codifications, Parsegian, Rau, and co-workers elucidated that water/hydration dominates the magnitude and form of DNA–DNA forces in the last nm of surface separation [120–122] (see Figure 10 for illustration of osmotic stress measurements and representative DNA force-extension curves). Specifically, it was observed that, upon the approaching of opposing surfaces, the DNA–DNA force–distance relations show a universal exponential form with a decay length of ~0.3 nm independent of the ionic species or concentration. More generally, such universal behavior for dsDNA forces was shown to hold for a large variety of molecular surfaces, charged (e.g., collagen and DNA) or non-charged (e.g., polysaccharides), cylindrical (e.g., DNA) or planar (e.g., lipid bilayers) [123,124]. This evidence culminates with the establishment of universal hydration forces for charged and polar molecular surfaces. Note that charged surfaces additionally make electrostatic contributions to the force, but the electrostatic interaction is dominated by hydration forces at close surface–surface separations.

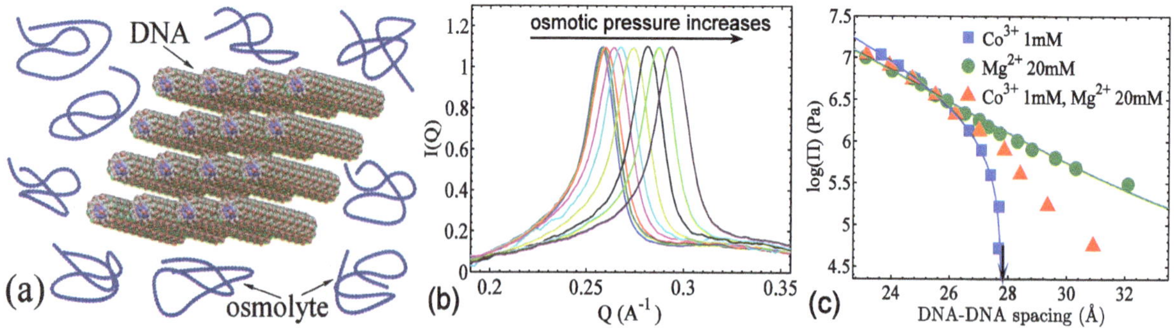

Figure 10. (**a**) Illustration of DNA arrays under osmotic stress; (**b**) Demonstration of shifts of the DNA–DNA X-ray diffraction peak as the osmotic pressure increases under DNA-condensing conditions. The peaks are scaled to assist visualization; (**c**) The force-spacing curves of dsDNA in varied salts as annotated in the legend. The *x*-axis shows the inter-axial spacing, whereas the surface-to-surface spacing is <2 nm. Symbols are experimental data and lines are the fits using exponential forces, as described in the text. The arrow indicates the DNA spacings at zero osmotic pressure.

Hydration forces can be either repulsive or attractive, in ways similar to "like charges repel and opposite charges attract" [123]. As the hydration shell is polar and can be viewed as pointing away or towards the molecular surface depending on the surface charge, the directionality of the hydration shell causes the approaching surfaces to have either conflicting or conforming hydration structures. Conflicting hydration structures lead to destructive dehydration and consequently repulsive hydration force, while conforming hydration structures lead to attractive hydration force. Hydration forces are generally reported in terms of the osmotic pressure $\Pi(d) = f_r e^{-d/\lambda_r} - f_a e^{-d/\lambda_a}$, where d is the inter-axial distance or surface separation, f_r and f_a are amplitudes of repulsive and attractive forces respectively, and λ_a and λ_r are the decay lengths. Figure 10 briefly illustrates the osmotic stress measurements and representative data in different salt conditions. In the case of counterion-mediated DNA–DNA forces in the last nm of surface separation, the absence of counterions would lead to two similarly homogenously hydrated surfaces and these two "like hydration" structures will collide and give rise to repulsive hydration forces. Addition of monovalent counterions weakens hydration but does not change the directionality of hydration, consequently decreasing the repulsive hydration force. However, for multivalent counterions, they can reverse the charge and hydration locally and create undulating hydration structures of both directions. This results in attractive hydration forces between opposing surfaces, noting that the hydration repulsion term still remains due to imperfect matches and/or incomplete charge neutralization near the surfaces. Divalent cations present an interesting border case for hydration forces because they can reverse the charge locally if confined to a single phosphate group. However, no attractive DNA–DNA hydration force has been observed for non-specifically interacting divalent counterions. It is likely that a divalent counterion is localized in between two phosphate groups and no inversion of the hydration direction is incurred.

3.3. The Role of DNA's Structure

The double helical structure of dsDNA is a critical element of several proposed models of DNA–DNA interactions. One possible role is providing discrete charge groups, regularly spaced charge patterns, and groove structures for counterion habitation. Varying DNA structure thus offers an avenue to test different theoretical models. However, studies of non-B-form DNA structures are relatively few [125]. One example of recent studies is the report of divalent counterion mediated attraction between triple-strand DNA (tsDNA) [92]. tsDNA is more highly charged than dsDNA, with ~40% larger linear charge density and ~32% higher surface charge density. tsDNA also has more regularly spaced charges and has narrower grooves. While shedding new light on the physical origin of DNA condensation, the observation fails, however, to differentiate between existing theoretical models, partly due to the dominant effect of substantially higher charge densities. More recently, it was shown that dsRNA of A-form helical structure is more resistant to condensation than B-form dsDNA [126], strongly suggesting that helical structures play non-negligible roles in nucleic acid interactions. Nonetheless, it is difficult to attribute the difference to a unique structural feature, e.g., counterion penetration toward central axis [127] or less evenly spaced helical charges [128]. This is further complicated by the additional hydroxyl group of RNA compared to DNA.

Electrostatic and hydration forces between DNA helices have proven to be fertile ground for biophysical approaches. The intricate interplay between the participating ions, solvent, and DNA makes it difficult to dissect the individual role of each component. It is also likely that different physical mechanisms may be invoked for DNA–DNA interactions in different settings. In order to parse the roles of ions, solvent, and DNA structures, systematic studies of the pertinent variables will be needed. Their strong interdependence further calls for new experimental probes and new theoretical treatments. While DNA–DNA forces can be measured, little experimental knowledge exists of the interstitial space filled with ions and solvent. Important unanswered questions include the ion hydration and distribution, DNA hydration structure, and how the interstitial space varies with ionic condition and DNA structure. Experimentally, the ability to directly probe ions and solvents will be very useful, as exemplified by a recent study to measure the number of ions within dsDNA arrays [102]. However, given the likely amorphous nature of the interstitial ions and solvent, inclusive theoretical models are probably better positioned to capture the relevant physics of nucleic acid interactions. An important step towards this direction was reported by He and Chen [129], including ion–ion correlation and dynamics, solvent polarization, and atomic DNA structure in a consistent framework. It was shown that ion-induced restructuring of the solvent polarization contributes significantly to DNA–DNA interactions. Further developments of such theoretical models will be essential to elucidate the intricate interplays between ions, solvent, and DNA structure.

4. Conclusions

The importance of mechanics, electrostatics, and hydration reveals the fundamental DNA molecular forces. Advances in instrumentation allow us to explore the interaction forces in detail and to further predict the behavior of DNA, which will benefit both our understanding of living systems and the creation of smart biomaterials. The DNA molecular force is a key determinant in the operation of cellular machinery in living organisms. An example is the tightly packaged DNA in bacteriophages, where both the mechanical and the electrostatic forces of DNA work together to create a large internal pressure that propel the initial DNA translocation when infecting bacteria. Outside the living system, the quantitative force information of DNA should help us to better design systems that apply DNA in nanotechnologies such as programmable self-assembly, DNA computing, material assembly, and nanomedicine.

Acknowledgments

We thank NSF DMR-0907676, Welch Foundation C-1632, and the Hamill Innovation Fund for support.

Author Contributions

Jingqiang Li and Sithara S. Wijeratn wrote Section 2; Xiangyun Qiu wrote Section 3. Ching-Hwa Kiang designed the outline and wrote the abstract. All authors contributed to the abstract; Section 1 and Section 4. Xiangyun Qiu and Ching-Hwa Kiang edited the entire manuscript.

Conflicts of Interest

The authors declare no conflict of interest.

References

1. Smith, S.B.; Cui, Y.J.; Bustamante, C. Overstretching B-DNA: The elastic response of individual double-stranded and single-stranded DNA molecules. *Science* **1996**, *271*, 795–799.
2. Rief, M.; Clausen-Schaumann, H.; Gaub, H.E. Sequence-dependent mechanics of single DNA molecules. *Nat. Struct. Biol.* **1999**, *6*, 346–349.
3. Chen, W.S.; Chen, W.H.; Chen, Z.P.; Gooding, A.A.; Lin, K.J.; Kiang, C.H. Direct observation of multiple pathways of single-stranded DNA stretching. *Phys. Rev. Lett.* **2010**, *105*, 218104:1–218104:5.
4. Livolant, F.; Leforestier, A. Condensed phases of DNA: Structures and phase transitions. *Prog. Polym. Sci.* **1996**, *21*, 1115–1164.
5. Parsegian, V.A.; Rand, R.P.; Rau, D.C. Macromolecules and water: Probing with osmotic stress. *Meth. Enzymol.* **1995**, *259*, 43–94.
6. Bloomfield, V.A. DNA condensation. *Curr. Opin. Struct. Biol.* **1996**, *6*, 334–341.
7. Gelbart, W.M.; Bruinsma, R.F.; Pincus, P.A.; Parsegian, V.A. DNA-inspired electrostatics. *Phys. Today* **2000**, *53*, 38–44.
8. Solis, F.J.; de la Cruz, M.O. Flexible polymers also counterattract. *Phys. Today* **2001**, *54*, 71–72.
9. Parsegian, V.A.; Zemb, T. Hydration forces: Observations, explanations, expectations, questions. *Curr. Opin. Colloid Interface Sci.* **2011**, *16*, 618–624.
10. Modi, S.; Bhatia, D.; Simmel, F.C.; Krishnan, Y. Structural DNA nanotechnology: From bases to bricks, from structure to function. *J. Phys. Chem. Lett.* **2010**, *1*, 1994–2005.
11. Seeman, N.C. DNA in a material world. *Nature* **2003**, *421*, 427–431.
12. Zhang, F.; Nangreave, J.; Liu, Y.; Yan, H. Structural DNA nanotechnology: State of the art and future perspective. *J. Am. Chem. Soc.* **2014**, *136*, 11198–11211.
13. Shih, W.M.; Douglas, S.M.; Dietz, H.; Liedl, T.; Hogberg, B.; Graf, F. Self-assembly of DNA into nanoscale three-dimensional shapes. *J. Biomol. Struct. Dyn.* **2009**, *26*, 414–418.

14. Liu, Q.H.; Wang, L.M.; Frutos, A.G.; Condon, A.E.; Corn, R.M.; Smith, L.M. DNA computing on surfaces. *Nature* **2000**, *403*, 175–179.
15. Adleman, L.M. Molecular computation of solutions to combinatorial problems. *Science* **1994**, *266*, 1021–1024.
16. Fu, J.L.; Yang, Y.R.; Johnson-Buck, A.; Liu, M.H.; Liu, Y.; Walter, N.G.; Woodbury, N.W.; Yan, H. Multi-enzyme complexes on DNA scaffolds capable of substrate channelling with an artificial swinging arm. *Nat. Nanotechnol.* **2014**, *9*, 531–536.
17. Han, D.R.; Pal, S.; Yang, Y.; Jiang, S.X.; Nangreave, J.; Liu, Y.; Yan, H. DNA gridiron nanostructures based on four-arm junctions. *Science* **2013**, *339*, 1412–1415.
18. Bustamante, C.; Marko, J.F.; Siggia, E.D.; Smith, S. Entropic elasticity of lambda-phage DNA. *Science* **1994**, *265*, 1599–1600.
19. Liphardt, J.; Onoa, B.; Smith, S.B.; Tinoco, I.; Bustamante, C. Reversible unfolding of single RNA molecules by mechanical force. *Science* **2001**, *292*, 733–737.
20. Wijeratne, S.S.; Botello, E.; Yeh, H.C.; Zhou, Z.; Bergeron, A.L.; Frey, E.W.; Patel, J.M.; Nolasco, L.; Turner, N.A.; Moake, J.L.; *et al*. Mechanical activation of a multimeric adhesive protein through domain conformational change. *Phys. Rev. Lett.* **2013**, *110*, 108102:1–108102:9.
21. Moy, V.T.; Florin, E.L.; Gaub, H.E. Intermolecular forces and energies between ligands and receptors. *Science* **1994**, *266*, 257–259.
22. Harris, N.C.; Song, Y.; Kiang, C.H. Experimental free energy surface reconstruction from single-molecule force spectroscopy using Jarzynski's equality. *Phys. Rev. Lett.* **2007**, *99*, 068101:1–068101:4.
23. Wang, M.D.; Yin, H.; Landick, R.; Gelles, J.; Block, S.M. Stretching DNA with optical tweezers. *Biophys. J.* **1997**, *72*, 1335–1346.
24. Smith, S.B.; Finzi, L.; Bustamante, C. Direct mechanical measurements of the elasticity of single DNA-molecules by using magnetic beads. *Science* **1992**, *258*, 1122–1126.
25. Cluzel, P.; Lebrun, A.; Heller, C.; Lavery, R.; Viovy, J.L.; Chatenay, D.; Caron, F. DNA: An extensible molecule. *Science* **1996**, *271*, 792–794.
26. Evans, E.; Ritchie, K.; Merkel, R. Sensitive force technique to probe molecular adhesion and structural linkages at biological interfaces. *Biophys. J.* **1995**, *68*, 2580–2587.
27. Calderon, C.P.; Chen, W.H.; Lin, K.J.; Harris, N.C.; Kiang, C.H. Quantifying DNA melting transitions using single-molecule force spectroscopy. *J. Phys.: Condens. Matter* **2009**, *21*, 034114:1–034114:12.
28. Bryant, Z.; Stone, M.D.; Gore, J.; Smith, S.B.; Cozzarelli, N.R.; Bustamante, C. Structural transitions and elasticity from torque measurements on DNA. *Nature* **2003**, *424*, 338–341.
29. Gore, J.; Bryant, Z.; Nollmann, M.; Le, M.U.; Cozzarelli, N.R.; Bustamante, C. DNA overwinds when stretched. *Nature* **2006**, *442*, 836–839.
30. Celedon, A.; Nodelman, I.M.; Wildt, B.; Dewan, R.; Searson, P.; Wirtz, D.; Bowman, G.D.; Sun, S.X. Magnetic tweezers measurement of single molecule torque. *Nano Lett.* **2009**, *9*, 1720–1725.
31. Mosconi, F.; Allemand, J.F.; Bensimon, D.; Croquette, V. Measurement of the torque on a single stretched and twisted DNA using magnetic tweezers. *Phys. Rev. Lett.* **2009**, *102*, doi:http://dx.doi.org/10.1103/PhysRevLett.102.078301.

32. Gross, P.; Laurens, N.; Oddershede, L.B.; Bockelmann, U.; Peterman, E.J.G.; Wuite, G.J.L. Quantifying how DNA stretches, melts and changes twist under tension. *Nat. Phys.* **2011**, *7*, 731–736.
33. Wijeratne, S.S.; Harris, N.C.; Kiang, C.-H. Principles Involved in Interpreting Single-Molecule Force Measurement of Biomolecules. In *MEMS and Nanotechnology*; Shaw, G.A., Prorok, B., Starman, L.A., Eds.; Springer: New York, NY, USA, 2012; Volume 6, pp. 19–23.
34. Frey, E.W.; Gooding, A.A.; Wijeratne, S.; Kiang, C.H. Understanding the physics of DNA using nanoscale single-molecule manipulation. *Front. Phys.* **2012**, *7*, 576–581.
35. Flory, P.J. *Statistical Mechanics of Chain Molecules*; Interscience Publishers: New York, NY, USA, 1969.
36. Marko, J.F.; Siggia, E.D. Stretching DNA. *Macromolecules* **1995**, *28*, 8759–8770.
37. Odijk, T. Stiff chains and filaments under tension. *Macromolecules* **1995**, *28*, 7016–7018.
38. Cocco, S.; Yan, J.; Leger, J.F.; Chatenay, D.; Marko, J.F. Overstretching and force-driven strand separation of double-helix DNA. *Phys. Rev. E* **2004**, *70*, 011910:1–011910:18.
39. Hatch, K.; Danilowicz, C.; Coljee, V.; Prentiss, M. Demonstration that the shear force required to separate short double-stranded DNA does not increase significantly with sequence length for sequences longer than 25 base pairs. *Phys. Rev. E* **2008**, *78*, 011920.
40. Ho, D.; Zimmermann, J.L.; Dehmelt, F.A.; Steinbach, U.; Erdmann, M.; Severin, P.; Falter, K.; Gaub, H.E. Force-driven separation of short double-stranded DNA. *Biophys. J.* **2009**, *97*, 3158–3167.
41. Pope, L.H.; Davies, M.C.; Laughton, C.A.; Roberts, C.J.; Tendler, S.J.B.; Williams, P.M. Force-induced melting of a short DNA double helix. *Eur. Biophys. J.* **2001**, *30*, 53–62.
42. Sattin, B.D.; Pelling, A.E.; Goh, M.C. DNA base pair resolution by single molecule force spectroscopy. *Nucleic Acids Res.* **2004**, *32*, 4876–4883.
43. Lee, G.U.; Chrisey, L.A.; Colton, R.J. Direct measurement of the forces between complementary strands of DNA. *Science* **1994**, *266*, 771–773.
44. Clausen-Schaumann, H.; Rief, M.; Tolksdorf, C.; Gaub, H.E. Mechanical stability of single DNA molecules. *Biophys. J.* **2000**, *78*, 1997–2007.
45. Scholl, Z.N.; Rabbi, M.; Lee, D.; Manson, L.; S-Gracz, H.; Marszalek, P.E. Origin of overstretching transitions in single-stranded nucleic acids. *Phys. Rev. Lett.* **2013**, *111*, 188302.
46. Bustamante, C.; Smith, S.B.; Liphardt, J.; Smith, D. Single-molecule studies of DNA mechanics. *Curr. Opin. Struct. Biol.* **2000**, *10*, 279–285.
47. Bustamante, C.; Bryant, Z.; Smith, S.B. Ten years of tension: Single-molecule DNA mechanics. *Nature* **2003**, *421*, 423–427.
48. Strunz, T.; Oroszlan, K.; Schafer, R.; Guntherodt, H.J. Dynamic force spectroscopy of single DNA molecules. *Proc. Natl. Acad. Sci. USA* **1999**, *96*, 11277–11282.
49. Grange, W.; Strunz, T.; Schumakovitch, I.; Guntherodt, H.J.; Hegner, M. Molecular recognition and adhesion of individual DNA strands studied by dynamic force microscopy. *Single Mol.* **2001**, *2*, 75–78.
50. Van Mameren, J.; Gross, P.; Farge, G.; Hooijman, P.; Modesti, M.; Falkenberg, M.; Wuite, G.J.L.; Peterman, E.J.G. Unraveling the structure of DNA during overstretching by using multicolor, single-molecule fluorescence imaging. *Proc. Natl. Acad. Sci. USA* **2009**, *106*, 18231–18236.

51. Bosaeus, N.; El-Sagheer, A.H.; Brown, T.; Smith, S.B.; Akerman, B.; Bustamante, C.; Norden, B. Tension induces a base-paired overstretched DNA conformation. *Proc. Natl. Acad. Sci. USA* **2012**, *109*, 15179–15184.
52. Bongini, L.; Melli, L.; Lombardi, V.; Bianco, P. Transient kinetics measured with force steps discriminate between double-stranded DNA elongation and melting and define the reaction energetics. *Nucleic Acids Res.* **2014**, *42*, 3436–3449.
53. Fu, H.X.; Chen, H.; Marko, J.F.; Yan, J. Two distinct overstretched DNA states. *Nucleic Acids Res.* **2010**, *38*, 5594–5600.
54. Zhang, X.; Chen, H.; Fu, H.; Doyle, P.S.; Yan, J. Two distinct overstretched DNA structures revealed by single-molecule thermodynamics measurements. *Proc. Natl. Acad. Sci. USA* **2012**, *109*, 8103–8108.
55. Zhang, X.H.; Chen, H.; Le, S.M.; Rouzina, I.; Doyle, P.S.; Yan, J. Revealing the competition between peeled ssDNA, melting bubbles, and s-DNA during DNA overstretching by single-molecule calorimetry. *Proc. Natl. Acad. Sci. USA* **2013**, *110*, 3865–3870.
56. Goddard, N.L.; Bonnet, G.; Krichevsky, O.; Libchaber, A. Sequence dependent rigidity of single stranded DNA. *Phys. Rev. Lett.* **2000**, *85*, 2400–2403.
57. Turner, N.A. Conformational Changes. In *Nucleic Acids: Structure Properties and Functions*; Bloomfield, V.A., Crothers, D.M., Tinoco, I., Eds.; University Science Books: Sausalito, CA, USA, 2000; pp. 259–334.
58. Buhot, A.; Halperin, A. Effects of stacking on the configurations and elasticity of single-stranded nucleic acids. *Phys. Rev. E* **2004**, *70*, 020902:1–020902:4.
59. Seol, Y.; Skinner, G.M.; Visscher, K.; Buhot, A.; Halperin, A. Stretching of homopolymeric RNA reveals single-stranded helices and base-stacking. *Phys. Rev. Lett.* **2007**, *98*, 158103.
60. Ke, C.; Humeniuk, M.; S-Gracz, H.; Marszalek, P.E. Direct measurements of base stacking interactions in DNA by single-molecule atomic-force spectroscopy. *Phys. Rev. Lett.* **2007**, *99*, doi:http://dx.doi.org/10.1103/PhysRevLett.99.018302.
61. Segal, E.; Widom, J. Poly(da:Dt) tracts: Major determinants of nucleosome organization. *Curr. Opin. Struct. Biol.* **2009**, *19*, 65–71.
62. Hopkins, J.C.; Dryden, D.M.; Ching, W.Y.; French, R.H.; Parsegian, V.A.; Podgornik, R. Dielectric response variation and the strength of van der waals interactions. *J. Colloid Interface Sci.* **2014**, *417*, 278–284.
63. Lorman, V.; Podgornik, R.; Zeks, B. Positional, reorientational, and bond orientational order in DNA mesophases. *Phys. Rev. Lett.* **2001**, *87*, 218101:1–218101:4.
64. Strey, H.H.; Wang, J.; Podgornik, R.; Rupprecht, A.; Yu, L.; Parsegian, V.A.; Sirota, E.B. Refusing to twist: Demonstration of a line hexatic phase in DNA liquid crystals. *Phys. Rev. Lett.* **2000**, *84*, 3105–3108.
65. Van Dam, L.; Korolev, N.; Nordenskiold, L. Polyamine-nucleic acid interactions and the effects on structure in oriented DNA fibers. *Nucleic Acids Res.* **2002**, *30*, 419–428.
66. Strey, H.H.; Podgornik, R.; Rau, D.C.; Parsegian, V.A. DNA–DNA interactions. *Curr. Opin. Struct. Biol.* **1998**, *8*, 309–313.

67. Wong, G.C.; Pollack, L. Electrostatics of strongly charged biological polymers: Ion-mediated interactions and self-organization in nucleic acids and proteins. *Annu. Rev. Phys. Chem.* **2010**, *61*, 171–189.
68. Tan, Z.J.; Chen, S.J. Predicting electrostatic forces in RNA folding. *Method Enzymol.* **2009**, *469*, 465–487.
69. Kanduc, M.; Dobnikar, J.; Podgornik, R. Counterion-mediated electrostatic interactions between helical molecules. *Soft Matter* **2009**, *5*, 868–877.
70. Dong, F.; Olsen, B.; Baker, N.A. Computational methods for biomolecular electrostatics. *Methods Cell Biol.* **2008**, *84*, 843–870.
71. Wong, G.C.L. Electrostatics of rigid polyelectrolytes. *Curr. Opin. Colloid Interface Sci.* **2006**, *11*, 310–315.
72. Baker, N.A.; McCammon, J.A. Electrostatic interactions. *Methods Biochem. Anal.* **2003**, *44*, 427–440.
73. Levin, Y. Electrostatic correlations: From plasma to biology. *Rep. Prog. Phys.* **2002**, *65*, 1577–1632.
74. Belloni, L. Colloidal interactions. *J. Phys. Condens. Matter.* **2000**, *12*, 549–587.
75. Lof, D.; Schillen, K.; Jonsson, B.; Evilevitch, A. Dynamic and static light scattering analysis of DNA ejection from the phage lambda. *Phys. Rev. E* **2007**, *76*, 011914:1–011914:11.
76. Fuller, D.N.; Raymer, D.M.; Rickgauer, J.P.; Robertson, R.M.; Catalano, C.E.; Anderson, D.L.; Grimes, S.; Smith, D.E. Measurements of single DNA molecule packaging dynamics in bacteriophage lambda reveal high forces, high motor processivity, and capsid transformations. *J. Mol. Biol.* **2007**, *373*, 1113–1122.
77. Bloomfield, V.A. DNA condensation by multivalent cations. *Biopolymers* **1997**, *44*, 269–282.
78. Li, G.-Y.; Guan, R.-L.; Ji, L.-N.; Chao, H. DNA condensation induced by metal complexes. *Coord. Chem. Rev.* **2014**, *281*, 100–113.
79. Verwey, E.J.W.; Overbeek, J.T.G. *Theory of the Stability of Lyophobic Colloids*; Elsevier: Amsterdam, The Netherland, 1948.
80. Deserno, M.; Holm, C. Theory and simulations of rigid polyelectrolytes. *Mol. Phys.* **2002**, *100*, 2941–2956.
81. Manning, G.S.; Ray, J. Counterion condensation revisited. *J. Biomol. Struct. Dyn.* **1998**, *16*, 461–476.
82. Manning, G.S. Limiting laws and counterion condensation in polyelectrolyte solutions I. Colligative properties. *J. Chem. Phys.* **1969**, *51*, 924–933.
83. Borukhov, I. Charge renormalization of cylinders and spheres: Ion size effects. *J. Polym. Sci. Polym. Phys.* **2004**, *42*, 3598–3615.
84. Hsin, W.L.; Wang, T.Y.; Sheng, Y.J.; Tsao, H.K. Charge renormalization of charged spheres based on thermodynamic properties. *J. Chem. Phys.* **2004**, *121*, 5494–5504.
85. Trizac, E.; Bocquet, L.; Aubouy, M.; von Grunberg, H.H. Alexander's prescription for colloidal charge renormalization. *Langmuir* **2003**, *19*, 4027–4033.
86. Alexander, S.; Chaikin, P.M.; Grant, P.; Morales, G.J.; Pincus, P.; Hone, D. Charge renormalization, osmotic-pressure, and bulk modulus of colloidal crystals—Theory. *J. Chem. Phys.* **1984**, *80*, 5776–5781.

87. Podgornik, R.; Hansen, P.L.; Parsegian, V.A. Elastic moduli renormalization in self-interacting stretchable polyelectrolytes. *J. Chem. Phys.* **2000**, *113*, 9343–9350.
88. Savelyev, A. Do monovalent mobile ions affect DNA's flexibility at high salt content? *Phys. Chem. Chem. Phys.* **2012**, *14*, 2250–2254.
89. Raspaud, E.; da Conceicao, M.; Livolant, F. Do free DNA counterions control the osmotic pressure? *Phys. Rev. Lett.* **2000**, *84*, 2533–2536.
90. Qiu, X.; Andresen, K.; Kwok, L.W.; Lamb, J.S.; Park, H.Y.; Pollack, L. Inter-DNA attraction mediated by divalent counterions. *Phys. Rev. Lett.* **2007**, *99*, 038104:1–038104:4.
91. Qiu, X.; Kwok, L.W.; Park, H.Y.; Lamb, J.S.; Andresen, K.; Pollack, L. Measuring inter-DNA potentials in solution. *Phys. Rev. Lett.* **2006**, *96*, 138101:1–138101:4.
92. Qiu, X.Y.; Parsegian, V.A.; Rau, D.C. Divalent counterion-induced condensation of triple-strand DNA. *Proc. Natl. Acad. Sci. USA* **2010**, *107*, 21482–21486.
93. Li, L.; Pabit, S.A.; Lamb, J.S.; Park, H.Y.; Pollack, L. Closing the lid on DNA end-to-end stacking interactions. *Appl. Phys. Lett.* **2008**, *92*, 22390:1–22390:13.
94. Pelta, J.; Livolant, F.; Sikorav, J.L. DNA aggregation induced by polyamines and cobalthexamine. *J. Biol. Chem.* **1996**, *271*, 5656–5662.
95. Podgornik, R.; Strey, H.H.; Parsegian, V.A. Colloidal DNA. *Curr. Opin. Colloid Interface Sci.* **1998**, *3*, 534–539.
96. Saminathan, M.; Antony, T.; Shirahata, A.; Sigal, L.H.; Thomas, T.; Thomas, T.J. Ionic and structural specificity effects of natural and synthetic polyamines on the aggregation and resolubilization of single-, double-, and triple-stranded DNA. *Biochemistry* **1999**, *38*, 3821–3830.
97. Korolev, N.; Lyubartsev, A.P.; Rupprecht, A.; Nordenskiold, L. Competitive substitution of hexammine cobalt(III) for Na^+ and K^+ ions in oriented DNA fibers. *Biopolymers* **2001**, *58*, 268–278.
98. Dai, L.; Mu, Y.G.; Nordenskiold, L.; van der Maarel, J.R.C. Molecular dynamics simulation of multivalent-ion mediated attraction between DNA molecules. *Phys. Rev. Lett.* **2008**, *100*, 118301:1–118301:4.
99. Raspaud, E.; de la Cruz, M.O.; Sikorav, J.L.; Livolant, F. Precipitation of DNA by polyamines: A polyelectrolyte behavior. *Biophys. J.* **1998**, *74*, 381–393.
100. Raspaud, E.; Durand, D.; Livolant, F. Interhelical spacing in liquid crystalline spermine and spermidine-DNA precipitates. *Biophys. J.* **2005**, *88*, 392–403.
101. Yang, J.; Rau, D.C. Incomplete ion dissociation underlies the weakened attraction between DNA helices at high spermidine concentrations. *Biophys. J.* **2005**, *89*, 1932–1940.
102. Qiu, X.Y.; Giannini, J.; Howell, S.C.; Xia, Q.; Ke, F.Y.; Andresen, K. Ion competition in condensed DNA arrays in the attractive regime. *Biophys. J.* **2013**, *105*, 984–992.
103. Liu, A.J.; Ha, B.Y. When like-charged chains attract—Physical questions posed by DNA condensation. *Abstr. Pap. Am. Chem. Soc.* **1999**, *217*, U295–U295.
104. Ha, B.Y.; Liu, A.J. Effect of non-pairwise-additive interactions on bundles of rodlike polyelectrolytes. *Phys. Rev. Lett.* **1998**, *81*, 1011–1014.
105. Ha, B.Y.; Liu, A.J. Counterion-mediated attraction between two like-charged rods. *Phys. Rev. Lett.* **1997**, *79*, 1289–1292.

106. Angelini, T.E.; Liang, H.; Wriggers, W.; Wong, G.C.L. Like-charge attraction between polyelectrolytes induced by counterion charge density waves. *Proc. Natl. Acad. Sci. USA* **2003**, *100*, 8634–8637.
107. Nguyen, T.T.; Rouzina, I.; Shklovskii, B.I. Reentrant condensation of DNA induced by multivalent counterions. *J. Chem. Phys.* **2000**, *112*, 2562–2568.
108. Shklovskii, B.I. Wigner crystal model of counterion induced bundle formation of rodlike polyelectrolytes. *Phys. Rev. Lett.* **1999**, *82*, 3268–3271.
109. Rouzina, I.; Bloomfield, V.A. Competitive electrostatic binding of charged ligands to polyelectrolytes: Practical approach using the non-linear poisson-boltzmann equation. *Biophys. Chem.* **1997**, *64*, 139–155.
110. Rouzina, I.; Bloomfield, V.A. Influence of ligand spatial organization on competitive electrostatic binding to DNA. *J. Phys. Chem.* **1996**, *100*, 4305–4313.
111. Grosberg, A.Y.; Nguyen, T.T.; Shklovskii, B.I. Colloquium: The physics of charge inversion in chemical and biological systems. *Rev. Mod. Phys.* **2002**, *74*, 329–345.
112. Naji, A.; Netz, R.R. Attraction of like-charged macroions in the strong-coupling limit. *Eur. Phys. J. E* **2004**, *13*, 43–59.
113. Kanduc, M.; Naji, A.; Podgornik, R. Counterion-mediated weak and strong coupling electrostatic interaction between like-charged cylindrical dielectrics. *J. Chem. Phys.* **2010**, *132*, 224703:1–224703:19.
114. Chen, Y.G.; Weeks, J.D. Local molecular field theory for effective attractions between like charged objects in systems with strong coulomb interactions. *Proc. Natl. Acad. Sci. USA* **2006**, *103*, 7560–7565.
115. Rodgers, J.M.; Kaur, C.; Chen, Y.G.; Weeks, J.D. Attraction between like-charged walls: Short-ranged simulations using local molecular field theory. *Phys. Rev. Lett.* **2006**, *97*, 097801:1–097801:5.
116. Tan, Z.J.; Chen, S.J. Nucleic acid helix stability: Effects of salt concentration, cation valence and size, and chain length. *Biophys. J.* **2006**, *90*, 1175–1190.
117. Tan, Z.J.; Chen, S.J. Ion-mediated nucleic acid helix-helix interactions. *Biophys. J.* **2006**, *91*, 518–536.
118. Kornyshev, A.A.; Leikin, S. Electrostatic zipper motif for DNA aggregation. *Phys. Rev. Lett.* **1999**, *82*, 4138–4141.
119. Kornyshev, A.A.; Wynveen, A. The homology recognition well as an innate property of DNA structure. *Proc. Natl. Acad. Sci. USA* **2009**, *106*, 4683–4688.
120. Rau, D.C.; Parsegian, V.A. Direct measurement of the intermolecular forces between counterion-condensed DNA double helices—Evidence for long-range attractive hydration forces. *Biophys. J.* **1992**, *61*, 246–259.
121. Rau, D.C.; Lee, B.; Parsegian, V.A. Measurement of the repulsive force between poly-electrolyte molecules in ionic solution—Hydration forces between parallel DNA double helices. *Proc. Natl. Acad. Sci. USA* **1984**, *81*, 2621–2625.
122. Podgornik, R.; Strey, H.H.; Rau, D.C.; Parsegian, V.A. Watching molecules crowd—DNA double helices under osmotic-stress. *Biophys. Chem.* **1995**, *57*, 111–121.

123. Leikin, S.; Parsegian, V.A.; Rau, D.C.; Rand, R.P. Hydration forces. *Annu. Rev. Phys. Chem.* **1993**, *44*, 369–395.
124. Stanley, C.; Rau, D.C. Evidence for water structuring forces between surfaces. *Curr. Opin. Colloid Interface Sci.* **2011**, *16*, 551–556.
125. Li, W.F.; Nordenskiold, L.; Zhou, R.H.; Mu, Y.G. Conformation-dependent DNA attraction. *Nanoscale* **2014**, *6*, 7085–7092.
126. Li, L.; Pabit, S.A.; Meisburger, S.P.; Pollack, L. Double-stranded RNA resists condensation. *Phys. Rev. Lett.* **2011**, *106*, 108101:1–108101:9.
127. Tolokh, I.S.; Pabit, S.A.; Katz, A.M.; Chen, Y.; Drozdetski, A.; Baker, N.; Pollack, L.; Onufriev, A.V. Why double-stranded RNA resists condensation. *Nucleic Acids Res.* **2014**, *42*, 10823–10831.
128. Kornyshev, A.A.; Leikin, S. Helical structure determines different susceptibilities of dsDNA, dsRNA, and tsDNA to counterion-induced condensation. *Biophys. J.* **2013**, *104*, 2031–2041.
129. He, Z.J.; Chen, S.J. Quantifying coulombic and solvent polarization-mediated forces between DNA helices. *J. Phys. Chem. B* **2013**, *117*, 7221–7227.

3

DNA/Ag Nanoparticles as Antibacterial Agents against Gram-Negative Bacteria

Tomomi Takeshima [1,2,†,*], Yuya Tada [2], Norihito Sakaguchi [3], Fumio Watari [4] and Bunshi Fugetsu [1,5]

1 Graduate School of Environmental Science, Hokkaido University, Hokkaido 060-0810, Japan; E-Mail: bunshifugetsu@pari.u-tokyo.ac.jp
2 Nissei Bio Co. Ltd., Hokkaido 061-1374, Japan; E-Mail: tada@nisseibio.co.jp
3 Division of Materials Science and Engineering, Faculty of Engineering, Hokkaido University, Hokkaido 060-8628, Japan; E-Mail: sakaguchi@eng.hokudai.ac.jp
4 Graduate School of Dental Medicine, Hokkaido University, Hokkaido 060-8586, Japan; E-Mail: watari@den.hokudai.ac.jp
5 Policy Alternative Research Institute, The University of Tokyo, Tokyo 113-0032, Japan

† The present address of this author is: Department of Biochemistry, University of Texas Southwestern Medical Center at Dallas, Dallas, TX 75390, USA.

* Author to whom correspondence should be addressed; E-Mail: tomomi.takeshima@utsouthwestern.edu

Academic Editor: Thomas Nann

Abstract: Silver (Ag) nanoparticles were produced using DNA extracted from salmon milt as templates. Particles spherical in shape with an average diameter smaller than 10 nm were obtained. The nanoparticles consisted of Ag as the core with an outermost thin layer of DNA. The DNA/Ag hybrid nanoparticles were immobilized over the surface of cotton based fabrics and their antibacterial efficiency was evaluated using *E. coli* as the typical Gram-negative bacteria. The antibacterial experiments were performed according to the Antibacterial Standard of Japanese Association for the Functional Evaluation of Textiles. The fabrics modified with DNA/Ag nanoparticles showed a high enough inhibitory and killing efficiency against *E. coli* at a concentration of Ag $\geq$ 10 ppm.

Keywords: antibacterial; DNA/Ag nanoparticles; electrostatic interaction; Gram-negative bacteria; nanoparticle immobilization; salmon milt DNA

---

1. Introduction

Metal nanoparticles, due to their extraordinary physical, chemical, and biological properties together with their unique morphologies, have been the key materials epitomizing the field of nanotechnology [1–5]. Silver (Ag) nanoparticles typify the common metal based nanoparticles and have been already used in some commercially available products, such as refrigerators, mobile phones, clothes, plasters, toothbrushes, cosmetics, catheters, bandages, scalpels, and needles [6]. Ag nanoparticles in these particular products function mainly as antibacterial agents against bacteria, fungi and/or viruses [6–12]. Ag nanoparticles release free Ag(I) ions, the free Ag(I) ions are capable of interacting with the negatively charged bacterial cell wall, deactivating cellular enzymes, disrupting membrane permeability, and as a result, leading to cell lysis and cell death [13,14]. The particular antibacterial efficiency of Ag nanoparticles depends on both their size [15] and shape [16] and higher antibacterial efficiency was obtainable for particles with larger surface areas. In addition, Ag nanoparticles can inhibit biofilm formation for *E. coli*, *P. aeruginosa* and *S. proteamaculans* [17] and the possible toxicity to the environment for the Ag nanoparticles is considered to be much lower than that of the toxicity of other metals and/or organic based antibacterial agents [18–20].

Many methods have been developed for preparing the metal based nanoparticles. The so-called wet-chemical reduction method has been the cornerstone approach, due to its low cost, being easier to operate and its high suitability for a wide range of metal based nanoparticle preparations. Precise control of the rate of particle growth during the reduction of the precursors (metal ions being commonly used) in the presence of suitable templates is the key for metal based nanoparticle preparations via the wet-chemical reduction method. In addition, adding a desirable stabilizer to prevent the metal nanoparticles forming aggregates is also an important approach in preparing metal based nanoparticles as highly stable suspensions. Polysaccharides [21–23], proteins [24], and nucleic acids [25–30] have been used as the so-called biopolymer based stabilizers and metal nanoparticle suspensions of high stability are obtainable in an easier manner.

In a previous study, we reported a novel approach to large scale production of Ag nanoparticles using DNA extracted from salmon milt as both the template and stabilizer [31]. Ag nanoparticles with average diameters smaller than 10 nm and highly stable in water were obtained. The DNA/Ag nanoparticles were found to carry highly negative potentials, providing a simple yet desirable approach to immobilization of the DNA/Ag nanoparticles through electrostatic interactions for practical applications. In this study, the DNA/Ag nanoparticles were immobilized onto the surface of cotton based fabrics through electrostatic interactions and their potential as antibacterial substances was evaluated using *E. coli* as the typical Gram-negative bacteria. New insights into the mechanism of antibiotic effectiveness of the DNA/Ag/cotton-fabrics against the targeting bacteria were observed.

2. Results

2.1. Characteristics of Size Distribution and Crystal Structure of the DNA/Ag Nanoparticles

An aqueous suspension containing Ag nanoparticles at a higher concentration (5.3×10^{-2} mol/L·Ag) was obtained by reducing Ag(I) ions using tetrahydroborate ($NaBH_4$) as the reductant in the presence of salmon milt DNA (approximately 20,000 Da, *i.e.*, 60 nucleotides, single-stranded). Figure 1A,B shows typical images observed using field emission transmission electron microscope (FE-TEM), of particles having diameters ranging from 2 to 20 nm. The nano-beam electron diffraction pattern (Inset b1 of Figure 1B) of the particle having a size of about 20 nm (b in Figure 1B) showed a single-crystal spot pattern with other faint patterns suggesting that the larger sized particles are polycrystals. The lattice spacing was calculated as follows:

$$L\lambda = dR \tag{1}$$

where L is the camera length, λ is the wavelength of the TEM beam, d is the lattice spacing and R is the measured diffraction ring radius or spot distance. The lattice spacing corresponding to the nearest spot was 0.24 nm, close to the (111) plane of face-centered cubic (FCC) Ag. A broad ring pattern (Inset a1 of Figure 1A) with the same diffraction ring radius as the spot in Figure 1B is observed in the selected area electron diffraction pattern from the wider area of the distributed particles (Figure 1A), indicating that the particles consist of Ag crystals. The lattice fringes with a spacing of 0.23 nm are shown in the enlarged image of the Ag nanoparticle both with a size of around 20 nm (Figure 1C) and a size of around 2 nm (d in Figure 1C,D), in good agreement with the lattice spacing calculated from the diffraction patterns.

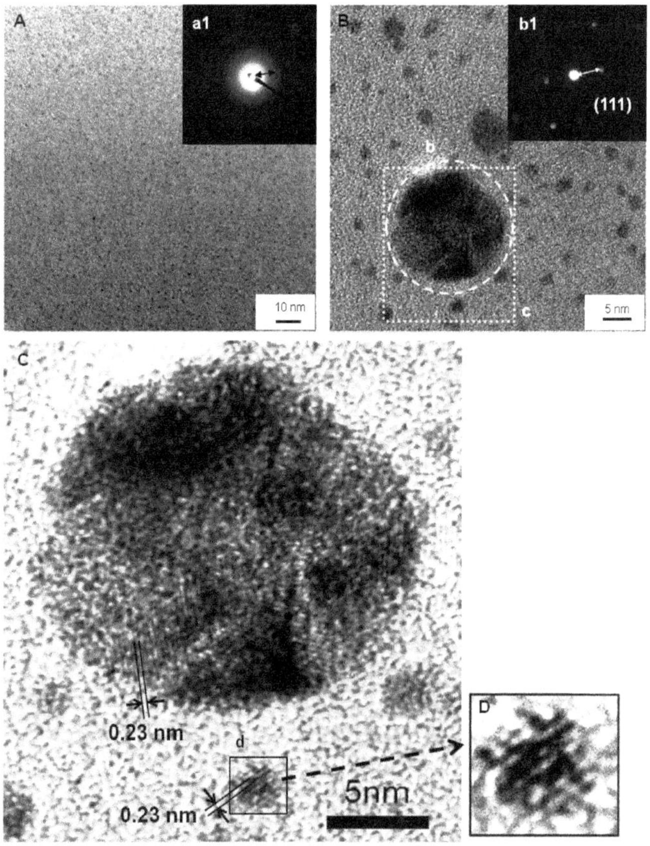

Figure 1. High-resolution field emission transmission electron microscope (FE-TEM) images of the DNA/Ag nanoparticles. The insets in (**A**) and (**B**) show the corresponding selected area electron diffraction pattern (a1) recorded from the whole area (**A**) and the nano-beam diffraction pattern (b1) from the area marked with the dashed line circle (b), respectively. (**C**) and (**D**) are the enlarged images from the areas indicated with the dotted square (c) in (**B**) and with the solid square (d) in (**C**), respectively.

X-ray powder diffraction (XRD) analysis was performed to identify the crystal phase of the DNA/Ag nanoparticles. Four peaks, each attributed to the (111), (200), (220) and (311) planes of FCC·Ag (Figure 2), respectively, were observed. The peak shape suggests a complex overlapped with a sharp and a broader peak, which indicating the existence of very fine Ag nanoparticles. Assuming that the peak attributed to the (111) consisted of sharp and broader peaks, the peak separation was done using the Lorentz function as peak shape, and the crystal size was estimated from the half-peak width of the broader peak with the equation as follows;

$$D = K\lambda/(\beta\cdot\cos\theta) \quad (2)$$

where D is the crystal size, K is a constant number, λ is the X-ray wavelength from CuKα, β is the half-peak width after correction of the equipment error and θ is the Bragg angle of the diffracted peak. The crystal size was estimated to be less than 10 nm, which was very consistent with the size as indicated from the FE-TEM images (Figure 1).

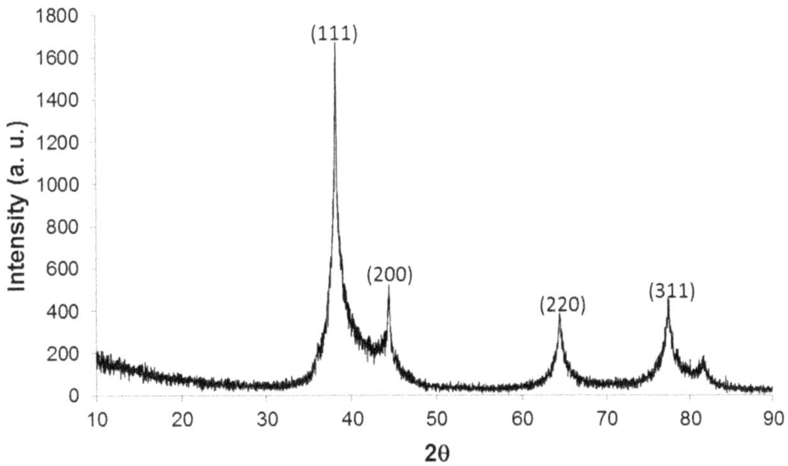

Figure 2. X-ray powder diffraction (XRD) patterns of the DNA/Ag nanoparticles.

2.2. Use of the DNA/Ag Nanoparticles as Antibacterial Materials

2.2.1. Evaluation of the Antibacterial Efficiency of the DNA/Ag Nanoparticles

Antibacterial efficiency of the DNA/Ag nanoparticles was evaluated by using both Gram-positive *S. aureus* and Gram-negative *E. coli*. Figure 3 shows the number of bacterial colonies on nutrient agar plates after incubation with and without the DNA/Ag nanoparticles. The growth inhibition effect was observed against both kinds of bacteria in a concentration-dependent manner. For the 10^6 and 10^7 colony-forming units (CFU)/mL of bacteria, the concentrations of DNA/Ag nanoparticles to completely prevent bacterial growth were 5 and 10 ppm, respectively, against *E. coli* (Figure 3A,B). The concentrations found were 50 and 100 ppm, respectively, to completely inhibit against *S. aureus* (Figure 3C,D).

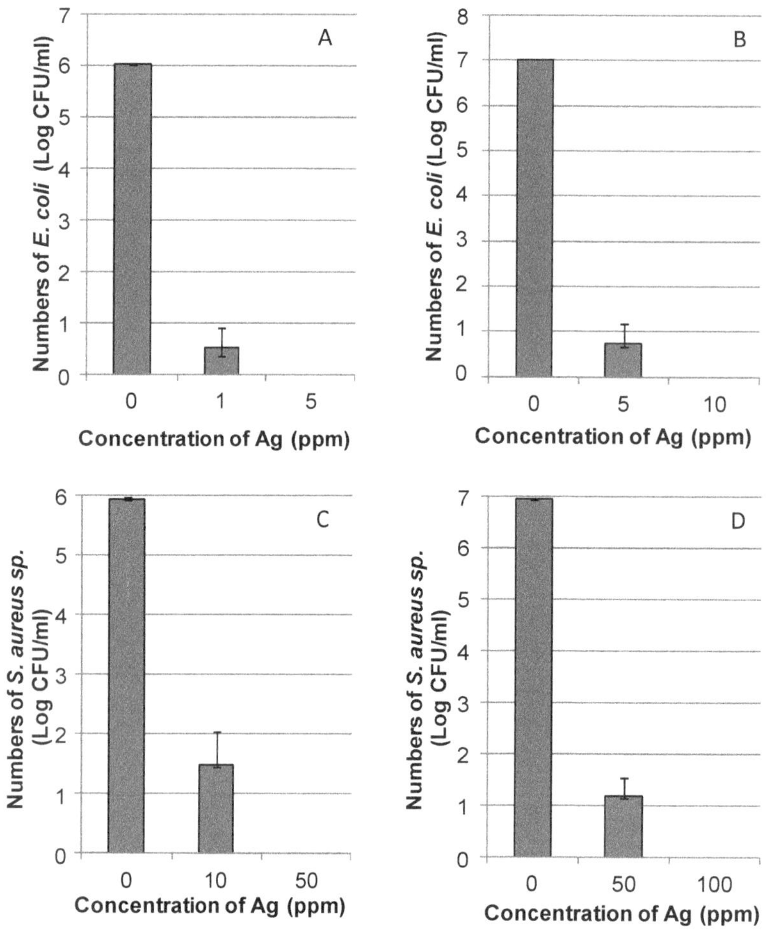

Figure 3. Number of colonies after incubation with a DNA/Ag nanoparticle aqueous suspension at different concentrations. (**A**) 1.04×10^6 colony-forming units (CFU)/mL; (**B**) 1.04×10^7 CFU/mL, of *E. coli*; (**C**) 8.65×10^5 CFU/mL; (**D**) 8.65×10^6 CFU/mL, of *S. aureus* were applied (without DNA/Ag nanoparticles, $n = 2$; with DNA/Ag nanoparticles, $n = 4$).

2.2.2. Preparation of the DNA/Ag Nanoparticle Immobilized Fabrics

The DNA/Ag nanoparticles were immobilized on the surfaces of cotton fabrics based on electrostatic interaction (Figure 4A). The cotton fabric samples were first modified with a cationic polymer, consisting of diallyl(3-chloro-2-hydroxypropyl)amine hydrochloride-diallyldimethylammonium chloride copolymer. The amount of immobilized Ag on the fabric was estimated by using inductively coupled plasma atomic emission spectrometry (ICP-AES). Cotton fabrics with DNA/Ag nanoparticles at concentrations such as Ag at 10, 30, 50, 3800 and 34,000 ppm were prepared (Figure 4B).

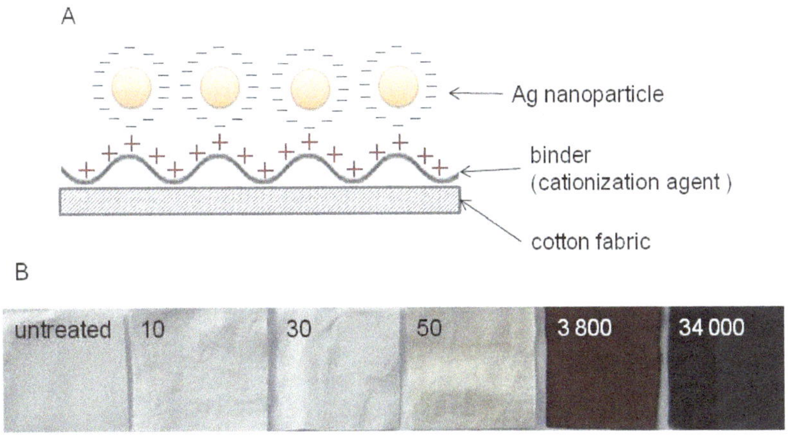

Figure 4. (**A**) Schematic description of the procedure for immobilizing DNA/Ag nanoparticles on to the cotton fabric; (**B**) Photographs of the DNA/Ag nanoparticle immobilized fabric samples. The amounts of immobilized Ag nanoparticles are shown as Ag (ppm).

2.2.3. Ag(I) Ions Release Ratio from the DNA/Ag Nanoparticle Immobilized Fabrics

Ag(I) ions release behavior was tested for the fabric with the immobilized DNA/Ag nanoparticles of 34,000 ppm as Ag. The fabric sample (approximately 200 mg) was immersed in a 50 mL glass vial containing 20 g water and then placed under shaking at 37 °C. The typical peak for Ag nanoparticles at around 410 nm was not observed in the UV-Vis absorption spectrum of the collected water samples, indicating a fact that the DNA/Ag nanoparticles had been firmly immobilized with the cotton fabrics. The water samples were analyzed using ICP-AES and Ag(I) ions were identified. Concentrations of Ag(I) ions in the water samples were found to be 1.9 ppm after 1 h and 6.4 ppm after 72 h of the release test, respectively. Ag(I) ion release ratio (Figure 5) was calculated as follows:

$$Q = N_a/N_o \times 100 \quad (3)$$

where Q is the release ratio, N_a is the Ag(I) ion content in the water sample at each time point of sample collecting, and N_o is the amount of the total immobilized Ag on the fabric (approximately 6.8 mg per sample). The amount of Ag(I) ions released from the fabric was found to be at around 1% and 2%, after 8 h and 72 h of the testing, respectively. The increase in the release ratio is predicted to reach a constant value according to Fick's law of diffusion.

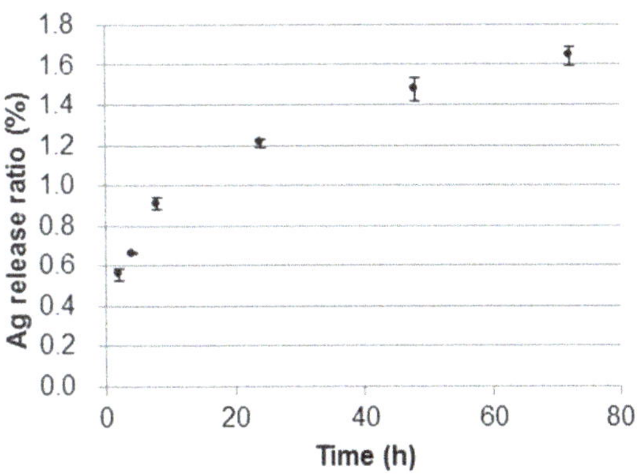

Figure 5. Ag(I) ions release profiles for the DNA/Ag nanoparticle immobilized cotton fabric (34,000 ppm as Ag). Error bars represent maximum and minimum value ($n = 2$).

2.2.4. Evaluation of the Antibacterial Efficiency of the DNA/Ag Nanoparticle Immobilized Fabrics

Antimicrobial efficiencies of the DNA/Ag nanoparticle immobilized fabrics were evaluated according to the Japanese Industrial Standard (JIS L 1902/2008) [32] using *E. coli*; the initial concentration of the bacteria seeding was chosen at 2.6×10^5 CFU/mL. Numbers of the bacteria at the beginning and at the time after incubation for 18 h are shown in Figure 6. For the controls (samples without DNA/Ag nanoparticles), the numbers of the bacteria increased to 5.40×10^6 after 18 h of incubation. In contrast to the controls, the number of bacteria inoculated on the DNA/Ag nanoparticle immobilized fabrics with 10 ppm Ag decreased to 1.82×10^2; for the fabrics with 30 ppm Ag, colonies were not observed after 18 h incubation.

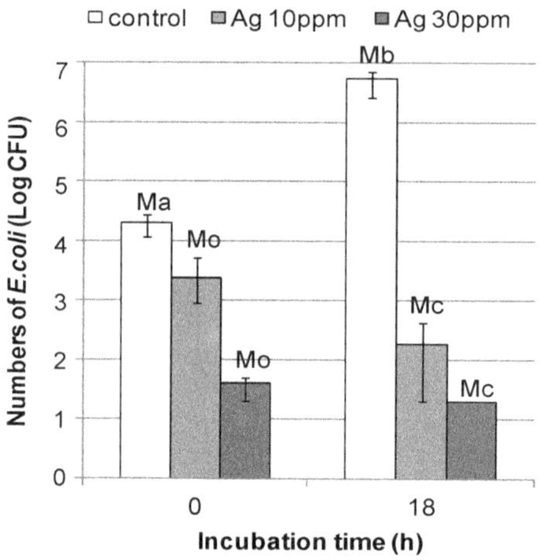

Figure 6. Numbers of colonies found for the untreated fabrics (controls) and for the DNA/Ag nanoparticles immobilized fabrics (10 and 30 ppm as Ag) at the starting point (0 h) and after 18 h incubation (18 h). If no colonies were observed, the number was regarded as less than 20 CFU/mL and was shown as 20 CFU/mL Error bars represent maximum and minimum value ($n = 3$).

Values of bacteria growth activity (BG), bacteriostatic value (BS) and bactericidal value (BC), were calculated; Table 1 summarizes the data. As denoted in the Antibacterial Standard of Japanese Association for the Functional Evaluation of Textiles (JAFET), when the BS value is found to be larger than 2.2, the test materials are classified as materials considerably capable of inhibiting bacteria; when the BC value is found to be greater than 0, the test materials are materials considerably capable of killing bacteria [33]. In this study, both the DNA/Ag nanoparticle immobilized fabrics (10 and 30 ppm·Ag) provided a BS value > 2.2 and a BC value > 0, indicating that their bacteria inhibitory efficiency and bacteria-killing effectiveness are high enough as antibacterial materials.

Table 1. Antibacterial evaluation values.

Testing Item	Test Results (Colony Numbers)			Reference [33]
	Controls	DNA/Ag Nanoparticle Immobilized Fabrics		
		10 ppm	30 ppm	
Ma	2.05×10^4	-	-	
Mb	5.40×10^6	-	-	
Mo	-	2.35×10^3	<40	
Mc	-	1.82×10^2	<20	
BG	2.4	-	-	
BS		>3.6	>2.7	>2.2 Inhibitory effect
BC		>2.3	>3.0	>0 Killing effect

Notes: Ma, Mb, Mo, Mc: corresponds to Figure 6; BG: the values of bacteria growth activity = log(Mb/Ma); BS: bacteriostatic value = log(Mb/Ma) − log(Mc/Mo); BC: bactericidal value = log(Ma/Mc); The mean value was used as the number of the bacteria (n = 3). These values were calculated according to JIS L 1902/2008 [32]. If no colonies appeared, the number was regarded as less than 20 CFU/mL.

3. Discussion

3.1. Formation Mechanism of DNA/Ag Nanoparticles

The zeta-potential of the DNA/Ag nanoparticles in water was found to be −73.8 mV, which was the value identical to the zeta-potential for the entire DNA. This fact suggests that the surfaces of the Ag nanoparticles were covered with DNA [31]. The molar ratio of nucleotide to Ag of the DNA/Ag nanoparticle was calculated to be 1:18 from the weight ratio of 14.3/85.2, measured using ICP-AES [31]. The number of Ag atoms combined to one molecule of the 60 nucleotides based DNA was calculated to be around 1080. The size of the Ag core, consisting of this number of Ag atoms covered with one molecule of the 60 nucleotides based DNA, was calculated to be 3.7 nm, which was in good agreement with the images in Figure 1 as well as the estimated crystal size calculated from the peak attributed to (111) in the XRD spectra.

Figure 7 shows a schematic image for describing a possible mechanism of the DNA/Ag nanoparticles formation: (i) A certain number of the free Ag(I) ions interacted with the nucleotide base of DNA to form a Ag(I)/DNA complex; (ii) The Ag(I)/DNA complex was densely packed with a certain number of the free Ag(I) ions being caged within this restricted range; (iii) Once $NaBH_4$, the reductant was introduced, these caged Ag(I) ions were reduced to Ag(0) atoms and this could be seen as the Ag core of the nanoparticles; (iv) DNA attached on to the core Ag(0) nanoparticles and formied the high natively charged DNA/Ag nano-sized hybrids.

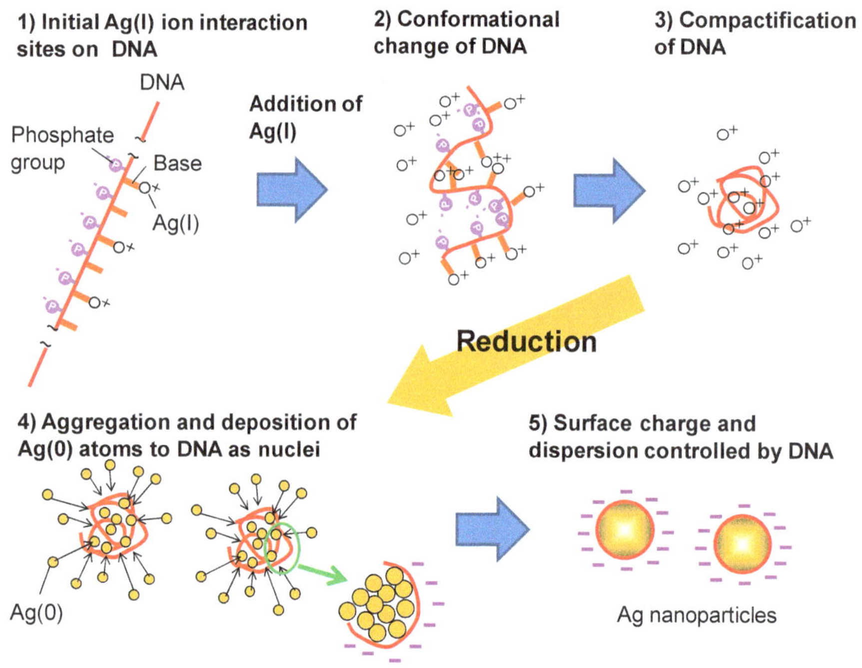

Figure 7. Proposed formation mechanism of the DNA/Ag nanoparticles.

3.2. Antibacterial Effect of the DNA/Ag Nanoparticles

Inhibitory concentrations of Ag nanoparticles reported in previous articles were in the range of 0.5–20 ppm against *E. coli* and 7.5–33.7 ppm against *S. aureus* for 10^3–10^5 CFU/mL [6], respectively. In this study, the inhibitory concentrations of the DNA/Ag nanoparticles were 5 ppm against *E. coli* and 50 ppm against *S. aureus* for the bacteria at 10^6 CFU/mL. In other words, the antibacterial efficiency of the DNA/Ag nanoparticles prepared in this study was found to be comparable or even better than that of other types of Ag nanoparticles reported in previous studies. In addition, the DNA/Ag nanoparticles showed higher antibacterial efficiency against *E. coli*, compared with *S. aureus*. Gram-positive bacteria are unique to the Gram-negative bacteria; the cell wall of Gram-positive bacteria is about a 30 nm thick peptidoglycan layer, while for the Gram-negative bacteria it is about 2–3 nm [8]. Feng *et al.* [13] reported that Ag ions are capable of stopping the DNA replication function and/or making proteins become inactivated; morphological changes of *S. aureus* after the treatment with Ag ions were found to be smaller than that for *E. coli*. Sondi *et al.* [4] reported that after the treatment with Ag nanoparticles, *E. coli* cells were damaged and a so-called formation of "pits" in the cell wall of the bacteria was observed. Ag nanoparticle accumulation on to the bacterial membrane leads to a significant increase in permeability and thereby is the cause of cell death [4], although so-called free-radical generation caused by the free Ag(I) ions is the widely accepted explanation for understanding the mechanism of growth-inhibition by Ag nanoparticles [5].

For the DNA/Ag nanoparticles, free Ag(I) ions released from the core Ag nanoparticle also play important roles in determining antibacterial efficiency. The DNA/Ag nanoparticles carry highly negative charges; this can lead to preventing microorganisms from contacting the Ag nanoparticle surface. In other words, the DNA/Ag nanoparticles function mainly as nano-sized reservoirs providing free Ag(I) ions in a continuous manner in to the media. The free Ag(I) ions, once reached to the bacteria function as antibacterial reagent in the manner similar as the mechanism proposed for the Ag(I) free ions provided by using other kinds of Ag nanoparticles as the providers.

4. Experimental Section

4.1. Materials

DNA was obtained from Nisseibio Co. Ltd. (Eniwa, Japan). The molecular weight of the DNA was estimated by HPLC, and the determination of double-stranded and/or single stranded of DNA was made by measuring the increase in absorbance at 260 nm upon heating the DNA solution at 95 °C. Silver nitrate ($AgNO_3$ special grade), potassium dihydrogen phosphate (KH_2PO_4, special grade) and polyoxyethylene sorbitan mono-oleate (Tween 80®) were purchased from Kishida Chemical Co. Ltd. (Osaka, Japan). Ammonia solution (10%) and sodium borohydride ($NaBH_4$ chemical grade) were purchased from Wako Pure Chemicals Industries Ltd. (Osaka, Japan). Potassium peroxydisulfate (for N and P analysis grade) and sodium carbonate (Na_2CO_3, special grade) were purchased from Kanto Chemical Co. Inc. (Tokyo, Japan). Sodium hydroxide (ACS grade) was purchased from Sigma-Aldrich (St. Louis, MO, USA). Ultra-pure water (supplied from Milli-Q water purification systems, Merck Millipore, Billerica, MA, USA, >18 MΩ) was used in all experiments.

S. aureus (AHU1142) was obtained from the Faculty of Agriculture, Hokkaido University (Sapporo, Japan). *E. coli* (NBRC3301) was purchased from the National Institute of Technology and Evaluation (Chiba, Japan). Peptone (Bacto Peotone) and beef extract (desiccated), which were purchased from BD (Franklin Lakes, NJ, USA), Agar (BA-10, high quality) from Ina Food Industry Co. Ltd., (Ina, Japan), were used to prepare the culture medium.

Cotton broad fabric, fabric mass of 122.5/m^2, was purchased from Shikisensya (Osaka, Japan). Cationic polymer solution (DANSHADE®) consists of diallyl(3-chloro-2-hydroxypropyl)amine hydrochloride-diallyldimethylammonium chloride copolymer which was used as a cationic agent to make the cotton fabric, carrying positive charges, was provided from Nittobo Medical (Tokyo, Japan).

4.2. Preparation and Characterization of the DNA/Ag Nanoparticles

The DNA/Ag nanoparticles were prepared as described in previous report [31]. An aqueous $AgNO_3$ (10^{-1} mol/L) was mixed with a DNA solution (10^{-2} mol/L nucleotide in 1% aqueous ammonia) at a volume ratio of 2:1, then a $NaBH_4$ solution (1 mol/L in 10% aqueous ammonia) was added at one tenth of the volume of the aqueous $AgNO_3$ to the mixture as a reductant at around 0 °C. The supernatant, which contained the DNA/Ag nanoparticles, was dialyzed using a dialysis membrane (molecular weight cut-off: 14,000, Viskase Companies Inc., Darien, IL, USA) in water for approximately 24 h. A detailed observation of the DNA/Ag nanoparticles was performed by high resolution observation using FE-TEM (JEM-2010F, JEOL, Akishima, Japan). The crystal structure and size of the DNA/Ag nanoparticles were analyzed by XRD measurement with an X-ray diffractometer (RINT 2000/PC, Rigaku Corporation, Akishima, Japan) with CuKα radiation operated at 40 kV and 40 mA.

4.3. Evaluation of the DNA/Ag Nanoparticles as Antibacterial Materials

4.3.1. Antibacterial Evaluation of the DNA/Ag Nanoparticles

Antibacterial efficiency of the DNA/Ag nanoparticles was examined against *S. aureus*, AHU1142, as Gram-positive bacteria, and *E. coli*, NBRC3301, as Gram-negative bacteria. The DNA/Ag nanoparticle aqueous dispersion was sterilized by filtration through a 0.45 μm syringe filter (Minisart, Sartorius Stedim Biotech, Goettingen, Germany), and then diluted (2–200 ppm as Ag) with phosphate buffer solution (0.25 M KH_2PO_4, pH 7.2). The bacteria were pre-cultured in nutrient broth (5% peptone, 3% beef extract) for 16 h at 37 °C. The culture solutions were diluted with phosphate buffer solution (0.25 M KH_2PO_4, pH 7.2), and mixed with the DNA/Ag nanoparticle dispersion at a volume ratio of 1:1. The mixture (0.1 mL) was put into bacterial petri dish (ϕ 90 mm) with 20 mL of nutrient agar medium (5% peptone, 3% beef extract, 15% agar), and incubated for 24–48 h at 37 °C.

4.3.2. Preparation of the DNA/Ag Nanoparticle Immobilized Fabrics

Cotton fabric was treated with the cationic agent diluted with water (1–10,000 ppm) for 20 min at 80 °C by using a hot plate stirrer, and then Na_2CO_3 was added to make the solution up to 1 wt.% Na_2CO_3. The mixture was kept for 30 min at 80 °C. The fabric was washed with deionized water and was then soaked in the DNA/Ag nanoparticles dispersion (5–5900 ppm as Ag) for 30 min. After careful washing with deionized water, the fabric was dried in air. The amount of the immobilized Ag was measured using ICP-AES (ICPE-9000, Shimadzu, Kyoto, Japan) after hydrolysis in the presence of aqueous potassium peroxydisulfate (4.0%) at 120 °C.

4.3.3. Ag(I) Ions Release from the DNA/Ag Nanoparticle Immobilized Fabrics

A small piece (approximately 0.2 g) of the DNA/Ag nanoparticles immobilized fabric (34,000 ppm as Ag) was immersed into water (20 mL) in 50 mL glass vial container with screw cap. The container was shaken at 37 °C in a swing shaker (around 110 rpm). The water (1 mL) was sampled at 2, 4, 8, 12, 24, and 48 h, and the Ag concentrations were measured with ICP-AES.

4.3.4. Antibacterial Evaluation of the DNA/Ag Nanoparticle Immobilized Fabrics

Antibacterial assay of the DNA/Ag nanoparticle immobilized fabrics (with 10 and 30 ppm as Ag) was performed according to JIS L 1902/2008—Adsorption test method [32] using *E. coli.* as follows: the bacteria were pre-cultured in nutrient broth for 16 h at 37 °C, and were then diluted with 20-fold diluted nutrient broth (pH 6.8) as bacteria suspension containing 1–3 × 10^5 CFU/mL. The bacteria suspension (0.2 mL) was inoculated to each specimen (0.4 g). Three untreated specimens and test specimens were washed immediately with 0.2% Tween 80-containing saline (20 mL), and the washed bacterial suspension (1 mL) was taken into a bacterial petri dish (ϕ 90 mm) with 20 mL of nutrient agar medium and then incubated at 37 °C for 24–48 h to measure the number of bacteria. The other three untreated specimens and the treated specimens were incubated at 37 °C for 18 h, and the number of bacteria was determined in the same way as at the beginning of the contacting experiment. The values of

BG, BS and BC of the DNA/Ag nanoparticle immobilized fabrics were then calculated according to JIS L 1902/2008 [32].

5. Conclusions

DNA modified Ag nanoparticles with average diameters smaller than 10 nm were successfully produced by using natural DNA extracted from salmon milt as the template. The resultant DNA/Ag nanoparticles carry highly negative charges; this provides a desirable approach to immobilization of Ag nanoparticles onto fabric surfaces via electrostatic interactions. Cotton based fabrics modified with the DNA/Ag nanoparticles showed a sufficiently high antibacterial efficiency against *E. coli.*

Acknowledgments

Analysis of the DNA/Ag nanoparticles was conducted using ICP-AES at the OPEN FACILITY, Sousei Hall, Hokkaido University; FE-TEM at the Center for Advanced Research of Energy and Materials, Faculty of Engineering, Hokkaido University. We sincerely appreciate Tetsu Yonezawa, Hokkaido University, for useful advice for the preparation of nanoparticles and their applications. We would like to thank for Norio Nishi, Masahito Sugi, Masao Nishihara and Yoshihisa Yamada, Nisseibio Co. Ltd., for helpful discussions.

This study was supported in part by Grant-in-Aid Regional R&D Proposal-Based Program from Northern Advancement Center for Science & Technology of Hokkaido Japan; this study was also supported in part by Grant-in-Aid "Development of high-performance-materials through cooperation with engineering technologies by using materials derived from agriculture, forestry and fishery products as the functioning element" program from the Ministry of Agriculture, Forestry and Fishery of Japan.

Author Contributions

The manuscript was written by contributions of all authors. Tomomi Takeshima designed the project, conducted the preparation of Ag nanoparticles, the characterization, the antibacterial test, immobilization to fabric and its antibacterial test, analyzed the results and was responsible for manuscript writing. Yuya Tada developed the methods for the antibacterial test and conducted the experiments. Norihito Sakaguchi conducted FE-TEM observation. Fumio Watari and Bunshi Fugetsu analyzed the results and contributed to the manuscript writing. All authors have given approval to the final version of the manuscript.

Conflicts of Interest

Tomomi Takeshima and Yuya Tada were employees of Nissei Bio Co., Ltd., provided DNA in this study, giving a potential conflict of interest.

References

1. Mulvaney, P. Surface plasmon spectroscopy of nanosized metal particles. *Langmuir* **1996**, *12*, 788–800.

2. Toshima, N.; Yonezawa, T. Bimetallic nanoparticles—Novel materials for chemical and physical applications. *New J. Chem.* **1998**, *22*, 1179–1201.
3. Burda, C.; Chen, X.B.; Narayanan, R.; El-Sayed, M.A. Chemistry and properties of nanocrystals of different shapes. *Chem. Rev.* **2005**, *105*, 1025–1102.
4. Sondi, I.; Salopek-Sondi, B. Silver nanoparticles as antimicrobial agent: A case study on *E. coli* as a model for Gram-negative bacteria. *J. Colloid Interface Sci.* **2004**, *275*, 177–182.
5. Kim, J.S.; Kuk, E.; Yu, K.N.; Kim, J.H.; Park, S.J.; Lee, H.J.; Kim, S.H.; Park, Y.K.; Park, Y.H.; Hwang, C.Y.; *et al*. Antimicrobial effects of silver nanoparticles. *Nanomed. Nanotechnol. Biol. Med.* **2007**, *3*, 95–101.
6. Chernousova, S.; Epple, M. Silver as antibacterial agent: Ion, nanoparticle, and metal. *Angew. Chem. Int. Ed.* **2012**, *51*, 2–20.
7. Rai, M.; Yadav, A.; Gade, A. Silver nanoparticles as a new generation of antimicrobials. *Biotechnol. Adv*. **2009**, *27*, 76–83.
8. Rai, M.K.; Deshmukh, S.D.; Ingle, A.P.; Gade, A.K. Silver nanoparticles: The powerful nanoweapon against multidrug-resistant bacteria. *J. Appl. Microbiol.* **2012**, *112*, 841–852.
9. Galdiero, S.; Falanga, A.; Vitiello, M.; Cantisani, M.; Marra, V.; Galdiero, M. Silver nanoparticles as potential antiviral agents. *Molecules* **2011**, *16*, 8894–8918.
10. Zhang, Y.; Peng, H.; Huang, W.; Zhou, Y.; Yan, D. Facile preparation and characterization of highly antimicrobial colloid Ag or Au nanoparticles. *J. Colloid Interface Sci.* **2008**, *325*, 371–376.
11. Lok, C.N.; Ho, C.M.; Chen, R.; He, Q.Y.; Yu, W.Y.; Sun, H.Z.; Tam, P.K. H.; Chiu, J.F.; Che, C.M. Proteomic analysis of the mode of antibacterial action of silver nanoparticles. *J. Proteome. Res*. **2006**, *5*, 916–924.
12. Silver, S. Bacterial silver resistance: Molecular biology and uses and misuses of silver compounds. *FEMS Microbiol. Rev*. **2003**, *27*, 341–353.
13. Feng, Q.L.; Wu, J.; Chen, G.Q.; Cui, F.Z.; Kim, T.N.; Kim, J.O. A mechanistic study of the antibacterial effect of silver ions on *Escherichia coli* and *Staphylococcus aureus*. *J. Biomed. Mater. Res.* **2000**, *52*, 662–668.
14. Choi, O.; Deng, K.K.; Kim, N.J.; Ross, L.; Surampalli, R.Y.; Hu, Z.Q. The inhibitory effects of silver nanoparticles, silver ions, and silver chloride colloids on microbial growth. *Water Res.* **2008**, *42*, 3066–3074.
15. Martinez-Castanon, G.A.; Nino-Martinez, N.; Martinez-Gutierrez, F.; Martinez-Mendoza, J.R.; Ruiz, F. Synthesis and antibacterial activity of silver nanoparticles with different sizes. *J. Nanopart. Res.* **2008**, *10*, 1343–1348.
16. Pal, S.; Tak, Y.K.; Song, J.M. Does the antibacterial activity of silver nanoparticles depend on the shape of the nanoparticle? A study of the Gram-negative bacterium *Escherichia coli*. *Appl. Environ. Microbiol.* **2007**, *73*, 1712–1720.
17. Radzig, M.A.; Nadtochenko, V.A.; Koksharova, O.A.; Kiwi, J.; Lipasova, V.A.; Khmel, I.A. Antibacterial effects of silver nanoparticles on Gram-negative bacteria: Influence on the growth and biofilms formation, mechanisms of action. *Colloids Surf. B* **2013**, *102*, 300–306.
18. Inoue, Y.; Hoshino, M.; Takahashi, H.; Noguchi, T.; Murata, T.; Kanzaki, Y.; Hamashima, H.; Sasatsu, M. Bactericidal activity of Ag-zeolite mediated by reactive oxygen species under aerated conditions. *J. Inorg. Biochem.* **2002**, *92*, 37–42.

19. Duran, N.; Marcato, P.D.; De Souza, G.I.H.; Alves, O.L.; Esposito, E. Antibacterial effect of silver nanoparticles produced by fungal process on textile fabrics and their effluent treatment. *J. Biomed. Nanotechnol.* **2007**, *3*, 203–208.
20. Tankhiwale, R.; Bajpai, S.K. Graft copolymerization onto cellulose-based filter paper and its further development as silver nanoparticles loaded antibacterial food-packaging material. *Colloids Surf. B* **2009**, *69*, 164–168.
21. Sharma, V.K.; Yngard, R.A.; Lin, Y. Silver nanoparticles: Green synthesis and their antimicrobial activities. *Adv. Colloid Interface Sci.* **2009**, *145*, 83–96.
22. Huang, H.Z.; Yuan, Q.; Yang, X.R. Preparation and characterization of metal-chitosan nanocomposites. *Colloids Surf. B* **2004**, *39*, 31–37.
23. Liu, Y.S.; Chen, S.M.; Zhong, L.; Wu, G.Z. Preparation of high-stable silver nanoparticle dispersion by using sodium alginate as a stabilizer under gamma radiation. *Radiat. Phys. Chem.* **2009**, *78*, 251–255.
24. Kasyutich, O.; Ilari, A.; Fiorillo, A.; Tatchev, D.; Hoell, A.; Ceci, P. Silver ion incorporation and nanoparticle formation inside the cavity of *Pyrococcus furiosus* ferritin: Structural and size-distribution analyses. *J. Am. Chem. Soc.* **2010**, *132*, 3621–3627.
25. Zheng, L.; Zhang, R.C.; Ni, Y.X.; Du, Q.A.; Wang, X.; Zhang, J.L.; Li, W. Catalytic performance of Ag nanoparticles templated by polymorphic DNA. *Catal. Lett.* **2010**, *139*, 145–150.
26. Petty, J.T.; Zheng, J.; Hud, N.V.; Dickson, R.M. DNA-templated Ag nanocluster formation. *J. Am. Chem. Soc.* **2004**, *126*, 5207–5212.
27. Molotsky, T.; Tamarin, T.; Ben Moshe, A.; Markovich, G.; Kotlyar, A.B. Synthesis of chiral silver clusters on a DNA template. *J. Phys. Chem. C* **2010**, *114*, 15951–15954.
28. Lan, G.Y.; Chen, W.Y.; Chang, H.T. Control of synthesis and optical properties of DNA templated silver nanoclusters by varying DNA length and sequence. *RSC Adv.* **2011**, *1*, 802–807.
29. Zhang, X.L.; Yu, M.; Liu, J.H.; Li, S.M. Fabrication and characterization of Ag nanoparticles based on plasmid DNA as templates. *Mater. Lett.* **2011**, *65*, 719–721.
30. Zinchenko, A.A.; Baigl, D.M.; Chen, N.; Pyshkina, O.; Endo, K.; Sergeyev, V.G.; Yoshikawa, K. Conformational behavior of giant DNA through binding with Ag^+ and metallization. *Biomacromolecules* **2008**, *9*, 1981–1987.
31. Takeshima, T.; Sun, L.; Wang, Y.; Yamada, Y.; Nishi, N.; Yonezawa, T.; Fugetsu, B. Salmon milt DNA as a template for the mass production of Ag nanoparticles. *Polym. J.* **2014**, *46*, 36–41.
32. JIS L 1902:2008. *Testing for Antimicrobial Activity and Efficacy on Textile Products*; Japanese Industrial Standards; Japanese Industrial Standards Committee: Tokyo, Japan, 2008.
33. JEC301. *The Certification Standards of Antibacterial Finished Textile Products*; Japan Textile Evaluation Technology Council: Osaka, Japan, 2012.

4

The Nucleotide Capture Region of Alpha Hemolysin: Insights into Nanopore Design for DNA Sequencing from Molecular Dynamics Simulations

Richard M. A. Manara, Susana Tomasio and Syma Khalid *

School of Chemistry, University of Southampton, Highfield Campus, Southampton SO17 1BJ, UK; E-Mails: rm16g09@soton.ac.uk (R.M.A.M.); susana@cresset-group.com (S.T.)

* Author to whom correspondence should be addressed; E-Mail: s.khalid@soton.ac.uk

Academic Editor: Stephen Ralph

Abstract: Nanopore technology for DNA sequencing is constantly being refined and improved. In strand sequencing a single strand of DNA is fed through a nanopore and subsequent fluctuations in the current are measured. A major hurdle is that the DNA is translocated through the pore at a rate that is too fast for the current measurement systems. An alternative approach is "exonuclease sequencing", in which an exonuclease is attached to the nanopore that is able to process the strand, cleaving off one base at a time. The bases then flow through the nanopore and the current is measured. This method has the advantage of potentially solving the translocation rate problem, as the speed is controlled by the exonuclease. Here we consider the practical details of exonuclease attachment to the protein alpha hemolysin. We employ molecular dynamics simulations to determine the ideal (a) distance from alpha-hemolysin, and (b) the orientation of the monophosphate nucleotides upon release from the exonuclease such that they will enter the protein. Our results indicate an almost linear decrease in the probability of entry into the protein with increasing distance of nucleotide release. The nucleotide orientation is less significant for entry into the protein.

Keywords: alpha-hemolysin; exonuclease sequencing; molecular dynamics; nanopore sequencing

---

1. Introduction

DNA sequencing using nanopores is a method that allows for direct analysis of genomic DNA [1–8]. Benefits of nanopore-based sequencing include detection of epigenetic markers [9,10], which is not possible using chemical methods as well as reducing costs compared to more traditional methods. The early promise has been realized by Oxford Nanopore Technologies in 2011–2012, however, the initial devices continue to be upgraded and improved to push the boundaries of what this technology can achieve.

The pore forming toxin, alpha-hemolysin (αHL) from *S. aureus* [11] is perhaps the best studied example of a proteinaceous nanopore for DNA sequencing [12]. This protein has two domains, the vestibule or cap domain and the transmembrane pore domain (Figure 1) [11]. While αHL is a robust protein that is resistant to changes in pH and temperature over practical ranges, issues related to the high voltage thresholds required for DNA translocation render wild type αHL unsuitable for incorporation into a sequencing device. However, engineered mutants have shown considerable success for DNA sequencing, although they too are continuously being improved for more efficient DNA sequencing. For example, methods to improve the accuracy of the device by decreasing the rate of translocation of ssDNA [2,13] through the protein pores are being investigated. It has been proposed that incorporation of an exonuclease enzyme, which would split the ssDNA into individual mononucleotides before they enter the nanopore, may be an efficient way to reduce the translocation rate [10,14]. Selection of an exonuclease enzyme for incorporation into an alpha-hemolysin based sequencing device is dependent upon practical considerations such as temperature and pH dependence, size and conformational flexibility of the enzyme. Furthermore it is imperative that the enzyme is positioned relative to the αHL such that there is a practical level of confidence mononucleotides exiting the enzyme will enter the αHL and not simply diffuse away in solution. This information is vital for the practical design and construction of the chimera protein.

Molecular dynamics simulations provide a route to study the translocation of biological molecules through nanoscale pores at levels of detail that are difficult to achieve with experimental methods alone [15–19]. Here we employ atomistic molecular dynamics simulations to characterize the "nucleotide capture area" of the wildtype alpha-hemolysin protein. Capture of the cytosine monophosphate (CMP), was investigated by embedding alpha-hemolysin in a 1,2-dimyristoyl-sn-glycero-3-phosphocholine (DMPC) bilayer and solvating in 1 M NaCl, resulting in a simulation system composed of 273,000 atoms. Only one of the four mononucleotides was simulated due to the similarity in their masses with respect to the protein. The distance of CMP from the protein was varied. Specifically, CMP was positioned above the protein on the vestibule side such that it was located centrally, above the entrance to the protein (where central is defined here as the average co-ordinates of the C-alpha atoms of K8 residues which are in the vestibule domain of the protein) as shown in Figure 1. Simulations were performed with the nucleotide initially positioned at 10, 15, 20, 30 and 40 angstroms above the ring of K8 residues, along the principal axis of the protein. Additional systems were set up in which, at a distance of 30 angstroms from the K8 residues, the nucleotides were displaced in the plane parallel to the membrane, by 5, 10, 15 and 20 angstroms. Two sets of simulations were performed for each location of CMP, in which the CMP was orientated with either the phosphate moiety furthest or closest to the vestibule entrance, which we define as phosphate "up" or "down" orientations respectively, shown in the inset of Figure 1. For each location, 20 independent simulations of 5 ns duration were performed at 310 K, with an applied electric field of magnitude 0.1 V·nm^{-1} equivalent to approximately 350 mV across the membrane. Simulations were performed using the GROMACS [20] software package, version 4.5.5 [20,21] with the GROMOS53a6 [22] forcefield and the SPC model of water [23]. Further details of the methodology are provided in the Experimental Section.

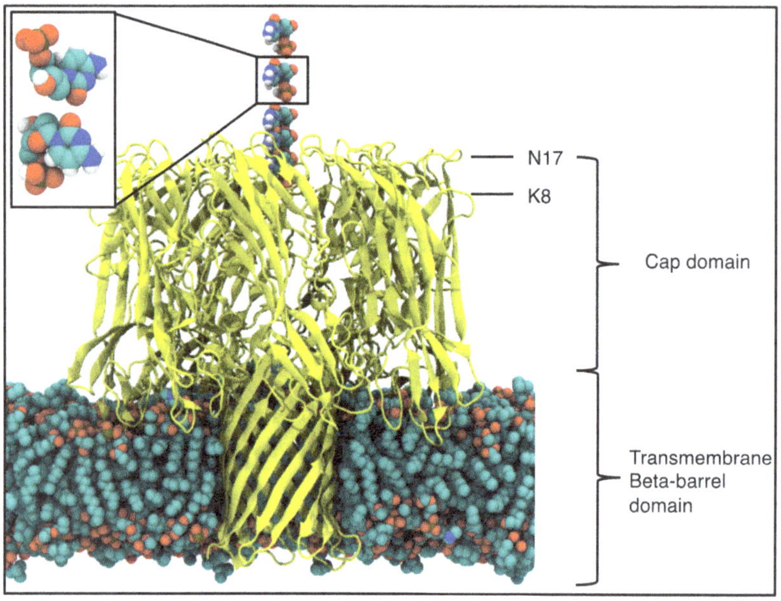

Figure 1. The alpha-hemolysin (αHL) protein in a 1,2-dimyristoyl-sn-glycero-3-phosphocholine (DMPC) bilayer, with the mononucleotide positioned above the vestibule entrance, where each mononucleotide represents an individual simulation. The protein is represented by yellow ribbons and for other atoms: Carbon is shown in cyan, oxygen in red, nitrogen in blue, phosphorus in brown and hydrogen in white. The waters and ions are excluded for clarity. (**Inset**) The phosphate orientations used, termed the "up" and "down" orientations, shown top and bottom respectively.

2. Results and Discussion

For the purposes of clarity when reporting our results, we use two terms for the behaviour of the nucleotide with respect to the protein: "capture" and "possible capture". These are defined as follows: When the entire CMP is below the ring of C-alpha atoms of N17, it is regarded as captured as over all simulations we observed no examples of exit from the vestibule entrance once interacting with the protein below this region. In contrast, if the CMP is entirely above this ring, it is deemed not captured. The final alternative is when simultaneously parts of the CMP are above and below the N17 ring, which we refer to as possible capture, as it was observed that CMP in this region, which we describe as the "edge" of the vestibule entrance, is capable of either entering the vestibule or diffusing into solution.

As the distance from which the mononucleotide is released from the protein is increased, the probability of entry into the vestibule decreases almost linearly, this reduction in the likelihood of entry occurs at approximately the same rate regardless of phosphate orientation (Figure 2). The probability of capture is higher with the phosphate in the down orientation, by an average of 10%. The phosphate orientation has negligible impact on the probability of "possible capture".

As the CMP is displaced away from the centre of the vestibule entrance at a distance of 30 angstroms, in general, the probability of entry decreases, however we note that there is an increased probability of entry with translation of 0.5 nm (Figure 2). We theorize that this is due to the CMP now being released above the edge of the vestibule entrance instead of above the centre; therefore, less lateral diffusion is required to interact with the protein. Here we note a higher correlation between the phosphate orientation and entry, with the phosphate down orientation demonstrating an average of 15% higher entry probability. The minimum probability of entry is generally higher by 10% with the CMP in the down orientation, and the rate of decrease in the capture probability as the translation away from the centre of the vestibule entrance increases, is lower than observed for the up orientation.

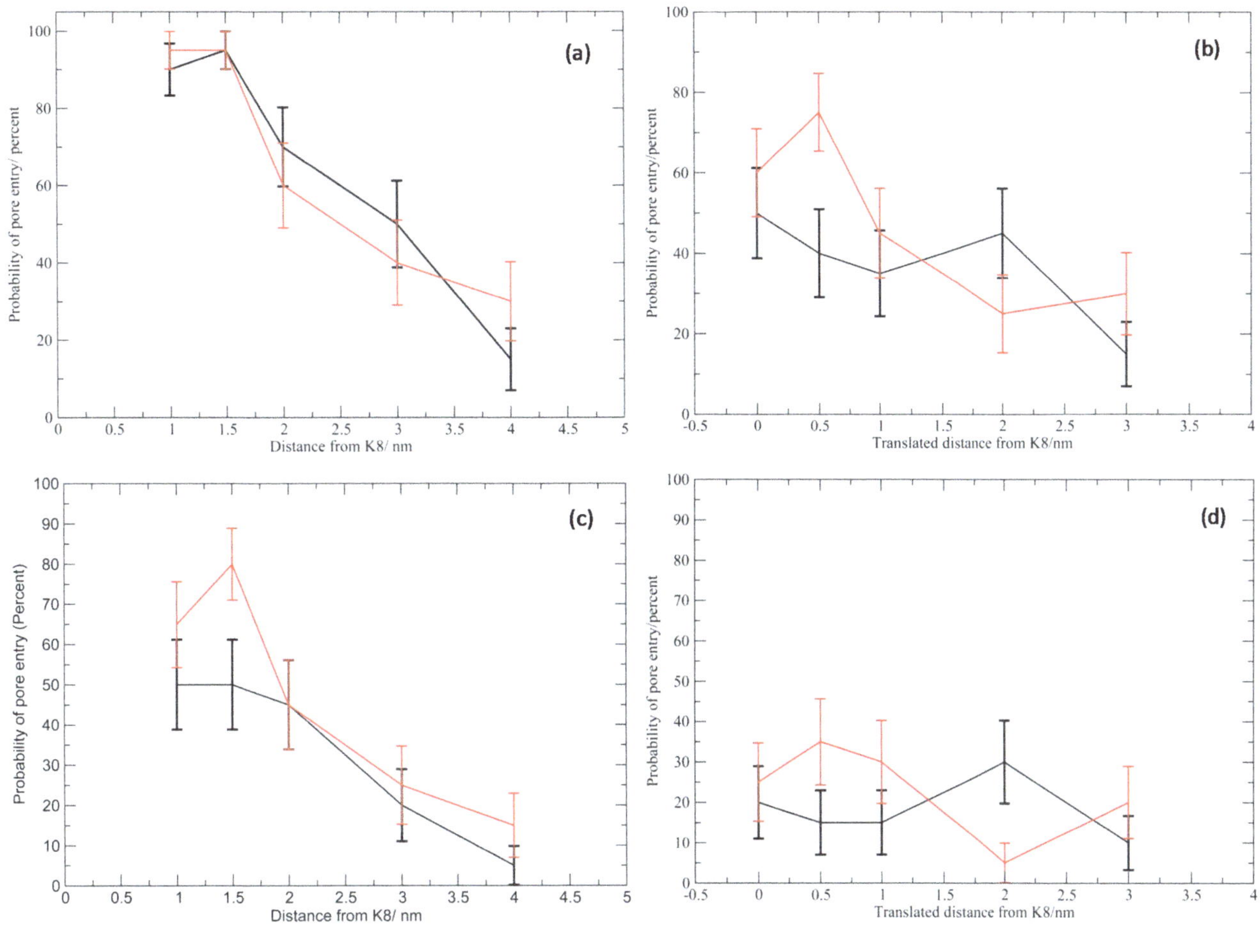

Figure 2. Plots showing thc relaionship between release loction and probability of capture. The highest possible probability of entry for (**a**) the distance study and (**b**) the displacement study, e.g., these are the simulations where either capture or possible capture is observed. The lowest possible probability of entry for (**c**) the distance study and (**d**) the displacement study, e.g., these are the simulations where capture is observed. The black lines are for the phosphate up orientation and the red lines are the phosphate down orientation. Error bars were calculated by using the standard error, treating the data as a binomial distribution.

We also note that unlike OccD1 [17] or OprP [24] the orientation of the molecule, beyond release, does not have a noticeable impact on entry into the vestibule entrance or the translocation process, as the protein freely interacts with the various moieties on the substrate, e.g., the nucleobase, the hydroxyl group on the sugar and the phosphate group.

An in-house script was used with the graphics software Visual Molecular Dynamics (VMD) [25] to calculate the frequency and duration of Van der Waals contacts between protein residues and CMP. For simulations in which entry to the vestibule was observed, the CMP had extended interactions (an average of more than 5% over all simulations) with: D2, S3, K8, T9, D13 and N293 residues, which are located predominately in the vestibule. Whilst for simulations in which possible entry was observed, interactions were noted with residues A1, N6, K8, T9, G10, T11, D13, G15, S16, D17, T18, T19, V20, D45, K46 and N47, these residues predominately lie on the edge of the entrance to the protein. For those simulations where entry to the vestibule was not observed we noticed the majority simply didn't interact with the protein at all, or when they did, it was with residues K8, I16, N17, T18, K46 and D47. This data and the location of these residues is summarized in Figure 3. Based on these observations we propose that the D45, K46 and N47 residues reduce the probability of entry to the vestibule via alternative favourable interactions with CMP outside the vestibule, on the surface of the cap region. The protein–nucleobase interactions that dominate here are largely hydrogen bonding of the amino acid side chains to the base and the sugar hydroxyl groups. The phosphate moiety is typically excluded from these interactions with the only exception being transient interactions with the side chain of K46 (for durations of less than 100 ps). An example interaction between CMP and the 3 residues is shown inset in Figure 3.

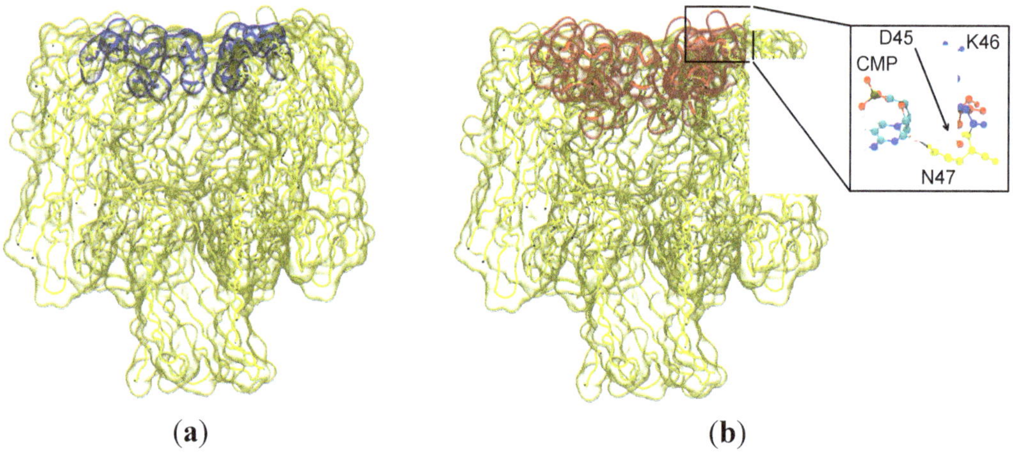

Figure 3. The residues with the highest propensity to interact with the nucleotide for: failed capture (**a**) and possible capture (**b**). The blue highlighted residues corresponding to failed capture are on the edge of the vestibule entrance and on the surface of the vestibule. The red highlighted residues associated with possible capture are predominantly on the edge of the entrance to the vestibule. (**Inset**) The most frequently observed binding mode between cytosine monophosphate (CMP) and the amino acid triplet, D45, K46 and N47. Hydrogen bonds (dashed lines) are observed between the hydroxyl group of the sugar and the side-chain of D45, as well as the nucleobase and the side-chain of N47. These hydrogen bonds are stable and are present for extended periods of time (greater than 1 ns). Residues are coloured for clarity; D45 in red K46 in blue, and N47 in yellow.

Over the 360 simulations performed, translocation through the entire alpha-hemolysin protein was observed in only two simulations, as the remainder of captured CMP substrates found various binding sites in the vestibule or beta barrel. Figure 4 provides a schematic summary of the probability of entry into the vestibule as a function of distance of release.

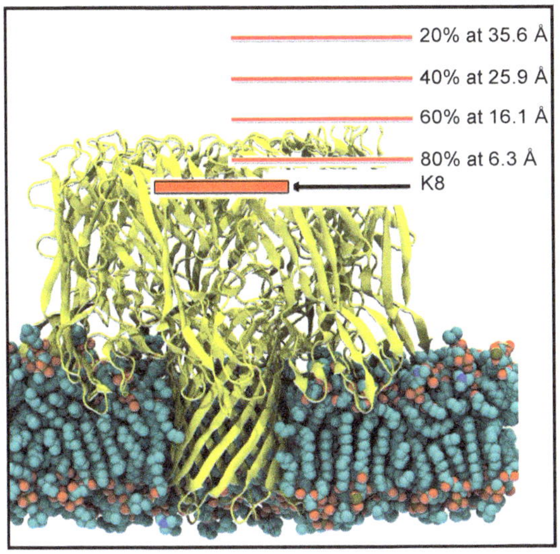

(**a**)

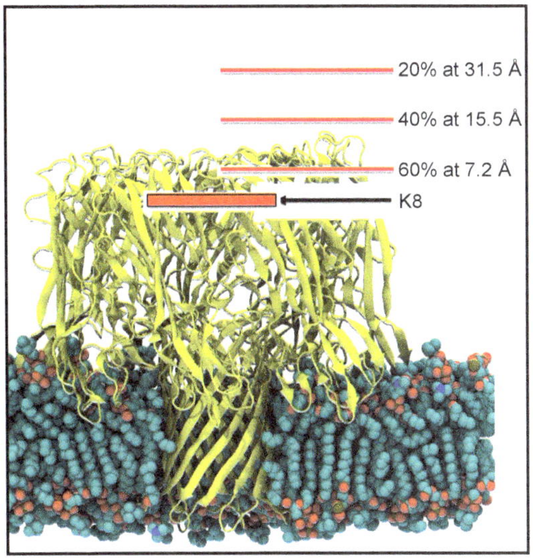

(**b**)

Figure 4. The probabilities of entry for given release points relative to the ring of K8 residues for (**a**) the phosphate down orientation and (**b**) the phosphate up orientation. Simulations performed with the phosphate in the down orientation are closer than the corresponding phosphate up simulations for the same capture probability.

Sum Effects to CMP Entry into the Vestibule

Root mean squares linear regression was used to calculate the equations of the lines for the relationship between distance away from the vestibule (variable labelled "height" in equations) and probability of entry into the vestibule, shown below in Equation (1), phosphate up and Equation (2), phosphate down. It was observed that with the nucleotide at the position of K8, or zero distance, the probability of capture is 72% to 93%, demonstrating that nucleotides can diffuse rapidly under the conditions used, failing to remain captured even when released below residue N17.

$$\text{Probability of possible capture} = 72\% - (16.5\%\ \text{nm height}^{-1}) \tag{1}$$

$$Probability\ of\ possible\ capture = 93\% - (20.5\%\ \text{nm height}^{-1}) \tag{2}$$

The data from the simulations in which the CMP had been translated in the *xy* plane, had the effect of distance removed by setting the probability at zero translation equal to the predicted probability from the previous equations. From this data an equation was calculated using the previous method, which provides the total effect of nucleotide release location on the probability of entry into the vestibule, see Equations (3), phosphate up, and (4), phosphate down, below.

$$\begin{aligned}&\text{Probability of possible capture}\\&\quad = 72\% - (16.5\%\ \text{nm}^{-1}\ \text{height}) - (2.3\%\ \text{nm translated}^{-1})\end{aligned} \tag{3}$$

$$\text{Probability of possible capture} = 93\% - (20.5\%\ \text{nm}^{-1}\ \text{height}) - (6.2\%\ \text{nm translated}^{-1}) \quad (4)$$

3. Experimental Section

Energy minimization was carried out by the method of steepest descents for either 2000 steps or a force of 1000 kJ·mol^{-1}. Equilibration was using the canonical (NVT) ensemble for 1 ns, and NPT for 1 ns, using the Berendsen thermostat and barostat [26] prior to electric field equilibration for 1 ns and a field strength of 0.1 V·nm^{-1}. The electric field equilibration and production molecular dynamics (MD) runs were run using the V-rescale thermostat [27] and the Parrinello-Rahman barostat [28]. All simulations were run at 310 K and 1 Bar.

The trajectories were visually analyzed using VMD to determine if the mononucleotide had drifted into the bulk water. A script was then used to compare of co-ordinates of the CMP to the c-alphas of N17, with successful capture defined as all atoms below this ring, possible capture as some above, some below and failed capture as all atoms above. The visualization step was required to prevent simulations where the nucleotide was below N17 but outside the pore being counted.

4. Conclusions

In conclusion, our simulations predict that optimisation of alpha-hemolysin for nanopore sequencing, which incorporates an exonuclease enzyme for cleaving nucleotides from a strand of DNA, must consider the protein-exonuclease distance of nucleotide release. We show that for successful capture of the nucleotide by the protein, the point of nucleotide release above the protein is more important than the lateral displacement of the nucleotide with respect to the dimensions of the entrance to the protein. In other words it is more important to release the nucleotide closer to the mouth of the vestibule, than it is to ensure that it is released directly above the centre of the mouth. Furthermore, our simulations reveal that the orientation of the nucleotide is also only likely to have negligible impact on the probability of entry into the protein.

Closer inspection of the wildtype alpha hemolysin revealed that residues D45, K46 and N47 play a role in interacting with CMP in instances of failed capture, therefore we recommend mutational studies to optimise this region. Of course, it must be noted that due to the number of simulations appropriate for this study, we have been unable to vary the effects of forces on the nucleotides due to the electro-osmotic flow of ions through the pore such as have been described in [29]. For a fuller picture of nucleotide dynamics in and around nanopores, in future work we plan longer simulations to include these effects and also to study all four nucleotides.

Acknowledgments

We acknowledge the use of both University of Southampton's computing cluster IRIDIS 4 and the ARCHER national facility.

Author Contributions

Syma Khalid conceived the experiments. Susanna Tomasio performed preliminary simulations. Richard M. A. Manara performed the simulations. Richard M. A. Manara and Syma Khalid analysed the simulations. Richard M. A. Manara and Syma Khalid wrote the manuscript.

Conflicts of Interest

The authors declare no conflict of interest.

References

1. Schneider, G.F.; Dekker, C. DNA sequencing with nanopores. *Nat. Biotechnol.* **2012**, *30*, 326–328.
2. Rincon-Restrepo, M.; Milthallova, E.; Bayley, H.; Maglia, G. Controlled translocation of individual DNA molecules through protein nanopores with engineered molecular brakes. *Nano Lett.* **2011**, *11*, 746–750.
3. Derrington, I.M.; Butler, T.Z.; Collins, M.D.; Manrao, E.; Pavlenok, M.; Niederweis, M.; Gundlach, J.H. Nanopore DNA sequencing with mspa. *Proc. Natl. Acad. Sci. USA* **2010**, *107*, 16060–16065.
4. Wanunu, M. Nanopores: A journey towards DNA sequencing. *Phys. Life Rev.* **2012**, *9*, 125–158.
5. Dekker, C. Solid-state nanopores. *Nat. Nanotechnol.* **2007**, *2*, 209–215.
6. Venkatesan, B.M.; Bashir, R. Nanopore sensors for nucleic acid analysis. *Nat. Nanotechnol.* **2011**, *6*, 615–624.
7. Fyta, M.; Melchionna, S.; Succi, S. Translocation of biomolecules through solid-state nanopores: Theory meets experiments. *J. Polym. Sci. Polym. Phys.* **2011**, *49*, 985–1011.
8. Scheicher, R.H.; Grigoriev, A.; Ahuja, R. DNA sequencing with nanopores from an ab initio perspective. *J. Mater. Sci.* **2012**, *47*, 7439–7446.
9. Laszlo, A.H.; Derrington, I.M.; Brinkerhoff, H.; Langford, K.W.; Nova, I.C.; Samson, J.M.; Bartlett, J.J.; Pavlenok, M.; Gundlach, J.H. Detection and mapping of 5-methylcytosine and 5-hydroxymethylcytosine with nanopore mspa. *Proc. Natl. Acad. Sci. USA* **2013**, *110*, 18904–18909.
10. Clarke, J.; Wu, H.C.; Jayasinghe, L.; Patel, A.; Reid, S.; Bayley, H. Continuous base identification for single-molecule nanopore DNA sequencing. *Nat. Nanotechnol.* **2009**, *4*, 265–270.
11. Song, L.Z.; Hobaugh, M.R.; Shustak, C.; Cheley, S.; Bayley, H.; Gouaux, J.E. Structure of staphylococcal alpha-hemolysin, a heptameric transmembrane pore. *Science* **1996**, *274*, 1859–1866.
12. Stoddart, D.; Maglia, G.; Mikhailova, E.; Heron, A.J.; Bayley, H. Multiple base-recognition sites in a biological nanopore: Two heads are better than one. *Angew. Chem. Int. Ed.* **2010**, *49*, 556–559.
13. Branton, D.; Deamer, D.W.; Marziali, A.; Bayley, H.; Benner, S.A.; Butler, T.; di Ventra, M.; Garaj, S.; Hibbs, A.; Huang, X.H.; *et al.* The potential and challenges of nanopore sequencing. *Nat. Biotechnol.* **2008**, *26*, 1146–1153.
14. Reiner, J.E.; Balijepalli, A.; Robertson, J.W.F.; Drown, B.S.; Burden, D.L.; Kasianowicz, J.J. The effects of diffusion on an exonuclease/nanopore-based DNA sequencing engine. *J. Chem. Phys.* **2012**, *137*, doi:10.1063/1.4766363.

15. Bond, P.J.; Guy, A.T.; Heron, A.J.; Bayley, H.; Khalid, S. Molecular dynamics simulations of DNA within a nanopore: Arginine-phosphate tethering and a binding/sliding mechanism for translocation. *Biochemistry* **2011**, *50*, 3777–3783.
16. Guy, A.T.; Piggot, T.J.; Khalid, S. Single-stranded DNA within nanopores: Conformational dynamics and implications for sequencing; a molecular dynamics simulation study. *Biophys. J.* **2012**, *103*, 1028–1036.
17. Parkin, J.; Khalid, S. Atomistic molecular-dynamics simulations enable prediction of the arginine permeation pathway through occd1/oprd from pseudomonas aeruginosa. *Biophys. J.* **2014**, *107*, 1853–1861.
18. Bhattacharya, S.; Derrington, I.M.; Pavlenok, M.; Niederweis, M.; Gundlach, J.H.; Aksimentiev, A. Molecular dynamics study of mspa arginine mutants predicts slow DNA translocations and ion current blockades indicative of DNA sequence. *ACS Nano* **2012**, *6*, 6960–6968.
19. Haider, S.; Neidle, S. Molecular modeling and simulation of g-quadruplexes and quadruplex-ligand complexes. *Methods Mol. Biol.* **2010**, *608*, 17–37.
20. Pronk, S.; Pall, S.; Schulz, R.; Larsson, P.; Bjelkmar, P.; Apostolov, R.; Shirts, M.R.; Smith, J.C.; Kasson, P.M.; van der Spoel, D.; *et al.* Gromacs 4.5: A high-throughput and highly parallel open source molecular simulation toolkit. *Bioinformatics* **2013**, *29*, 845–854.
21. Hess, B.; Kutzner, C.; van der Spoel, D.; Lindahl, E. Gromacs 4: Algorithms for highly efficient, load-balanced, and scalable molecular simulation. *J. Chem. Theory Comput.* **2008**, *4*, 435–447.
22. Oostenbrink, C.; Villa, A.; Mark, A.E.; van Gunsteren, W.F. A biomolecular force field based on the free enthalpy of hydration and solvation: The gromos force-field parameter sets 53a5 and 53a6. *J. Comput. Chem.* **2004**, *25*, 1656–1676.
23. Berendsen, H.J.C.; Postma, J.P.M.; van Gunsteren, W.F.; Hermans, J. Interaction Models for Water in Relation to Protein Hydration. In *Intermol. Forces*; Pullman, B., Ed.; Springer: Berlin, Germany, 1981; pp. 331–342.
24. Pongprayoon, P.; Beckstein, O.; Wee, C.L.; Sansom, M.S.P. Simulations of anion transport through oprp reveal the molecular basis for high affinity and selectivity for phosphate. *Proc. Natl. Acad. Sci. USA* **2009**, *106*, 21614–21618.
25. Humphrey, W.; Dalke, A.; Schulten, K. Vmd: Visual molecular dynamics. *J. Mol. Graph.* **1996**, *14*, 33–38.
26. Berendsen, H.J.C.; Postma, J.P.M.; Vangunsteren, W.F.; Dinola, A.; Haak, J.R. Molecular-dynamics with coupling to an external bath. *J. Chem. Phys.* **1984**, *81*, 3684–3690.
27. Bussi, G.; Donadio, D.; Parrinello, M. Canonical sampling through velocity rescaling. *J. Chem. Phys.* **2007**, *126*, doi:10.1063/1.2408420.
28. Nosé, S.; Klein, M.L. Constant pressure molecular dynamics for molecular systems. *Mol. Phys.* **1983**, *50*, 1055–1076.

29. He, Y.H.; Tsutsui, M.; Scheicher, R.H.; Fan, C.; Taniguchi, M.; Kawai, T. Mechanism of how salt-gradient-induced charges affect the translocation of DNA molecules through a nanopore. *Biophys. J.* **2013**, *105*, 776–782.

5

Hybrids of Nucleic Acids and Carbon Nanotubes for Nanobiotechnology

Kazuo Umemura

Biophysics Section, Department of Physics, Tokyo University of Science, 1-3 Kagurazaka, Shinjuku, Tokyo 1628601, Japan; E-Mail: meicun2006@163.com

Academic Editor: Stephen Ralph

Abstract: Recent progress in the combination of nucleic acids and carbon nanotubes (CNTs) has been briefly reviewed here. Since discovering the hybridization phenomenon of DNA molecules and CNTs in 2003, a large amount of fundamental and applied research has been carried out. Among thousands of papers published since 2003, approximately 240 papers focused on biological applications were selected and categorized based on the types of nucleic acids used, but not the types of CNTs. This survey revealed that the hybridization phenomenon is strongly affected by various factors, such as DNA sequences, and for this reason, fundamental studies on the hybridization phenomenon are important. Additionally, many research groups have proposed numerous practical applications, such as nanobiosensors. The goal of this review is to provide perspective on biological applications using hybrids of nucleic acids and CNTs.

Keywords: single-stranded DNA; double-stranded DNA; RNA; carbon nanotubes (CNTs); functionalization

---

1. Introduction

The hybridization of DNA and carbon nanotubes (CNTs) was first reported in 2003 when Zheng *et al.* presented the hybridization of single-stranded DNA (ssDNA) and single-walled carbon nanotubes (SWNTs) [1]. They not only presented experimental data, but also introduced a theoretical model of the adsorption of DNA onto SWNT surfaces (Figure 1). Furthermore, this work suggested there is a correlation between specific bases of DNA molecules and SWNT chirality. Nakashima *et al.* independently described the solubility of SWNTs with DNA molecules in 2003 [2] using double-stranded

DNA (dsDNA) molecules. Thereafter, many research groups have reported on the hybridization of DNA and CNTs, especially SWNTs.

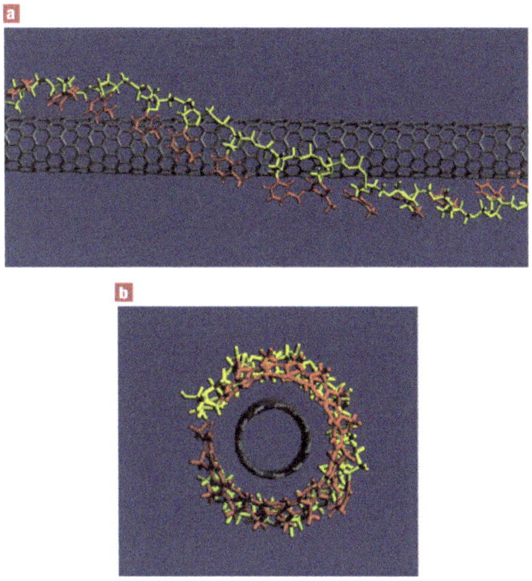

Figure 1. A theoretical model of structure of single-stranded DNA (ssDNA) and single-walled carbon nanotubes (SWNTs) hybrids, proposed by Zheng *et al.* (10,0) SWNT and poly(T) were assumed for the calculation (Reprinted from reference [1] with permission).

There have been two central hybridization research goals since 2003. The first is to improve the solubility of CNTs. Although the unique structural and physicochemical properties of CNTs are appealing for fundamental and industrial applications, uncoated CNTs typically form insoluble bundles in aqueous solutions. To take full advantage of CNTs, they must be prepared in a mono-dispersed form, which is especially true for SWNTs. Mono-dispersed CNTs can be prepared with attached hydrophilic organic molecules, such as sodium dodecyl sulfate (SDS), peptides, proteins, or DNA molecules, which can enhance CNT solubility. Another goal of hybridization is to purify SWNTs. SWNTs exhibit chirality, which is determined by the chirality vector, $C_h = na_1 + ma_2$. The a_1 and a_2 variables represent unit vectors on a graphene surface, and n and m are integers. Physical properties of SWNTs, such as electrical and optical characteristics, can vary depending on chirality. Because the chirality of SWNTs cannot be well controlled during synthesis, post-synthesis purification is required for some SWNTs for industrial applications. These two objectives were the main rationales for initial studies on hybridization.

While the method for DNA adsorption onto CNTs was originally developed to solubilize and purify CNTs, DNA-CNT hybrids also have great potential as DNA nanodevices. In this review, the fundamental studies of DNA and CNT hybridization for biological applications, especially in solutions, will be presented. Modification of electrode surfaces with DNA and CNTs is another major area of research; however, this topic will be explored elsewhere.

2. Non-Covalent Hybridization of DNA Molecules onto SWNT Surfaces

The solubility of CNTs is a major area of CNT research, with the aim of preparing mono-dispersed CNT suspensions. Prior to using DNA, various organic molecules were employed to hybridize with CNTs [3–18]. Although this article focuses on DNA technology, several examples of hybridization using other molecules are introduced for comparison. Various organic molecules, such as polyvinyl pyrrolidone (PVP) and polystyrene sulfonate (PSS) [3], poly(propionylethylenimine-co-ethylenimine) [5], hyperbranched polymers [6], 12-membered cyclodextrins [7], SDS [8], Nafion [10], poly(*N*-cetyl-4-vinylpyridinium bromide-co-*N*-ethyl-4-vinylpyridinium bromideco-4-vinylpyridine) [14], and electroactive methylene blue (MB) dyes [15] have been employed. Polysaccharides, such as lactose-appended schizophyllan [12] and *N*-trimethyl-chitosan [17], have also been used to control the electrical properties of prepared hybrids. Furthermore, proteins such as bovine serum albumin (BSA) [4] have been employed to solubilize both SWNTs and multiple-walled CNTs (MWNTs). Early reports suggested the possibility of separating specific CNTs using surface modification.

Numerous review articles have been published on the subject of solubilization [19–38]. At present, the review titled "*Advances toward Bioapplications of Carbon Nanotubes*" and prepared by Lin *et al.* in 2004 has been cited over 500 times [24]. Although this review described examples of solubilization studies using various organic molecules, only a few studies illustrating the relationship between DNA and CNTs existed before 2004 because the review was written at the early stages of research on DNA and CNT hybridization. Since 2005, several review articles that have focused on DNA applications with CNTs have been published [25,27,29–32,34–36,38].

In 2014, The Web of Science database contained approximately 4000 studies with both the keywords "DNA" and "carbon nanotube". Among these, approximately 1000 were reports on suspensions or solutions. In this review, approximately 240 papers were selected that focus on biological applications. First, the fundamental studies of hybridization are reviewed [39–79]. The articles are divided into categories based solely on the types of nucleic acid molecules used.

Interestingly, although some of the articles provided detailed information on the DNA molecules that were used in the experiments, some of them did not include details on the DNA molecules. If the types of the employed DNA molecules were important factors for the research, the authors would likely have described them in detail. However, if the DNA molecules were merely one of the organic molecules used to solubilize CNTs, detailed descriptions might not be necessary. A researcher's stance on DNA-CNT hybrids is apparent based on the description of DNA in an article.

3. Hybridization with dsDNA

A large portion of hybridization research uses double-stranded DNA molecules, including DNA from salmon testes [2,39,54,67,70,72], calf thymus [74,76,79], or fish sperm [75]. These commercially available DNA molecules are suitable for industrial applications because their cost is less than that of synthesized ones. At the initial stages of hybridization research, finding optimal hybridization conditions was one of the major aims. Barisci *et al.* proposed that a 1% SWNT suspension could be prepared with DNA molecules [39]. Characterization of the prepared DNA-CNT hybrids was the next area of study. Badaire *et al.* measured the average length and diameter of the hybrids using dynamic light scattering [40]. Cathcart *et al.* reported an increase in the photoluminescence intensity of SWNTs with decreasing concentrations of SWNTs [54].

Several applications for DNA-CNT hybrids have been reported. Asada *et al.* fabricated thin-film transistors (TFTs) using dsDNA-SWNT hybrids [67]. Wang *et al.* prepared transparent conductive films on PET substrates [70]. Wang *et al.* modified a glassy carbon electrode with dsDNA-SWNT hybrids and Nafion composites [75].

Although the hybridization method is essentially well established, the search for optimal hybridization conditions is ongoing. Chhikara *et al.* established that cationic cholesterol induces higher stability in DNA-SWNT hybrids [74]. Ao *et al.* reported cholesteric and nematic liquid crystalline phase behavior of SWNTs with dsDNA [72]. Primo *et al.* dispersed bamboo-like multi-walled carbon nanotubes (bCNTs) for the modification of glassy carbon electrodes [76]. Recently, Tardani *et al.* demonstrated the hybridization in a nematic phase of dsDNA-water-NaCl [79]. These reports suggest that optimal conditions of the hybridization are easily changed according to types of DNA/CNT and concentrations of DNA/CNT mixtures.

As described above, various fundamental and applied studies have been reported using natural dsDNA molecules. Variation of dsDNA molecules revealed huge potential for industrial applications. However, the mechanisms of hybridization between dsDNA and CNTs are currently not well understood. Fundamental research clarifying the details of dsDNA-CNT hybrids is necessary. Furthermore, many studies do not identify the sequences of dsDNA molecules used. Examining the effects of the sequence of dsDNA used could be a focus of future study.

4. Hybridization with ssDNA

Hybridization of ssDNA and CNTs has also been demonstrated. In most cases, synthesized oligonucleotides were utilized. Because Zheng *et al.* reported that the 30-m of thymine (T_{30}) resulted in the highest yield of hybridization, oligonucleotides containing thymine have been widely employed.

Hu *et al.* used $(GT)_{15}$ and achieved selective and sensitive detection in the form of quasi-reversible voltammetric responses [43]. Bertoncini *et al.* employed a poly $(GT)_{15}$ sequence to confirm the debundling of SWNTs [53]. Haggenmueller *et al.* compared $(GT)_{15}$, $(GT)_{10}$, $(AC)_{15}$, $(AC)_{10}$, and C_{10-30} as well as surfactants and other biomolecules in hybridization [61]. Hughes *et al.* used A_{15}, G_{15}, C_{15}, and T_{15} to disperse CNTs and found that the order of dispersion efficiency was T > C > G >> A, and that the adsorption speed of cytosine was the fastest [68].

The use of synthesized ssDNA molecules enabled detailed characterization of the physical and chemical properties of the prepared hybrids. Fagan *et al.* used the 30-m 5'-GT(GT)13GT-3' to investigate the effects of pH on hybrid properties [46]. Jin *et al.* used $(GT)_{15}$, $(GC)_{15}$, and 5'-TAG CTA TGG AAT TCC TCG TAG GCA-3' to detect the transition phenomenon between 0 and 50 °C [58]. Cooper *et al.* debundled double-walled CNTs (DWNTs) using ssDNA and evaluated the DWNTs using cyclic voltammograms [63]. Ghosh *et al.* utilized ssDNA, such as $(GT)_{40}$, to disperse MWNTs while studying thermal ablation [64]. Jovanovic *et al.* used gamma-irradiation for hybridization instead of sonication [65]. Qiu *et al.* used $(GT)_{20}$ and investigated electrostatically driven interactions using force-distance curves [77].

Throughout, technical progress has been reported. Chen *et al.* demonstrated controlled precipitation using T_{30} [45]. Onoa *et al.* fabricated a field effect transistor (FET) with poly(CT) and SWNTs [51]. Zhao *et al.* hybridized SWNTs with long ssDNA molecules synthesized by rolling circle amplification [52]. Han *et al.* used $(GT)_{29}$SH for hybridization with SWNTs and attached Au nanoparticles to the prepared hybrids [55]. Hopkins *et al.* demonstrated the alignment of ssDNA-SWNT hybrids using an inkjet printing method [56]. Pease *et al.* investigated the length distribution of SWNTs using electrospray differential mobility analysis (ES-DMA).

Short ssDNA molecules with user-specified lengths and sequences can be easily synthesized. Therefore, comparative studies using multiple combinations have been reported for understanding the hybridization mechanism. In addition, examination of assorted combinations can be used to obtain clues for sorting and purifying CNTs. The focus of many of the studies using synthesized ssDNA molecules was closer to chemistry than to biology. Further work from the viewpoint of biology and biotechnology are anticipated in the near future.

5. Comparative Studies Using ssDNA, dsDNA, and Other Nucleic Acids

Investigations comparing ssDNA, dsDNA, and other nucleic acids are also popular approaches to studying hybridization. He *et al.* deposited calf thymus dsDNA, calf thymus ssDNA, poly(G), or poly(diallyldimethylammonium) (PDDA) on water-soluble oxidized SWNTs to generate DNA sensors [41]. Yan *et al.* compared the hybridization of MWNTs with either calf thymus ssDNA or dsDNA using cyclic voltammetry [44]. Gigliotti *et al.* found that long genomic ssDNA molecules formed tight helices around CNTs [47]. Gladchenko discussed the behaviors of double-stranded and single-stranded regions of ultrasonically examined dsDNA present in the hybrid [48]. Vogel *et al.* examined $(GT)_n$ and $(AC)_n$ with n = 2, 3, 5, 10, 20, or 40 to find the optimum conditions for hybridization [59]. Yamamoto *et al.* used short dsDNA molecules, A_{20}/T_{20}, as well as A_{20} and T_{20} to purify highly stable hybrids by chromatography [71].

RNA molecules were examined by several researchers. Ishibashi *et al.* employed RNA molecules, poly(A), poly(C), and poly(G) to prepare RNA-SWNT hybrids, and fabricated multilayer films using the layer-by-layer technique [50]. Sanz *et al.* hybridized RNA molecules from baker's yeast with SWNTs, DWNTs, and MWNTs for use in gene delivery systems [73].

There are unique approaches using specific molecules that are related to DNA or RNA. Ikeda *et al.* prepared SWNT hybrids with adenosine triphosphate (ATP) and other related molecules using a high-speed vibration milling technique [49]. Schmucker modified DNA molecules with pyrene and prepared DNA-SWNT hybrids in 2013, revealing that DNA modification affected the solubilization and photoluminescence of SWNTs [78].

Comparative studies using both ssDNA and dsDNA molecules are attractive to illuminate the differences of the nucleic acid structures on CNT surfaces. When naturally derived long dsDNA molecules are employed, preparation of ssDNA molecules that have similar length and similar sequence is not easy. When dsDNA molecules are synthesized using hybridization of two short synthesized ssDNA molecules, residual ssDNA molecules should be considered. The use of well-purified ssDNA and dsDNA molecules that are 500 to 1000 bp long and have similar sequences could be an interesting area of research. There are not many examples of hybridization of RNA and other nucleic acids with CNTs. The use of various nucleic acids with diverse physical properties is an attractive target in biophysics.

6. Selective Adsorption and Separation of CNTs

Zheng *et al.* described the selective adsorption of DNA molecules that was dependent on the type of DNA base, and they indicated the possibility of separating SWNTs by attaching DNA molecules [1]. This idea has been followed up in numerous recent papers. When a study uses more than one ssDNA sequence for hybridization with SWNTs, selective adhesion is expected [41,44,47,50,58,59,67,71]. Albertorio *et al.* specifically focused on selective adhesion, and they found that the binding strengths between nucleobases and SWNTs followed the order G > C > A > T [80].

Multiple studies have reported on the separation of specific SWNTs using selective adhesion of DNA molecules [50,59,61,68,80–83]. Currently, the chirality of SWNTs cannot be well controlled during synthesis. A technique for separating SWNTs is one of the major aims of CNT research [62,84–93]. For example, Vetcher *et al.* separated DNA/RNA-SWNT hybrids using agarose gel electrophoresis [84].

Chromatography techniques have also been used to purify or separate DNA-CNT hybrids. Bauer *et al.* prepared three types of hybrids: octadecylamine functionalization (SWNT-ODA) in tetrahydrofuran (THF), butyl group functionalization (SWNT-butyl) in THF, and DNA-SWNT hybrids. Then, the hybrids were separated by size-exclusion chromatography (SEC) in 2007 [85]. They also found that the molar mass of the hybrids was proportional to the rod length of the SWNTs [87]. Chun *et al.* used flow-field flow fractionation (flow-FFF) to separate DNA-SWNT hybrids [88]. Asada *et al.* reported the separation of DNA-SWNTs by length and the separation of DNA-DWNTs using size-exclusion high-performance liquid chromatography (HPLC) [90]. Much progress using chromatographic methods for separation was demonstrated by Tu *et al.* in 2009 [91] where they examined many combinations of ssDNA molecules with specific sequences and SWNTs with various chiralities. They found that 12 major single-chirality semiconducting SWNTs could be separated by ion-exchange chromatography, and the results of the separation are clearly indicated in the absorption spectra (Figure 2).

Although most separation studies focus on hybrids of SWNTs, Kim *et al.* reported DWNT hybrids as a function of nanotube diameter, suggesting that stronger van der Waals forces between large-diameter tubes were an important factor [89]. Shakhmaeva *et al.* proposed a unique method of removing DNA molecules from CNT surfaces. Using divalent metal ions, the state of plasmid DNA changed from supercoiled to relaxed form and regulated the adhesion of DNA molecules onto the CNTs [92]. Tardani *et al.* presented the effects of polymers and ionic strength on the phase separation of DNA-SWNT suspensions [93].

Separation and purification studies have been an area of intense focus because they are absolutely vital for progress in CNT nanotechnology. If separation techniques can be applied to DNA nanotechnology, this may open a new field of study.

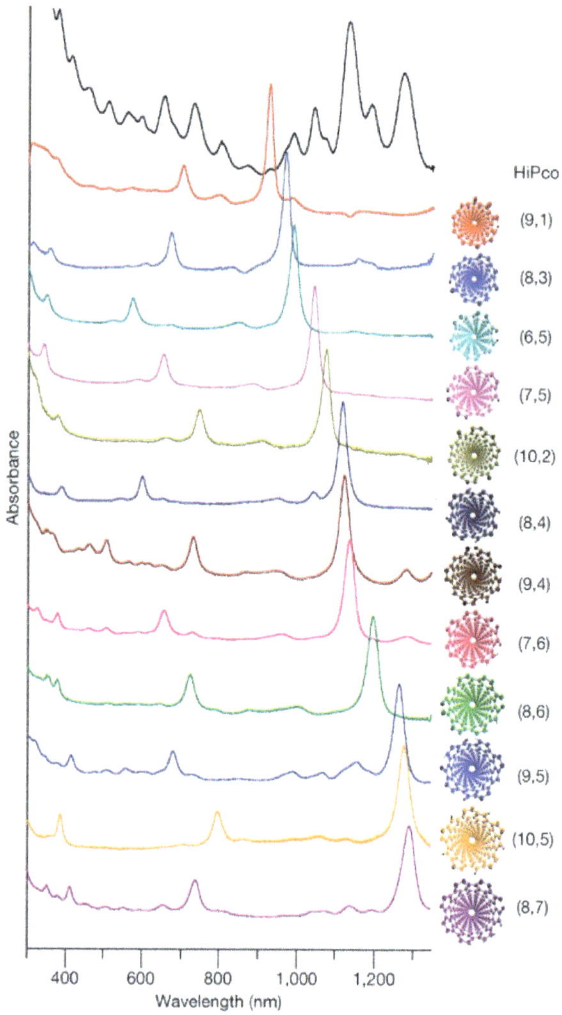

Figure 2. Optical absorption spectra of separated SWNTs by selective adsorption of ssDNA, reported by Tu *et al.* (Reprinted from reference [91] with permission).

7. Characterization of DNA-CNT Hybrids

The structural, physical, and chemical properties of DNA-CNT hybrids have been characterized by various methods, including UV-Vis spectroscopy (UV-Vis), near infrared (NIR) spectroscopy, Raman spectroscopy, photoluminescence (PL), transmission electron microscopy (TEM), scanning electron microscopy (SEM), scanning probe microscopy (SPM), and agarose gel electrophoresis. Because SWNTs have unique optical properties, spectroscopic studies are a prevalent approach to study hybrids [41,42,46,48,54,71,81,94–115].

Raman spectroscopy has been utilized since the establishment of hybridization research. Glamazda *et al.* described spectroscopic differences between DNA-SWNTs and SDS-SWNTs using Raman spectroscopy in 2006 [94]. Shoda *et al.* reported a decrease in the G-band in Raman spectra with DNA adsorption, only in the case of metallic tubes [106]. Simpson *et al.* performed Raman spectroscopy of length-separated fractions of ssDNA-SWNT hybrids to evaluate the role of defects in the fluorescence quantum yield of SWNTs [107]. Chen *et al.* attached silver nanoparticles onto DNA-SWNT hybrids to measure surface-enhanced Raman scattering (SERS). From the SERS measurements, they suggested that DNA-SWNT hybrids could be useful as nanoprobes for marking cells [108].

Jeng *et al.* suggested that an ssDNA-SWNT hybrid concentration of 6 nM could be detected using NIR band-gap fluorescence [95]. Asada *et al.* performed various types of spectroscopies involving NIR to assess charge transport on DNA-SWNT hybrids [98]. Karachevtsev *et al.* achieved glucose sensing by measuring the NIR spectra of DNA-SWNT hybrids using enzyme reactions [99]. Tu *et al.* reported a redox reaction of ssDNA-SWNT hybrids, suggesting that the NIR signals were tunable by the type and pH of the buffer solutions used [100]. Xu *et al.* described the potential of DNA-SWNT hybrids in immunoassays and glucose sensing as DNA-SWNT hybrids were optically sensitive to H_2O_2 and glucose in the NIR measurements [101]. Cao *et al.* suggested that red shifts in NIR absorption spectra were significant in hybrids with semiconducting SWNTs, but not in those with metallic SWNTs [102]. Noguchi *et al.* found that the optical properties of the dsDNA-SWNT hybrids could be modulated by changing the pH of the solution [81]. Ozturk *et al.* performed Raman spectroscopy on DNA-SWNT hybrids that were prepared with poly(A), poly(T), poly(G), and poly(C) at different lengths. Their results indicated that although the kinetics of DNA adsorption was affected by DNA length, the electronic properties of the DNA-SWNT hybrids were dependent on the species of DNA bases used [114]. Hobbie *et al.* examined the properties of G_{30} NIR [42], and Gladchenko *et al.* combined Raman spectroscopy with NIR to evaluate sonicated dsDNA that contained ssDNA regions [48]. Ishibashi *et al.* also used UV-Vis-NIR and Raman spectroscopy for comparative studies of poly(A), poly(C), and poly(G) [50].

Lacerda *et al.* demonstrated that nanotube aggregation and bundling led to decreased PL [97]. Cathcart *et al.* examined the aging effects of the DNA-SWNT hybrids on PL, observing that between 20 and 50 days after sample preparation, the PL intensity of the hybrids increased [103]. They also reported an increase in the PL intensity of SWNTs with decreasing concentrations of SWNTs, as mentioned in the previous section [54]. Schmucker discussed the PL spectra of hybrid DNA molecules modified with pyrene and SWNTs [78]. Ignatova *et al.* suggested fluorescence resonance energy transfer (FRET) could occur between two rare earth ions and the ionized phosphate groups of DNA [110]. Shao *et al.* combined a classical ruthenium (II) complex with guanine to yield tunable PL materials [111]. Hughes *et al.* used PL to compare 15-m of A, G, C, and T [68]. Figure 3 shows the changes in PL spectra due to solvent pH that were reported by Kim *et al.* [89]. Such phenomena are expected to apply to optical switches.

Other types of spectroscopic measurements were used by numerous investigators. Park *et al.* studied the kinetics of ssDNA adsorption onto SWNT surfaces using surface plasmon resonance imaging (SPRi) and demonstrated the effects of length and buffer using SPRi [109]. Badaire *et al.* used dynamic light scattering to estimate the length and diameter of the hybrids [40]. Tardani *et al.* used circular dichroism and rheology to study dsDNA-SWNT hybrids, as cited in the previous section [79]. He *et al.* systematically carried out various characterization methods, such as SEM, TEM, UV-Vis spectroscopy, Raman spectroscopy, and thermogravimetric analysis (TGA), to study positively and negatively charged hybrids [41].

Microscopic techniques are also powerful tools to directly characterize DNA-CNT hybrids [56,83,98,99,113,116–126]. Although TEM and SEM are major techniques, they are too broad to be adequately summarized in this section. A specific review article for using electron microscopy for studying DNA-CNT hybrids will be compiled. Here, several studies that used SPM and other specific microscopies as their main characterization methods are introduced.

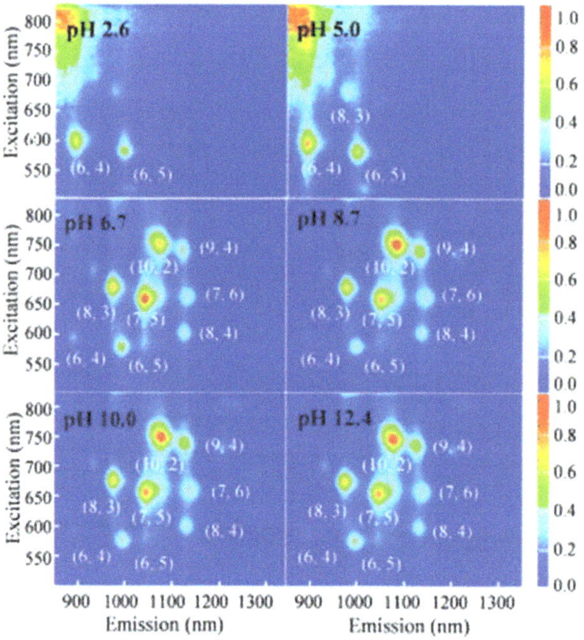

Figure 3. Differences in photoluminescence (PL) maps due to the pH values of solvents indicated by Kim *et al.* (Reprinted from reference [89] with permission).

Takahashi *et al.* published an article that specifically focused on atomic force microscopy (AFM) titled "*AFM Imaging of Wrapped Multiwall Carbon Nanotube in DNA*" in 2006. They examined dsDNA-MWNT hybrids using AFM on highly oriented pyrolytic graphite (HOPG) surfaces, and concluded that the diameters of the MWNTs strongly affected the adsorption of DNA [116]. Campbell *et al.* observed a regular pattern on the surface of CNT-DNA hybrids by AFM. Furthermore, they attached core/shell CdSe/ZnS quantum dots to the DNA-SWNTs using thiol-modified DNA molecules, and the attachment of the quantum dots was also confirmed by AFM [117]. Toita *et al.* prepared two types of SWNTs that were synthesized using high-pressure carbon monoxide (HiPco) and arc discharge (Arc) methods. Based on AFM measurements, they suggested that the adsorption mechanism of DNA molecules was affected by the diameter of SWNTs [118]. Hayashida *et al.* hybridized ssDNA or dsDNA using HiPco or chemical vapor deposition (CVD) methods and compared the thickness of the DNA-SWNT hybrids using cross-sections of AFM images. They found that the manner of adsorption was significantly different among the four combinations of hybridization [122]. Nii *et al.* studied the adsorption and desorption of dsDNA onto/away from SWNTs that were functionalized with polyethylene glycol (PEG), which was confirmed by cross-sectional analysis of AFM images and agarose gel electrophoresis [113]. Using a similar experimental approach, Nii *et al.* also reported the selective binding of ssDNA-binding (SSB) proteins to ssDNA-SWNTs, but not to dsDNA-SWNT hybrids [123].

Force spectroscopy is an advanced mode of AFM. Using force spectroscopy, Iliafar *et al.* found that the binding strength of ssDNA to curved SWNTs is much greater than that to flat graphite [83]. Figure 4 shows an example of their force spectroscopic data. Moreover, they estimated the free energy of DNA binding to SWNTs for every nucleic base. Qiu *et al.* carried out force spectroscopy to measure the electrostatic forces between GT_{20} and SWNTs [77].

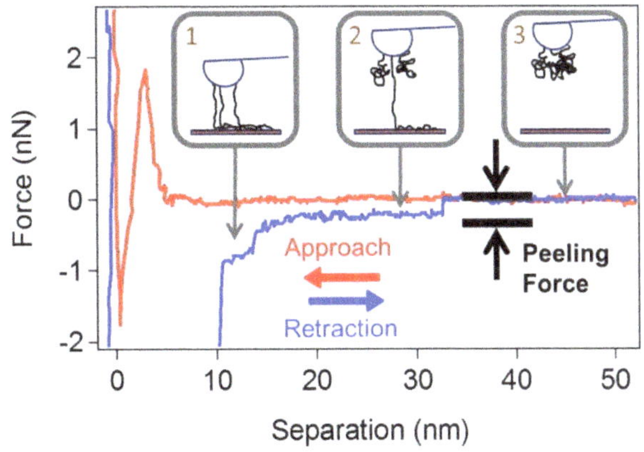

Figure 4. A typical force curve during the peeling of a T_{100} ssDNA molecule away from a SWNT surface, demonstrated by Iliafar *et al.* (Reprinted from reference [83] with permission).

Several unique approaches related to microscopic methods have been demonstrated. Xu *et al.* investigated the motion of SWNTs in solution using optical trapping techniques and scanning photocurrent microscopy [119]. For measurement, microbeads were attached to SWNTs via DNA molecules. The study demonstrated that the surface potential against Au substrate varied depending on the type of SWNTs. Arnett *et al.* reacted ssDNA ligase with DNA-CNT hybrids that were prepared with pyrimidine-functionalized CNTs and observed the formation of a 3D lattice-like CNT structure using wet-cell TEM [120]. Hayashida *et al.* performed Kelvin force microscopy to measure the surface potential of ssDNA-SWNT hybrids and other related samples [121].

Electrophoresis methods are popular in biology and have been applied to characterize DNA-CNT hybrids. For example, Vetcher *et al.* employed agarose gel electrophoresis to characterize separated DNA/RNA-SWNT hybrids [84]. Nii *et al.* combined staining of DNA molecules using ethidium bromide (EtBr), of proteins with Coomassie Brilliant Blue (CBB), and direct observation without staining for SWNTs to characterize DNA-SWNT hybrids after SSB reaction [123]. Figure 5 shows the results of the electrophoresis presented by Nii *et al.*, where significant differences between ssDNA and dsDNA were observed in the electrophoresis patterns.

Fluorescent spectroscopy and fluorescent microscopy revealed a peculiar quenching phenomenon [127–137]. Rao *et al.* confirmed the binding of poly(rU) molecules using single-molecule fluorescence microscopy and Raman spectroscopy [127]. Yang *et al.* described carbon nanotube-quenched fluorescent oligonucleotides and hairpin-structured fluorescent oligonucleotides [128]. Zhang *et al.* detected mercury ions (Hg^{2+}) in an aqueous solution using fluorescence quenching [129]. Zhao *et al.* prepared a reusable fluorescent sensor using DNA-SWNT hybrids for highly sensitive and selective detection of Ag^{+} and cysteine (Cys) in aqueous solutions [130]. D'Souza *et al.* used ssDNA hybrids to measure electron transfer in self-assemblies of ion-paired porphyrins employing (6,5) and (7,6) semiconductive SWNTs [131]. Martinez *et al.* found that streptavidin was an effective linker between SWNTs and oligonucleotides, and therefore, streptavidin efficiently prevented the non-specific adsorption of high-affinity molecules, such as DNA, onto CNTs [134]. Judkins *et al.* performed EPI-fluorescence microscopy (EFM) on DNA-SWNT hybrids to estimate their diffusion constants [137]. In this study, SWNTs were non-covalently coated with 1-pyrenebutanoic succinimidyl ester (PSE) as a fluorophore, and then DNA molecules were covalently attached to the PSE. This study suggested these tools could be used in the development of drug delivery systems.

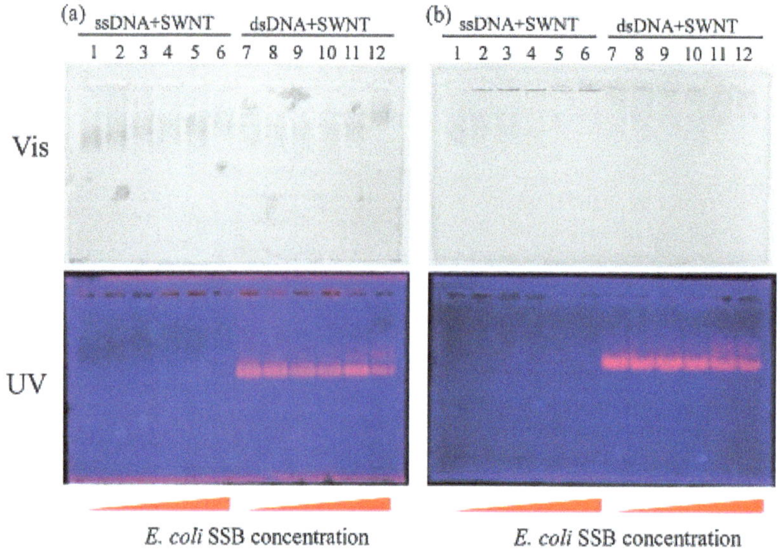

Figure 5. An example of agarose gel electrophoresis of DNA-SWNT hybrids with and without ssDNA-binding (SSB) proteins as performed by Nii *et al.* (Reprinted from reference [123] with permission).

There has been much work done in characterizing the physical and chemical properties of DNA-CNT hybrids. This is especially true for various nanomaterials where specific optical properties of CNTs seem to depend on CNT chirality and DNA sequence. A future application of this work might include development of optical sensors and switches. However, it is important to note that when a fluorescent dye is linked to DNA, quenching can occur on the CNT surface. If there are excess unbound DNA molecules nearby, this quenching can be difficult to observe.

These studies also indicate the importance of single-molecule measurements. In order to determine specific physical properties of CNTs, further progress using single-molecule techniques is needed. For example, if single-molecule electrophoresis were possible for DNA-CNT hybrids, this would be a powerful method to characterize DNA-CNT hybrids.

8. Functionalization of CNTs for DNA Attachment

The use of functionalized CNTs for DNA attachment is a prevalent approach in hybridization research [138–150]. In most cases of this approach, DNA molecules are covalently attached onto the functionalized CNTs. CNT structures should be chemically modified, and therefore, the physical and chemical properties of the CNTs should also be different in comparison with those of bare CNTs.

Hazani *et al.* attached ssDNA molecules onto SWNT surfaces using carbodiimide-assisted amidation in 2003 [138]. They characterized the samples using UV-Vis spectroscopy and confocal fluorescence microscopy and revealed hybridization of the ssDNA-SWNT hybrids with complementary ssDNA. Stevens *et al.* prepared fluorinated SWNTs (fluoronanotubes) to covalently attach various molecules such as DNA in 2003 [139]. Singh *et al.* synthesized conjugates of peptide nucleic acid (PNA, an artificial analogue of DNA) and SWNTs [141]. Wang *et al.* functionalized SWNTs with carboxylic groups and DNA molecules with amino groups and hybridized them using amide linkages [142]. Yang *et al.* covalently attached DNA molecules onto oxidized SWNT surfaces in organic and aqueous solutions. The attachment of the DNA molecules was confirmed by TEM using complementary DNA molecules that were functionalized using gold nanoparticles with ssDNA-SWNT conjugates [144]. Moghaddam *et al.* formed photo-adducts on graphitic surfaces using *N*-5-azido-nitrobenzoyloxy succinimide (ANB-NOS) and covalently attached DNA molecules that were terminated with amine groups [145]. Alidori *et al.* prepared ammonium-functionalized CNTs for DNA attachment [146]. Dolash *et al.* proposed a model in which covalent bonding was formed between DNA and CNTs using sonication [147]. Singh *et al.* developed the covalent double functionalization method using oxidized SWNTs with a combination of purine-pyrimidine and purine-purine nucleobases [148]. Canete-Rosales *et al.* fabricated hybrids of single-stranded deoxyoligonucleotide oligomers (DNO) and CNTs using both covalent and non-covalent attachment [149]. Kaufmann *et al.* functionalized MWNTs with hydroxyl groups [150], subsequently attaching a carrier strand oligodeoxynucleotide (CS-ODN) to the functionalized MWNTs and hybridizing a therapeutic antisense oligodeoxynucleotide (AS-ODN) with the conjugates. Finally, Yuan *et al.* reviewed the various functionalization methods [143].

Although not a chemical functionalization method of CNTs, the insertion of DNA molecules into CNTs is a unique idea [36,69,124,126,151–157]. For example, Gao *et al.* predicted that a DNA could possibly be spontaneously inserted into CNTs using molecular dynamics (MD) simulations [151]. Okada *et al.* demonstrated insertion of ssDNA molecules into SWNTs using ion irradiation in electrolyte plasma [152,153]. Lulevich *et al.* pulled a DNA molecule out of a SWNT using an AFM probe. Since the measured force was nearly constant, they suggested that the removal of the DNA molecule was frictionless [126].

Many new small companies that commercialize the functionalization of CNT products have been established. These companies are quite valuable for physicists and biologists who do not synthesize their own CNTs. Currently, since each company proposes their own protocols, qualities of some functionalized CNTs are dependent on manufacturers even though similar structural formulas are written in their catalogs. Further accumulation of the use of functionalized CNTs is necessary for the DNA nanotechnology field.

9. Theoretical Studies of Nucleic Acid Adsorption to CNTs

The combination of theoretical and experimental approaches is a trend in fundamental hybridization research of nucleic acids and CNTs [155,158–174]. Enyashin *et al.* employed a quantum mechanical density-functional tight-binding method (DFTB) to assess the adsorption of DNA molecules onto CNT surfaces [158]. Frischknecht *et al.* performed atomistic MD simulations to measure the binding energy of six different nucleotide monophosphates (NMPs) under two distinctive environmental conditions. They established that the binding energy between the NMPs and (6,0) SWNTs is a few kilocalories/mole [159]. Roxbury *et al.* studied sequence-specific self-stitching motifs of short ssDNA using replica exchange molecular dynamics (REMD) [164]. Umadevi *et al.* demonstrated that the binding energy between CNTs and DNA/RNA nucleobases could be controlled by the curvature of the CNTs using quantum chemical calculations [165]. Xiao *et al.* studied sequence- and base-dependent interactions using all atom MD simulations and thermodynamic analyses based on their fundamental interests [166]. Akdim *et al.* examined the selectivity of DNA-CNT hybridization using density functional theory calculations [167]. While ssDNA has mainly been studied from the theoretical aspect, Alegret *et al.* investigated dsDNA-SWNT hybrids using MD simulations [168]. Their results indicated that the disruption of short dsDNA helices was accelerated in the presence of SWNTs. Roxbury *et al.* reported that DNA formed ordered structures on SWNT surfaces, and the structures are strongly dependent on the sequence of the DNA and the types of SWNTs [169]. Figure 6 depicts their theoretical models of ssDNA-SWNT hybridization with distinct sequences and chiralities. Santosh *et al.* suggested that dsDNA undergoes much less unzipping and wrapping on SWNTs within 70 ns using MD simulations [170]. Wu *et al.* indicated that bundled CNTs can be dispersed by applying torsional energy based on MD simulations [174]. Iliafar verified their experimental data using REMD [112]. Makarucha *et al.* presented a review article to summarize theoretical research in this field in 2011 [163].

In the past ten years, theoretical modeling has proven useful for examining various conditions of DNA-CNT hybridization. Theoretical models of dsDNA-CNT hybrids, not only ssDNA-CNT hybrids, have also recently been proposed. At this stage, systematic combination of theoretical and experimental studies might be required.

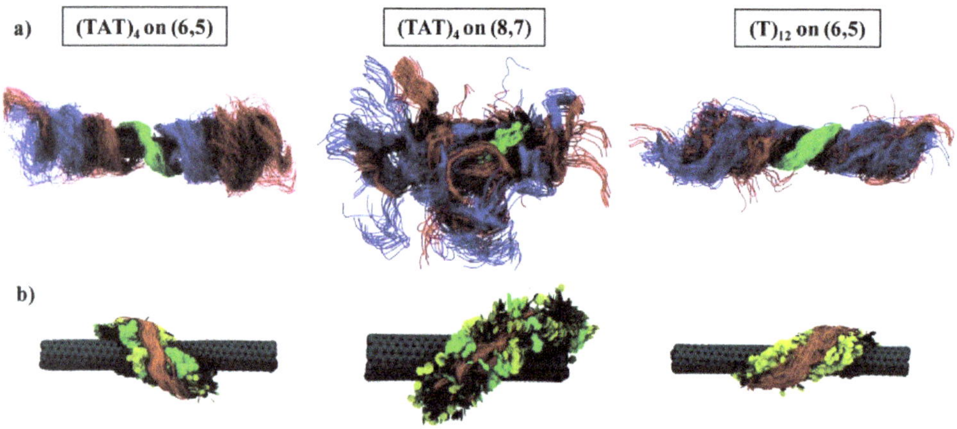

Figure 6. Recent theoretical models of several combinations of ssDNA and SWNTs, proposed by Roxbury *et al.* (Reprinted from reference [169] with permission).

10. Biological Applications

Although much work has focused on the fundamental aspects of DNA-CNT hybridization, many recent reports have presented ideas for and demonstrations of biological applications [40,43,50,55,56,63,64,67,74–76,82,99,101,108,112,117,130,133,175–200]. The following describes several examples.

Biosensors comprise a widespread field in the application of DNA-CNT hybrids. Zhang *et al.* reported the selective detection of single nitric oxide (NO) molecules. In this study, $(AT)_{15}$ ssDNA was adsorbed onto an array of near-infrared fluorescent semiconducting SWNTs, which resulted in AT_{15}-SWNT hybrids [133]. Zhang *et al.* achieved label-free detection of sequence-specific DNA using MWNTs, noting that the mismatch of a single base could be detected using light scattering signals [176]. Aravind *et al.* demonstrated the detection of dopamine using ssDNA-MWNTs [180]. Cao *et al.* developed a highly sensitive immunosensor for detecting human chorionic gonadotrophin (HCG) using dsDNA-MWNT hybrids and methylene blue (MB) on a glass carbon electrode [182]. Shi *et al.* fabricated highly sensitive glucose and ATP microbiosensors using ssDNA-SWNT hybrids [185]. Thuy *et al.* detected *Escherichia coli* O157:H7 using DNA-MWNT hybrids [189].

Su *et al.* fabricated gas sensors for H_2, H_2S, NH_3, and NO_2 using metal-DNA-SWNT hybrids on microfabricated electrodes [195].

Kawaguchi *et al.* used dsDNA-SWNT hybrids as heat delivery vehicles for the thermal ablation of tumors. Anti-human IgG was attached to the hybrids for specific binding with the targeted protein molecules [187]. The attachment of quantum dots to DNA-CNT hybrids was reported by several authors [117,190]. Gong *et al.* found that molecular structural switching of T_{15} adsorbed onto SWNTs was induced when the sample was exposed to Hg^{2+} [192]. Williams *et al.* described an inhibitory concentration of DNA-SWNTs in PCR reactions [200]. Mangalum *et al.* fabricated unique structures by combining ssDNA-SWNT hybrids and DNA origami techniques, as shown in Figure 7 [193].

Finally, the interaction between nucleic acids and graphene has becoming a major area of interest [201–219], and the safety of CNTs is an important area of research [220–245]. Further review articles that focus on these topics will be compiled.

The number of biological applications has increased drastically over the past several years. This suggests that biophysicists and biochemists can use DNA-CNT hybrids even though they do not specialize in CNTs. Although fundamental analysis of DNA-CNT hybridization is an important area of research, it appears that biological applications can be developed using our present understanding of DNA-CNT hybrids.

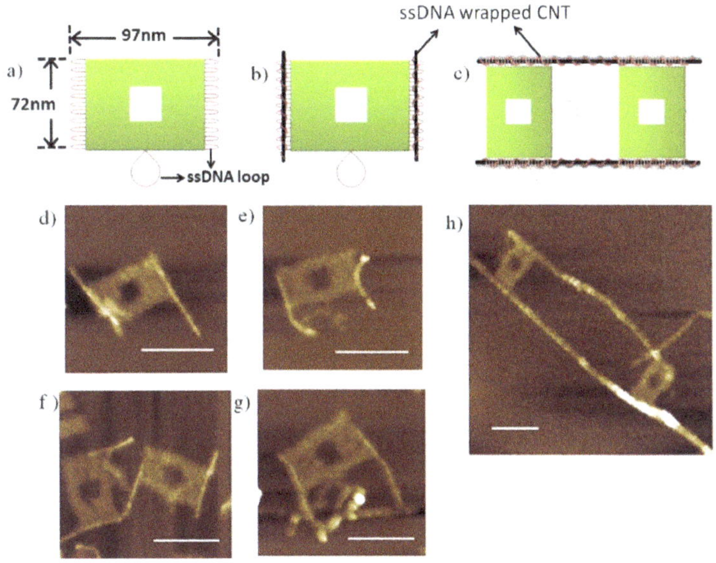

Figure 7. Schematic representations and AFM images of structures fabricated with ssDNA-SWNT hybrids using the DNA origami technique, generated by Mangalum *et al.* (Reprinted from reference [193] with permission).

11. Conclusions

In this review, biological applications of CNT hybrids with nucleic acids are discussed and categorized by several key factors. This survey suggests that fundamental research of the hybridization phenomenon of nucleic acids and CNTs has continued intensively from both theoretical and experimental viewpoints. Progress in practical applications, such as fabrication of nanosensors, has recently accelerated. Some topics cannot be covered in this article, and thus, more specific review articles will be written in the future.

Acknowledgments

This work was supported by Grant-in-Aid for Scientific Research (26400436) of the Japan Society for the Promotion of Science (JSPS).

Conflicts of Interest

The author declares no conflict of interest.

References

1. Zheng, M.; Jagota, A.; Semke, E.D.; Diner, B.A.; McLean, R.S.; Lustig, S.R.; Richardson, R.E.; Tassi, N.G. DNA-assisted dispersion and separation of carbon nanotubes. *Nat. Mater.* **2003**, *2*, 338–342.
2. Nakashima, N.; Okuzono, S.; Murakami, H.; Nakai, T.; Yoshikawa, K. DNA dissolves single-walled carbon nanotubes in water. *Chem. Lett.* **2003**, *32*, 456–457.
3. O'Connell, M.J.; Boul, P.; Ericson, L.M.; Huffman, C.; Wang, Y.H.; Haroz, E.; Kuper, C.; Tour, J.; Ausman, K.D.; Smalley, R.E. Reversible water-solubilization of single-walled carbon nanotubes by polymer wrapping. *Chem. Phys. Lett.* **2001**, *342*, 265–271.
4. Fu, K.F.; Huang, W.J.; Lin, Y.; Zhang, D.H.; Hanks, T.W.; Rao, A.M.; Sun, Y.P. Functionalization of carbon nanotubes with bovine serum albumin in homogeneous aqueous solution. *J. Nanosci. Nanotechnol.* **2002**, *2*, 457–461.
5. Huang, W.J.; Lin, Y.; Taylor, S.; Gaillard, J.; Rao, A.M.; Sun, Y.P. Sonication-assisted functionalization and solubilization of carbon nanotubes. *Nano Lett.* **2002**, *2*, 231–234.
6. Star, A.; Stoddart, J.F. Dispersion and solubilization of single-walled carbon nanotubes with a hyperbranched polymer. *Macromolecules* **2002**, *35*, 7516–7520.
7. Dodziuk, H.; Ejchart, A.; Anczewski, W.; Ueda, H.; Krinichnaya, E.; Dolgonos, G.; Kutner, W. Water solubilization, determination of the number of different types of single-wall carbon nanotubes and their partial separation with respect to diameters by complexation with eta-cyclodextrin. *Chem. Commun.* **2003**, *8*, 986–987.
8. Jiang, L.Q.; Gao, L.; Sun, J. Production of aqueous colloidal dispersions of carbon nanotubes. *J. Colloid Interface Sci.* **2003**, *260*, 89–94.
9. Petrov, P.; Stassin, F.; Pagnoulle, C.; Jerome, R. Noncovalent functionalization of multi-walled carbon nanotubes by pyrene containing polymers. *Chem. Commun.* **2003**, *23*, 2904–2905.
10. Wang, J.; Musameh, M.; Lin, Y.H. Solubilization of carbon nanotubes by nafion toward the preparation of amperometric biosensors. *J. Am. Chem. Soc.* **2003**, *125*, 2408–2409.
11. Hasegawa, T.; Fujisawa, T.; Numata, M.; Umeda, M.; Matsumoto, T.; Kimura, T.; Okumura, S.; Sakurai, K.; Shinkai, S. Single-walled carbon nanotubes acquire a specific lectin-affinity through supramolecular wrapping with lactose-appended schizophyllan. *Chem. Commun.* **2004**, *19*, 2150–2151.
12. Zhang, M.G.; Smith, A.; Gorski, W. Carbon nanotube-chitosan system for electrochemical sensing based on dehydrogenase enzymes. *Anal. Chem.* **2004**, *76*, 5045–5050.
13. Liu, P. Modifications of carbon nanotubes with polymers. *Eur. Polym. J.* **2005**, *41*, 2693–2703.

14. Sinani, V.A.; Gheith, M.K.; Yaroslavov, A.A.; Rakhnyanskaya, A.A.; Sun, K.; Mamedov, A.A.; Wicksted, J.P.; Kotov, N.A. Aqueous dispersions of single-wall and multiwall carbon nanotubes with designed amphiphilic polycations. *J. Am. Chem. Soc.* **2005**, *127*, 3463–3472.
15. Yan, Y.M.; Zhang, M.N.; Gong, K.P.; Su, L.; Guo, Z.X.; Mao, L.Q. Adsorption of methylene blue dye onto carbon nanotubes: A route to an electrochemically functional nanostructure and its layer-by-layer assembled nanocomposite. *Chem. Mater.* **2005**, *17*, 3457–3463.
16. Wang, D.; Chen, L.W. Temperature and ph-responsive single-walled carbon nanotube dispersions. *Nano Lett.* **2007**, *7*, 1480–1484.
17. Wise, A.J.; Smith, J.R.; Bouropoulos, N.; Yannopoulos, S.N.; van der Merwe, S.M.; Fatouros, D.G. Single-wall carbon nanotube dispersions stabilised with *N*-trimethyl-chitosan. *J. Biomed. Nanotechnol.* **2008**, *4*, 67–72.
18. Plisko, T.V.; Bildyukevich, A.V. Debundling of multiwalled carbon nanotubes in *N*, *N*-dimethylacetamide by polymers. *Colloid Polym. Sci.* **2014**, *292*, 2571–2580.
19. Niyogi, S.; Hamon, M.A.; Hu, H.; Zhao, B.; Bhowmik, P.; Sen, R.; Itkis, M.E.; Haddon, R.C. Chemistry of single-walled carbon nanotubes. *Accounts Chem. Res.* **2002**, *35*, 1105–1113.
20. Williams, K.A.; Veenhuizen, P.T.M.; de la Torre, B.G.; Eritja, R.; Dekker, C. Nanotechnology—Carbon nanotubes with DNA recognition. *Nature* **2002**, *420*, 761–761.
21. Fu, K.F.; Sun, Y.P. Dispersion and solubilization of carbon nanotubes. *J. Nanosci. Nanotechnol.* **2003**, *3*, 351–364.
22. Martin, C.R.; Kohli, P. The emerging field of nanotube biotechnology. *Nat. Rev. Drug Discov.* **2003**, *2*, 29–37.
23. Jin, W.J.; Sun, X.F.; Wang, Y. Solubilization and functionalization of carbon nanotubes. *New Carbon Mater.* **2004**, *19*, 312–318.
24. Lin, Y.; Taylor, S.; Li, H.P.; Fernando, K.A.S.; Qu, L.W.; Wang, W.; Gu, L.R.; Zhou, B.; Sun, Y.P. Advances toward bioapplications of carbon nanotubes. *J. Mater. Chem.* **2004**, *14*, 527–541.
25. Pividori, M.I.; Alegret, S. DNA Adsorption on Carbonaceous Materials. In *Immobilisation of DNA on Chips I*; Wittmann, C., Ed.; Springer-Verlag Berlin: Berlin, Germany, 2005; Volume 260, pp. 1–36.
26. Nakashima, N. Solubilization of single-walled carbon nanotubes with condensed aromatic compounds. *Sci. Technol. Adv. Mater.* **2006**, *7*, 609–616.
27. Daniel, S.; Rao, T.P.; Rao, K.S.; Rani, S.U.; Naidu, G.R.K.; Lee, H.Y.; Kawai, T. A review of DNA functionalized/grafted carbon nanotubes and their characterization. *Sens. Actuator B* **2007**, *122*, 672–682.
28. Wei, W.; Sethuraman, A.; Jin, C.; Monteiro-Riviere, N.A.; Narayan, R.J. Biological properties of carbon nanotubes. *J. Nanosci. Nanotechnol.* **2007**, *7*, 1284–1297.
29. Tu, X.M.; Zheng, M. A DNA-based approach to the carbon nanotube sorting problem. *Nano Res.* **2008**, *1*, 185–194.
30. Becerril, H.A.; Woolley, A.T. DNA-templated nanofabrication. *Chem. Soc. Rev.* **2009**, *38*, 329–337.
31. Cheung, W.; Chiu, P.L.; Parajuli, R.R.; Ma, Y.F.; Ali, S.R.; He, H.X. Fabrication of high performance conducting polymer nanocomposites for biosensors and flexible electronics: Summary of the multiple roles of DNA dispersed and functionalized single walled carbon nanotubes. *J. Mater. Chem.* **2009**, *19*, 6465–6480.

32. Sanchez-Pomales, G.; Santiago-Rodriguez, L.; Cabrera, C.R. DNA-functionalized carbon nanotubes for biosensing applications. *J. Nanosci. Nanotechnol.* **2009**, *9*, 2175–2188.
33. Zhao, Y.L.; Stoddart, J.F. Noncovalent functionalization of single-walled carbon nanotubes. *Accounts Chem. Res.* **2009**, *42*, 1161–1171.
34. Fujigaya, T.; Tanaka, Y.; Nakashima, N. Soluble carbon nanotubes and application to electrochemistry. *Electrochemistry* **2010**, *78*, 2–15.
35. Lei, J.P.; Ju, H.X. Nanotubes in biosensing. *Nanomed. Nanobiotechnol.* **2010**, *2*, 496–509.
36. Contreras-Torres, F.F.; Martinez-Loran, E. DNA insertion in and wrapping around carbon nanotubes. *Comput. Mol. Sci.* **2011**, *1*, 902–919.
37. Zhang, W.X.; Zhang, Z.Z.; Zhang, Y.G. The application of carbon nanotubes in target drug delivery systems for cancer therapies. *Nanoscale Res. Lett.* **2011**, *6*, doi:10.1186/1556-276X-6-555.
38. Zhang, X.Q.; Xu, X.Y.; Lam, R.; Giljohann, D.; Ho, D.; Mirkin, C.A. Strategy for increasing drug solubility and efficacy through covalent attachment to polyvalent DNA—Nanoparticle conjugates. *ACS Nano* **2011**, *5*, 6962–6970.
39. Barisci, J.N.; Tahhan, M.; Wallace, G.G.; Badaire, S.; Vaugien, T.; Maugey, M.; Poulin, P. Properties of carbon nanotube fibers spun from DNA-stabilized dispersions. *Adv. Funct. Mater.* **2004**, *14*, 133–138.
40. Badaire, S.; Zakri, C.; Maugey, M.; Derre, A.; Barisci, J.N.; Wallace, G.; Poulin, P. Liquid crystals of DNA-stabilized carbon nanotubes. *Adv. Mater.* **2005**, *17*, 1673–1676.
41. He, P.G.; Bayachou, M. Layer-by-layer fabrication and characterization of DNA-wrapped single-walled carbon nanotube particles. *Langmuir* **2005**, *21*, 6086–6092.
42. Hobbie, E.K.; Bauer, B.J.; Stephens, J.; Becker, M.L.; McGuiggan, P.; Hudson, S.D.; Wang, H. Colloidal particles coated and stabilized by DNA-wrapped carbon nanotubes. *Langmuir* **2005**, *21*, 10284–10287.
43. Hu, C.G.; Zhang, Y.Y.; Bao, G.; Zhang, Y.L.; Liu, M.L.; Wang, Z.L. DNA functionalized single-walled carbon nanotubes for electrochemical detection. *J. Phys. Chem. B* **2005**, *109*, 20072–20076.
44. Yan, W.; Shen, X.C.; Zhang, Z.L.; Chen, C.; Pang, D.W. Electrochemical behavior of daunorubicin at DNA-MWCNT bioconjugates modified glassy carbon electrodes. *Anal. Lett.* **2005**, *38*, 2579–2595.
45. Chen, R.J.; Zhang, Y.G. Controlled precipitation of solubilized carbon nanotubes by delamination of DNA. *J. Phys. Chem. B* **2006**, *110*, 54–57.
46. Fagan, J.A.; Landi, B.J.; Mandelbaum, I.; Simpson, J.R.; Bajpai, V.; Bauer, B.J.; Migler, K.; Walker, A.R.H.; Raffaelle, R.; Hobbie, E.K. Comparative measures of single-wall carbon nanotube dispersion. *J. Phys. Chem. B* **2006**, *110*, 23801–23805.
47. Gigliotti, B.; Sakizzie, B.; Bethune, D.S.; Shelby, R.M.; Cha, J.N. Sequence-independent helical wrapping of singles-walled carbon nanotubes by long genomic DNA. *Nano Lett.* **2006**, *6*, 159–164.
48. Gladchenko, G.O.; Karachevtsev, M.V.; Leontiev, V.S.; Valeev, V.A.; Glamazda, A.Y.; Plokhotnichenko, A.M.; Stepanian, S.G. Interaction of fragmented double-stranded DNA with carbon nanotubes in aqueous solution. *Mol. Phys.* **2006**, *104*, 3193–3201.

49. Ikeda, A.; Hamano, T.; Hayashi, K.; Kikuchi, J. Water-solubilization of nucleotides-coated single-walled carbon nanotubes using a high-speed vibration milling technique. *Org. Lett.* **2006**, *8*, 1153–1156.
50. Ishibashi, A.; Yamaguchi, Y.; Murakami, H.; Nakashima, N. Layer-by-layer assembly of RNA/single-walled carbon nanotube nanocomposites. *Chem. Phys. Lett.* **2006**, *419*, 574–577.
51. Onoa, B.; Zheng, M.; Dresselhaus, M.S.; Diner, B.A. Carbon nanotubes and nucleic acids: Tools and targets. *Phys. Status Solidi A* **2006**, *203*, 1124–1131.
52. Zhao, W.; Gao, Y.; Brook, M.A.; Li, Y.F. Wrapping single-walled carbon nanotubes with long single-stranded DNA molecules produced by rolling circle amplification. *Chem. Commun.* **2006**, 3582–3584.
53. Bertoncini, P.; Gresil, M.; Lardoux, J.; Riou, I.; Chauvet, O. Morphology of DNA/single walled nanotubes complexes. *Dig. J. Nanomater. Biostruct.* **2007**, *2*, 293–297.
54. Cathcart, H.; Quinn, S.; Nicolosi, V.; Kelly, J.M.; Blau, W.J.; Coleman, J.N. Spontaneous debundling of single-walled carbon nanotubes in DNA-based dispersions. *J. Phys. Chem. C* **2007**, *111*, 66–74.
55. Han, X.G.; Li, Y.L.; Deng, Z.X. DNA-wrapped single-walled carbon nanotubes as rigid templates for assembling linear gold nanoparticle arrays. *Adv. Mater.* **2007**, *19*, 1518–1522.
56. Hopkins, A.R.; Kruk, N.A.; Lipeles, R.A. Macroscopic alignment of single-walled carbon nanotubes (SWNTs). *Surf. Coat. Technol.* **2007**, *202*, 1282–1286.
57. Jeng, E.S.; Barone, P.W.; Nelson, J.D.; Strano, M.S. Hybridization kinetics and thermodynamics of DNA adsorbed to individually dispersed single-walled carbon nanotubes. *Small* **2007**, *3*, 1602–1609.
58. Jin, H.; Jeng, E.S.; Heller, D.A.; Jena, P.V.; Kirmse, R.; Langowski, J.; Strano, M.S. Divalent ion and thermally induced DNA conformational polymorphism on single-walled carbon nanotubes. *Macromolecules* **2007**, *40*, 6731–6739.
59. Vogel, S.R.; Kappes, M.M.; Hennrich, F.; Richert, C. An unexpected new optimum in the structure space of DNA solubilizing single-walled carbon nanotubes. *Chemistry* **2007**, *13*, 1815–1820.
60. Vogel, S.R.; Muller, K.; Plutowski, U.; Kappes, M.M.; Richert, C. DNA-carbon nanotube interactions and nanostructuring based on DNA. *Phys. Status Solidi B* **2007**, *244*, 4026–4029.
61. Haggenmueller, R.; Rahatekar, S.S.; Fagan, J.A.; Chun, J.H.; Becker, M.L.; Naik, R.R.; Krauss, T.; Carlson, L.; Kadla, J.F.; Trulove, P.C.; *et al.* Comparison of the quality of aqueous dispersions of single wall carbon nanotubes using surfactants and biomolecules. *Langmuir* **2008**, *24*, 5070–5078.
62. Noguchi, Y.; Fujigaya, T.; Niidome, Y.; Nakashima, N. Single-walled carbon nanotubes/DNA hybrids in water are highly stable. *Chem. Phys. Lett.* **2008**, *455*, 249–251.
63. Cooper, L.; Amano, H.; Hiraide, M.; Houkyou, S.; Jang, I.Y.; Kim, Y.J.; Muramatsu, H.; Kim, J.H.; Hayashi, T.; Kim, Y.A.; *et al.* Freestanding, bendable thin film for supercapacitors using DNA-dispersed double walled carbon nanotubes. *Appl. Phys. Lett.* **2009**, doi:10.1063/1.3271768.
64. Ghosh, S.; Dutta, S.; Gomes, E.; Carroll, D.; D'Agostino, R.; Olson, J.; Guthold, M.; Gmeiner, W.H. Increased heating efficiency and selective thermal ablation of malignant tissue with DNA-encased multiwalled carbon nanotubes. *ACS Nano* **2009**, *3*, 2667–2673.

65. Jovanovic, S.P.; Markovic, Z.M.; Kleut, D.N.; Romcevic, N.Z.; Trajkovic, V.S.; Dramicanin, M.D.; Markovic, B.M.T. A novel method for the functionalization of gamma-irradiated single wall carbon nanotubes with DNA. *Nanotechnology* **2009**, *20*, doi:10.1088/0957-4484/20/44/445602.
66. Pease, L.F.; Tsai, D.H.; Fagan, J.A.; Bauer, B.J.; Zangmeister, R.A.; Tarlov, M.J.; Zachariah, M.R. Length distribution of single-walled carbon nanotubes in aqueous suspension measured by electrospray differential mobility analysis. *Small* **2009**, *5*, 2894–2901.
67. Asada, Y.; Miyata, Y.; Ohno, Y.; Kitaura, R.; Sugai, T.; Mizutani, T.; Shinohara, H. High-performance thin-film transistors with DNA-assisted solution processing of isolated single-walled carbon nanotubes. *Adv. Mater.* **2010**, *22*, 2698–2701.
68. Hughes, J.M.; Cathcart, H.; Coleman, J.N. Dispersion and exfoliation of nanotubes with synthetic oligonucleotides: Variation of dispersion efficiency and oligo-nanotube interaction with base type. *J. Phys. Chem. C* **2010**, *114*, 11741–11747.
69. Li, Z.H.; Yang, W. Capture and manipulation of hybrid DNAs by carbon nanotube bundles. *Nanotechnology* **2010**, *21*, doi:10.1088/0957-4484/21/19/195301.
70. Wang, R.R.; Sun, J.; Gao, L.A.; Zhang, J. Dispersion of single-walled carbon nanotubes by DNA for preparing transparent conductive films. *J. Mater. Chem.* **2010**, *20*, 6903–6909.
71. Yamamoto, Y.; Fujigaya, T.; Niidome, Y.; Nakashima, N. Fundamental properties of oligo double-stranded DNA/single-walled carbon nanotube nanobiohybrids. *Nanoscale* **2010**, *2*, 1767–1772.
72. Ao, G.Y.; Nepal, D.; Aono, M.; Davis, V.A. Cholesteric and nematic liquid crystalline phase behavior of double-stranded DNA stabilized single-walled carbon nanotube dispersions. *ACS Nano* **2011**, *5*, 1450–1458.
73. Sanz, V.; Borowiak, E.; Lukanov, P.; Galibert, A.M.; Flahaut, E.; Coley, H.M.; Silva, S.R.P.; McFadden, J. Optimising DNA binding to carbon nanotubes by non-covalent methods. *Carbon* **2011**, *49*, 1775–1781.
74. Chhikara, B.S.; Misra, S.K.; Bhattacharya, S. Cnt loading into cationic cholesterol suspensions show improved DNA binding and serum stability and ability to internalize into cancer cells. *Nanotechnology* **2012**, *23*, doi:10.1088/0957-4484/23/6/065101.
75. Wang, Y.W.; Liu, H.; Wang, F.; Gao, Y.M. Electrochemical oxidation behavior of methotrexate at DNA/SWCNT/nafion composite film-modified glassy carbon electrode. *J. Solid State Electrochem.* **2012**, *16*, 3227–3235.
76. Primo, E.N.; Canete-Rosales, P.; Bollo, S.; Rubianes, M.D.; Rivas, G.A. Dispersion of bamboo type multi-wall carbon nanotubes in calf-thymus double stranded DNA. *Colloid Surf. B* **2013**, *108*, 329–336.
77. Qiu, X.Y.; Khripin, C.Y.; Ke, F.Y.; Howell, S.C.; Zheng, M. Electrostatically driven interactions between hybrid DNA-carbon nanotubes. *Phys. Rev. Lett.* **2013**, *111*, doi:10.1103/PhysRevLett.111.048301.
78. Schmucker, W.; Klumpp, S.; Hennrich, F.; Kappes, M.; Wagenknecht, H.A. A simple pyrene "click"-type modification of DNA affects solubilisation and photoluminescence of single-walled carbon nanotubes. *RSC Adv.* **2013**, *3*, 6331–6333.

79. Tardani, F.; Strobbia, P.; Scipioni, A.; La Mesa, C. Encapsulating carbon nanotubes in aqueous ds-DNA anisotropic phases: Shear orientation and rheological properties. *RSC Adv.* **2013**, *3*, 25917–25923.
80. Albertorio, F.; Hughes, M.E.; Golovchenko, J.A.; Branton, D. Base dependent DNA-carbon nanotube interactions: Activation enthalpies and assembly-disassembly control. *Nanotechnology* **2009**, *20*, doi:10.1088/0957-4484/20/39/395101.
81. Noguchi, Y.; Fujigaya, T.; Niidome, Y.; Nakashima, N. Regulation of the near-IR spectral properties of individually dissolved single-walled carbon nanotubes in aqueous solutions of dsDNA. *Chemistry* **2008**, *14*, 5966–5973.
82. Ao, G.Y.; Khripin, C.Y.; Zheng, M. DNA-controlled partition of carbon nanotubes in polymer aqueous two-phase systems. *J. Am. Chem. Soc.* **2014**, *136*, 10383–10392.
83. Iliafar, S.; Mittal, J.; Vezenov, D.; Jagota, A. Interaction of single-stranded DNA with curved carbon nanotube is much stronger than with flat graphite. *J. Am. Chem. Soc.* **2014**, *136*, 12947–12957.
84. Vetcher, A.A.; Srinivasan, S.; Vetcher, I.A.; Abramov, S.M.; Kozlov, M.; Baughman, R.H.; Levene, S.D. Fractionation of SWNT/nucleic acid complexes by agarose gel electrophoresis. *Nanotechnology* **2006**, *17*, 4263–4269.
85. Bauer, B.J.; Becker, M.L.; Bajpai, V.; Fagan, J.A.; Hobbie, E.K.; Migler, K.; Guttman, C.M.; Blair, W.R. Measurement of single-wall nanotube dispersion by size exclusion chromatography. *J. Phys. Chem. C* **2007**, *111*, 17914–17918.
86. Sanchez-Pomales, G.; Santiago-Rodriguez, L.; Rivera-Velez, N.E.; Cabrera, C.R. Characterization of the DNA-assisted purification of single-walled carbon nanotubes. *Phys. Status Solidi A* **2007**, *204*, 1791–1796.
87. Bauer, B.J.; Fagan, J.A.; Hobbie, E.K.; Chun, J.; Bajpai, V. Chromatographic fractionation of SWNT/DNA dispersions with on-line multi-angle light scattering. *J. Phys. Chem. C* **2008**, *112*, 1842–1850.
88. Chun, J.; Fagan, J.A.; Hobbie, E.K.; Bauer, B.J. Size separation of single-wall carbon nanotubes by flow-field flow fractionation. *Anal. Chem.* **2008**, *80*, 2514–2523.
89. Kim, J.H.; Kataoka, M.; Kim, Y.A.; Shimamoto, D.; Muramatsu, H.; Hayashi, T.; Endo, M.; Terrones, M.; Dresselhaus, M.S. Diameter-selective separation of double-walled carbon nanotubes. *Appl. Phys. Lett.* **2008**, *93*, doi:10.1063/1.3039790.
90. Asada, Y.; Sugai, T.; Kitaura, R.; Shinohara, H. Chromatographic length separation and photoluminescence study on DNA-wrapped single-wall and double-wall carbon nanotubes. *J. Nanomater.* **2009**, *2009*, doi:10.1155/2009/257892.
91. Tu, X.M.; Manohar, S.; Jagota, A.; Zheng, M. DNA sequence motifs for structure-specific recognition and separation of carbon nanotubes. *Nature* **2009**, *460*, 250–253.
92. Shakhmaeva, I.I.; Bulatov, E.R.; Bondar, O.V.; Saifullina, D.V.; Culha, M.; Rizvanov, A.A.; Abdullin, T.I. Binding and purification of plasmid DNA using multi-layered carbon nanotubes. *J. Biotechnol.* **2011**, *152*, 102–107.
93. Tardani, F.; la Mesa, C.; Poulin, P.; Maugey, M. Phase behavior of DNA-based dispersions containing carbon nanotubes: Effects of added polymers and ionic strength on excluded volume. *J. Phys. Chem. C* **2012**, *116*, 9888–9894.

94. Glamazda, A.Y.; Dettlaff-Weglikowska, U.; Leontiev, V.S.; Mateichenko, P.V.; Karachevtsev, V.A. Raman spectroscopy and sem study of swnts in aqueous solution and films with surfactant or polymer surroundings. *Fuller. Nanotub. Carbon Nanostruct.* **2006**, *14*, 221–225.
95. Jeng, E.S.; Moll, A.E.; Roy, A.C.; Gastala, J.B.; Strano, M.S. Detection of DNA hybridization using the near-infrared band-gap fluorescence of single-walled carbon nanotubes. *Nano Lett.* **2006**, *6*, 371–375.
96. Karachevtsev, V.A.; Glamazda, A.Y.; Dettlaff-Weglikowska, U.; Leontiev, V.S.; Mateichenko, P.V.; Roth, S.; Rao, A.M. Spectroscopic and SEM studies of SWNTs: Polymer solutions and films. *Carbon* **2006**, *44*, 1292–1297.
97. Lacerda, L.; Pastorin, G.; Wu, W.; Prato, M.; Bianco, A.; Kostarelos, K. Luminescence of functionalized carbon nanotubes as a tool to monitor bundle formation and dissociation in water: The effect of plasmid-DNA complexation. *Adv. Funct. Mater.* **2006**, *16*, 1839–1846.
98. Asada, Y.; Dohi, H.; Kuwahara, S.; Sugai, T.; Kitaura, R.; Shinohara, H. Synthesis and spectroscopic characterization of salmon DNA-wrapped single-wall carbon nanotubes. *Nano* **2007**, *2*, 295–299.
99. Karachevtsev, V.A.; Glamazda, A.Y.; Leontiev, V.S.; Lytvyn, O.S.; Dettlaff-Weglikowska, U. Glucose sensing based on nir fluorescence of DNA-wrapped single-walled carbon nanotubes. *Chem. Phys. Lett.* **2007**, *435*, 104–108.
100. Tu, X.M.; Pehrsson, P.E.; Zhao, W. Redox reaction of DNA-encased HiPCO carbon nanotubes with hydrogen peroxide: A near infrared optical sensitivity and kinetics study. *J. Phys. Chem. C* **2007**, *111*, 17227–17231.
101. Xu, Y.; Pehrsson, P.E.; Chen, L.W.; Zhang, R.; Zhao, W. Double-stranded DNA single-walled carbon nanotube hybrids for optical hydrogen peroxide and glucose sensing. *J. Phys. Chem. C* **2007**, *111*, 8638–8643.
102. Cao, C.F.; Kim, J.H.; Yoon, D.; Hwang, E.S.; Kim, Y.J.; Baik, S. Optical detection of DNA hybridization using absorption spectra of single-walled carbon nanotubes. *Mater. Chem. Phys.* **2008**, *112*, 738–741.
103. Cathcart, H.; Nicolosi, V.; Hughes, J.M.; Blau, W.J.; Kelly, J.M.; Quinn, S.J.; Coleman, J.N. Ordered DNA wrapping switches on luminescence in single-walled nanotube dispersions. *J. Am. Chem. Soc.* **2008**, *130*, 12734–12744.
104. Qian, H.; Araujo, P.T.; Georgi, C.; Gokus, T.; Hartmann, N.; Green, A.A.; Jorio, A.; Hersam, M.C.; Novotny, L.; Hartschuh, A. Visualizing the local optical response of semiconducting carbon nanotubes to DNA-wrapping. *Nano Lett.* **2008**, *8*, 2706–2711.
105. Yoon, D.; Cao, C.; Choi, J.B.; Kim, Y.J.; Baik, S. Optical characterization of DNA-wrapped single walled carbon nanotubes irradiated with ultraviolet light. *J. Nanosci. Nanotechnol.* **2008**, *8*, 5135–5138.
106. Shoda, M.; Bandow, S.; Maruyama, Y.; Iijima, S. Probing interaction between ssDNA and carbon nanotubes by raman scattering and electron microscopy. *J. Phys. Chem. C* **2009**, *113*, 6033–6036.
107. Simpson, J.R.; Fagan, J.A.; Becker, M.L.; Hobbie, E.K.; Walker, A.R.H. The effect of dispersant on defects in length-separated single-wall carbon nanotubes measured by raman spectroscopy. *Carbon* **2009**, *47*, 3238–3241.

108. Chen, Z.; Liu, R.; Wang, Y.L.; Zhu, H.R.; Sun, Z.P.; Zuo, T.S.; Chang, X.L.; Zhao, F.; Xing, G.M.; Yuan, H.; *et al.* Ag nanoparticles coated swcnt with surface enhanced raman scattering (SERS) signals. *J. Nanosci. Nanotechnol.* **2010**, *10*, 8538–8543.
109. Park, J.J.; Fagan, J.A.; Huh, J.Y.; Migler, K.B.; Karim, A.; Raghavan, D. Spr imaging study of DNA wrapped single wall carbon nanotube (ssDNA-SWCNT) adsorption on a model biological (collagen) substrate. *Soft Matter* **2010**, *6*, 5581–5588.
110. Ignatova, T.; Najafov, H.; Ryasnyanskiy, A.; Biaggio, I.; Zheng, M.; Rotkin, S.V. Significant fret between swnt/DNA and rare earth ions: A signature of their spatial correlations. *ACS Nano* **2011**, *5*, 6052–6059.
111. Shao, J.Y.; Sun, T.; Guo, Q.Y.; Li, H.; Lan, S. *In situ* electrochemically tuned photoluminescence of $[Ru(bpy)_2(dppz)]^{2+}$ aggregates with single-walled carbon nanotubes and DNA monitored by guanine oxidation. *Transit. Met. Chem.* **2011**, *36*, 499–504.
112. Ensafi, A.A.; Amini, M.; Rezaei, B. Biosensor based on ds-DNA decorated chitosan modified multiwall carbon nanotubes for voltammetric biodetection of herbicide amitrole. *Colloid Surf. B* **2013**, *109*, 45–51.
113. Nii, D.; Hayashida, T.; Umemura, K. Controlling the adsorption and desorption of double-stranded DNA on functionalized carbon nanotube surface. *Colloid Surf. B* **2013**, *106*, 234–239.
114. Ozturk, S.A.; Kelestemur, S.; Culha, M. Influence of oligonucleotide interaction on electronic properties of single walled carbon nanotubes. *J. Raman Spectrosc.* **2013**, *44*, 183–189.
115. Ito, M.; Kobayashi, T.; Ito, Y.; Hayashida, T.; Nii, D.; Umemura, K.; Homma, Y. Intense photoluminescence from dried double-stranded DNA and single-walled carbon nanotube hybrid. *Appl. Phys. Lett.* **2014**, *104*, doi:10.1063/1.4863272
116. Takahashi, H.; Numao, S.; Bandow, S.; Iijima, S. Afm imaging of wrapped multiwall carbon nanotube in DNA. *Chem. Phys. Lett.* **2006**, *418*, 535–539.
117. Campbell, J.F.; Tessmer, I.; Thorp, H.H.; Erie, D.A. Atomic force microscopy studies of DNA-wrapped carbon nanotube structure and binding to quantum dots. *J. Am. Chem. Soc.* **2008**, *130*, 10648–10655.
118. Toita, S.; Kang, D.; Kobayashi, K.; Kawamoto, H.; Kojima, K.; Tachibana, A. Atomic force microscopic study on DNA-wrapping for different diameter single-wall carbon nanotubes. *Diam. Relat. Mater.* **2008**, *17*, 1389–1393.
119. Xu, Y.Q.; Barnard, A.; McEuen, P.L. Bending and twisting of suspended single-walled carbon nanotubes in solution. *Nano Lett.* **2009**, *9*, 1609–1614.
120. Arnett, C.M.; Marsh, C.P.; Welch, C.R.; Strano, M.S.; Han, J.H.; Gray, J.H.; Carlson, T.A. Enzyme-mediated assimilation of DNA-functionalized single-walled carbon nanotubes. *Langmuir* **2010**, *26*, 613–617.
121. Hayashida, T.; Kawashima, T.; Nii, D.; Ozasa, K.; Umemura, K. Kelvin probe force microscopy of single-walled carbon nanotubes modified with DNA or poly(ethylene glycol). *Chem. Lett.* **2013**, *42*, 666–668.
122. Hayashida, T.; Umemura, K. Surface morphology of hybrids of double-stranded DNA and single-walled carbon nanotubes studied by atomic force microscopy. *Colloid Surf. B* **2013**, *101*, 49–54.

123. Nii, D.; Hayashida, T.; Yamaguchi, Y.; Ikawa, S.; Shibata, T.; Umemura, K. Selective binding of single-stranded DNA-binding proteins onto DNA molecules adsorbed on single-walled carbon nanotubes. *Colloid Surf. B* **2014**, *121*, 325–330.
124. Iijima, M.; Watabe, T.; Ishii, S.; Koshio, A.; Yamaguchi, T.; Bandow, S.; Iijima, S.; Suzuki, K.; Maruyama, Y. Fabrication and STM-characterization of novel hybrid materials of DNA/carbon nanotube. *Chem. Phys. Lett.* **2005**, *414*, 520–524.
125. Karachevtsev, M.V.; Lytvyn, O.S.; Stepanian, S.G.; Leontiev, V.S.; Adamowicz, L.; Karachevtsev, V.A. SWNT-DNA and SWNT-polyC hybrids: AFM study and computer modeling. *J. Nanosci. Nanotechnol.* **2008**, *8*, 1473–1480.
126. Lulevich, V.; Kim, S.; Grigoropoulos, C.P.; Noy, A. Frictionless sliding of single-stranded DNA in a carbon nanotube pore observed by single molecule force spectroscopy. *Nano Lett.* **2011**, *11*, 1171–1176.
127. Rao, R.; Lee, J.; Lu, Q.; Keskar, G.; Freedman, K.O.; Floyd, W.C.; Rao, A.M.; Ke, P.C. Single-molecule fluorescence microscopy and raman spectroscopy studies of RNA bound carbon nanotubes. *Appl. Phys. Lett.* **2004**, *85*, 4228–4230.
128. Yang, R.H.; Jin, J.Y.; Chen, Y.; Shao, N.; Kang, H.Z.; Xiao, Z.; Tang, Z.W.; Wu, Y.R.; Zhu, Z.; Tan, W.H. Carbon nanotube-quenched fluorescent oligonucleotides: Probes that fluoresce upon hybridization. *J. Am. Chem. Soc.* **2008**, *130*, 8351–8358.
129. Zhang, L.B.; Tao, L.; Li, B.L.; Jing, L.; Wang, E.K. Carbon nanotube-DNA hybrid fluorescent sensor for sensitive and selective detection of mercury(II) ion. *Chem. Commun.* **2010**, *46*, 1476–1478.
130. Zhao, C.; Qu, K.G.; Song, Y.J.; Xu, C.; Ren, J.S.; Qu, X.G. A reusable DNA single-walled carbon-nanotube-based fluorescent sensor for highly sensitive and selective detection of Ag+ and cysteine in aqueous solutions. *Chemistry* **2010**, *16*, 8147–8154.
131. D'Souza, F.; Das, S.K.; Zandler, M.E.; Sandanayaka, A.S.D.; Ito, O. Bionano donor-acceptor hybrids of porphyrin, ssDNA, and semiconductive single-wall carbon nanotubes for electron transfer via porphyrin excitation. *J. Am. Chem. Soc.* **2011**, *133*, 19922–19930.
132. Qin, W.L.; Yang, K.Q.; Tang, H.; Tan, L.A.; Xie, Q.J.; Ma, M.; Zhang, Y.Y.; Yao, S.Z. Improved GFP gene transfection mediated by polyamidoamine dendrimer-functionalized multi-walled carbon nanotubes with high biocompatibility. *Colloid Surf. B* **2011**, *84*, 206–213.
133. Zhang, J.Q.; Boghossian, A.A.; Barone, P.W.; Rwei, A.; Kim, J.H.; Lin, D.H.; Heller, D.A.; Hilmer, A.J.; Nair, N.; Reuel, N.F.; *et al.* Single molecule detection of nitric oxide enabled by d(AT)(15) DNA adsorbed to near infrared fluorescent single-walled carbon nanotubes. *J. Am. Chem. Soc.* **2011**, *133*, 567–581.
134. Martinez, M.T.; Tseng, Y.C.; Gonzalez, M.; Bokor, J. Streptavidin as cnts and DNA linker for the specific electronic and optical detection of DNA hybridization. *J. Phys. Chem. C* **2012**, *116*, 22579–22586.
135. Wang, L.Y.; Cheng, Y.Q.; Wang, H.; Li, Z.P. A homogeneous fluorescence sensing platform with water-soluble carbon nanoparticles for detection of microrna and nuclease activity. *Analyst* **2012**, *137*, 3667–3672.

136. Bergler, F.F.; Schoppler, F.; Brunecker, F.K.; Hailman, M.; Hertel, T. Fluorescence spectroscopy of gel-immobilized single-wall carbon nanotubes with microfluidic control of the surfactant environment. *J. Phys. Chem. C* **2013**, *117*, 13318–13323.
137. Judkins, J.; Lee, H.H.; Tung, S.; Kim, J.W. Diffusion of single-walled carbon nanotube under physiological conditions. *J. Biomed. Nanotechnol.* **2013**, *9*, 1065–1070.
138. Hazani, M.; Naaman, R.; Hennrich, F.; Kappes, M.M. Confocal fluorescence imaging of DNA-functionalized carbon nanotubes. *Nano Lett.* **2003**, *3*, 153–155.
139. Stevens, J.L.; Huang, A.Y.; Peng, H.Q.; Chiang, L.W.; Khabashesku, V.N.; Margrave, J.L. Sidewall amino-functionalization of single-walled carbon nanotubes through fluorination and subsequent reactions with terminal diamines. *Nano Lett.* **2003**, *3*, 331–336.
140. Jung, D.H.; Kim, B.H.; Ko, Y.K.; Jung, M.S.; Jung, S.; Lee, S.Y.; Jung, H.T. Covalent attachment and hybridization of DNA oligonucleotides on patterned single-walled carbon nanotube films. *Langmuir* **2004**, *20*, 8886–8891.
141. Singh, K.V.; Pandey, R.R.; Wang, X.; Lake, R.; Ozkan, C.S.; Wang, K.; Ozkan, M. Covalent functionalization of single walled carbon nanotubes with peptide nucleic acid: Nanocomponents for molecular level electronics. *Carbon* **2006**, *44*, 1730–1739.
142. Wang, X.; Liu, F.; Andavan, G.T.S.; Jing, X.Y.; Singh, K.; Yazdanpanah, V.R.; Bruque, N.; Pandey, R.R.; Lake, R.; Ozkan, M.; *et al.* Carbon nanotube-DNA nanoarchitectures and electronic functionality. *Small* **2006**, *2*, 1356–1365.
143. Yuan, W.K.; Wu, H.; Jiang, Z.Y.; Xu, S.W. Covalent functionalization of carbon nanotubes. *Chin. J. Org. Chem.* **2006**, *26*, 1508–1517.
144. Yang, W.R.; Moghaddam, M.J.; Taylor, S.; Bojarski, B.; Wieczorek, L.; Herrmann, J.; McCall, M.J. Single-walled carbon nanotubes with DNA recognition. *Chem. Phys. Lett.* **2007**, *443*, 169–172.
145. Moghaddam, M.J.; Yang, W.R.; Bojarski, B.; Gengenbach, T.R.; Gao, M.; Zareie, H.; McCall, M.J. Azide photochemistry for facile modification of graphitic surfaces: Preparation of DNA-coated carbon nanotubes for biosensing. *Nanotechnology* **2012**, doi:10.1088/0957-4484/23/42/425503.
146. Alidori, S.; Asqiriba, K.; Londero, P.; Bergkvist, M.; Leona, M.; Scheinberg, D.A.; McDevitt, M.R. Deploying rna and DNA with functionalized carbon nanotubes. *J. Phys. Chem. C* **2013**, *117*, 5982–5992.
147. Dolash, B.D.; Lahiji, R.R.; Zemlyanov, D.Y.; Drachev, V.P.; Reifenberger, R.; Bergstrom, D.E. Sonication mediated covalent cross-linking of DNA to single-walled carbon nanotubes. *Chem. Phys.* **2013**, *413*, 11–19.
148. Singh, P.; Menard-Moyon, C.; Battigelli, A.; Toma, F.M.; Raya, J.; Kumar, J.; Nidamanuri, N.; Verma, S.; Bianco, A. Double functionalization of carbon nanotubes with purine and pyrimidine derivatives. *Chemistry* **2013**, *8*, 1472–1481.
149. Canete-Rosales, P.; Gonzalez, M.; Anson, A.; Martinez, M.T.; Yanez, C.; Bollo, S. Electrochemical characterization of oligonucleotide-carbon nanotube functionalized using different strategies. *Electrochim. Acta* **2014**, *140*, 489–496.
150. Kaufmann, A.; Kunhardt, D.; Cirillo, G.; Hampel, S.; Schwenzer, B. Functionalized carbon nanotubes as transporters for antisense oligodeoxynucleotides. *J. Mater. Chem. B* **2014**, *2*, 7000–7008.

151. Gao, H.J.; Kong, Y.; Cui, D.X.; Ozkan, C.S. Spontaneous insertion of DNA oligonucleotides into carbon nanotubes. *Nano Lett.* **2003**, *3*, 471–473.
152. Okada, T.; Kaneko, T.; Hatakeyama, R. Single-stranded DNA insertion into single-walled carbon nanotubes by ion irradiation in an electrolyte plasma. *Jpn. J. Appl. Phys. Part 1 Regul. Pap. Brief Commun. Rev. Pap.* **2006**, *45*, 8335–8339.
153. Okada, T.; Kaneko, T.; Hatakeyama, R.; Tohji, K. Electrically triggered insertion of single-stranded DNA into single-walled carbon nanotubes. *Chem. Phys. Lett.* **2006**, *417*, 288–292.
154. Kaneko, T.; Okada, T.; Hatakeyama, R. DNA encapsulation inside carbon nanotubes using micro electrolyte plasmas. *Contrib. Plasma Phys.* **2007**, *47*, 57–63.
155. Pei, Q.X.; Lim, C.G.; Cheng, Y.; Gao, H.J. Molecular dynamics study on DNA oligonucleotide translocation through carbon nanotubes. *J. Chem. Phys.* **2008**, *129*, doi:10.1063/1.2981798.
156. Zimmerli, U.; Koumoutsakos, P. Simulations of electrophoretic rna transport through transmembrane carbon nanotubes. *Biophys. J.* **2008**, *94*, 2546–2557.
157. Kamiya, K.; Okada, S. Energetics and electronic structure of encapsulated single-stranded DNA in carbon nanotubes. *Phys. Rev. B* **2011**, *83*, doi:10.1103/PhysRevB.83.155444.
158. Enyashin, A.N.; Gemming, S.; Seifert, G. DNA-wrapped carbon nanotubes. *Nanotechnology* **2007**, *18*, doi:10.1088/0957-4484/18/24/245702.
159. Frischknecht, A.L.; Martin, M.G. Simulation of the adsorption of nucleotide monophosphates on carbon nanotubes in aqueous solution. *J. Phys. Chem. C* **2008**, *112*, 6271–6278.
160. Zheng, M.; Eom, K.; Ke, C.H. Calculations of the resonant response of carbon nanotubes to binding of DNA. *J. Phys. D* **2009**, *42*, 8.
161. Bobadilla, A.D.; Seminario, J.M. DNA-CNT interactions and gating mechanism using MD and DFT. *J. Phys. Chem. C* **2011**, *115*, 3466–3474.
162. Ding, Z.J.; Jiao, Y.; Meng, S. Quantum simulation of molecular interaction and dynamics at surfaces. *Front. Phys.* **2011**, *6*, 294–308.
163. Makarucha, A.J.; Todorova, N.; Yarovsky, I. Nanomaterials in biological environment: A review of computer modelling studies. *Eur. Biophys. J. Biophys. Lett.* **2011**, *40*, 103–115.
164. Roxbury, D.; Jagota, A.; Mittal, J. Sequence-specific self-stitching motif of short single-stranded DNA on a single-walled carbon nanotube. *J. Am. Chem. Soc.* **2011**, *133*, 13545–13550.
165. Umadevi, D.; Sastry, G.N. Quantum mechanical study of physisorption of nucleobases on carbon materials: Graphene *versus* carbon nanotubes. *J. Phys. Chem. Lett.* **2011**, *2*, 1572–1576.
166. Xiao, Z.T.; Wang, X.; Xu, X.; Zhang, H.; Li, Y.; Wang, Y.H. Base- and structure-dependent DNA dinucleotide-carbon nanotube interactions: Molecular dynamics simulations and thermodynamic analysis. *J. Phys. Chem. C* **2011**, *115*, 21546–21558.
167. Akdim, B.; Pachter, R.; Day, P.N.; Kim, S.S.; Naik, R.R. On modeling biomolecular-surface nonbonded interactions: Application to nucleobase adsorption on single-wall carbon nanotube surfaces. *Nanotechnology* **2012**, *23*, doi:10.1088/0957-4484/23/16/165703.
168. Alegret, N.; Santos, E.; Rodriguez-Fortea, A.; Rius, F.X.; Poblet, J.M. Disruption of small double stranded DNA molecules on carbon nanotubes: A molecular dynamics study. *Chem. Phys. Lett.* **2012**, *525–526*, 120–124.
169. Roxbury, D.; Mittal, J.; Jagota, A. Molecular-basis of single-walled carbon nanotube recognition by single-stranded DNA. *Nano Lett.* **2012**, *12*, 1464–1469.

170. Santosh, M.; Panigrahi, S.; Bhattacharyya, D.; Sood, A.K.; Maiti, P.K. Unzipping and binding of small interfering RNA with single walled carbon nanotube: A platform for small interfering RNA delivery. *J. Chem. Phys.* **2012**, *136*, doi:10.1063/1.3682780.
171. Yamazaki, T.; Fenniri, H. Imaging carbon nanotube interaction with nucleobases in water using the statistical mechanical theory of molecular liquids. *J. Phys. Chem. C* **2012**, *116*, 15087–15092.
172. Amirani, M.C.; Tang, T.; Cuervo, J. Quantum mechanical treatment of binding energy between DNA nucleobases and carbon nanotube: A DFT analysis. *Physica E* **2013**, *54*, 65–71.
173. Clavier, A.; Kraszewski, S.; Ramseyer, C.; Picaud, F. Insertion kinetics of small nucleotides through single walled carbon nanotube. *J. Biotechnol.* **2013**, *164*, 13–18.
174. Wu, N.; Wang, Q.; Pang, S.S. Dispersion of a bundle of carbon nanotubes by mechanical torsional energy. *Carbon* **2013**, *59*, 229–236.
175. Ostojic, G.N.; Ireland, J.R.; Hersam, M.C. Noncovalent functionalization of DNA-wrapped single-walled carbon nanotubes with platinum-based DNA cross-linkers. *Langmuir* **2008**, *24*, 9784–9789.
176. Zhang, L.; Huang, C.Z.; Li, Y.F.; Xiao, S.J.; Xie, J.P. Label-free detection of sequence-specific DNA with multiwalled carbon nanotubes and their light scattering signals. *J. Phys. Chem. B* **2008**, *112*, 7120–7122.
177. Zhou, Z.P.; Kang, H.; Clarke, M.L.; Lacerda, S.H.D.; Zhao, M.H.; Fagan, J.A.; Shapiro, A.; Nguyen, T.; Hwang, J. Water-soluble DNA-wrapped single-walled carbon-nanotuble/quantum-dot complexes. *Small* **2009**, *5*, 2149–2155.
178. Campbell, J.F.; Napier, M.E.; Feldberg, S.W.; Thorp, H.H. Metal-mediated electrochemical oxidation of DNA-wrapped carbon nanotubes. *J. Phys. Chem. B* **2010**, *114*, 8861–8870.
179. Weizmann, Y.; Chenoweth, D.M.; Swager, T.M. Addressable terminally linked DNA-CNT nanowires. *J. Am. Chem. Soc.* **2010**, *132*, 14009–14011.
180. Aravind, S.S.J.; Ramaprabhu, S. Noble metal dispersed multiwalled carbon nanotubes immobilized ss-DNA for selective detection of dopamine. *Sens. Actuator B* **2011**, *155*, 679–686.
181. Asada, Y.; Miyata, Y.; Shiozawa, K.; Ohno, Y.; Kitaura, R.; Mizutani, T.; Shinohara, H. Thin-film transistors with length-sorted DNA-wrapped single-wall carbon nanotubes. *J. Phys. Chem. C* **2011**, *115*, 270–273.
182. Cao, S.R.; Wang, J.F.; Yuan, R.; Chai, Y.Q. An amperometric immunoassay for human chorionic gonadotrophin based on DNA-multiwalled carbon nanotubes and methylene blue as sensing interface. *Sens. Lett.* **2011**, *9*, 1636–1642.
183. Fujigaya, T.; Yamamoto, Y.; Kano, A.; Maruyama, A.; Nakashima, N. Enhanced cell uptake via non-covalent decollation of a single-walled carbon nanotube-DNA hybrid with polyethylene glycol-grafted poly(l-lysine) labeled with an alexa-dye and its efficient uptake in a cancer cell. *Nanoscale* **2011**, *3*, 4352–4358.
184. Guo, Q.Y.; Shao, J.Y.; Sun, T.; Li, H.; Lan, S.; Xu, Z.H. Electrochemical fabrication and potential-enhanced luminescence of $[Ru(bpy)_2tatp]^{2+}$ incorporating DNA-stabilized single-wall carbon nanotubes on an indium tin oxide electrode. *Electrochim. Acta* **2011**, *56*, 1432–1438.
185. Shi, J.; Cha, T.G.; Claussen, J.C.; Diggs, A.R.; Choi, J.H.; Porterfield, D.M. Microbiosensors based on DNA modified single-walled carbon nanotube and Pt black nanocomposites. *Analyst* **2011**, *136*, 4916–4924.

186. Han, S.P.; Maune, H.T.; Barish, R.D.; Bockrath, M.; Goddard, W.A. DNA-linker-induced surface assembly of ultra dense parallel single walled carbon nanotube arrays. *Nano Lett.* **2012**, *12*, 1129–1135.
187. Kawaguchi, M.; Yamazaki, J.; Ohno, J.; Fukushima, T. Preparation and binding study of a complex made of DNA-treated single-walled carbon nanotubes and antibody for specific delivery of a "molecular heater" platform. *Int. J. Nanomed.* **2012**, *7*, 4363–4371.
188. Shin, S.R.; Lee, C.K.; Eom, T.W.; Lee, S.H.; Kwon, C.H.; So, I.; Kim, S.J. DNA-coated MWNT microfibers for electrochemical actuator. *Sens. Actuator B* **2012**, *162*, 173–177.
189. Thuy, N.T.; Tam, P.D.; Tuan, M.A.; Le, A.T.; Tam, L.T.; Thu, V.V.; Hieu, N.V.; Chien, N.D. Detection of pathogenic microorganisms using biosensor based on multi-walled carbon nanotubes dispersed in DNA solution. *Curr. Appl. Phys.* **2012**, *12*, 1553–1560.
190. Yu, T.; Gong, Y.X.; Lu, T.T.; Wei, L.; Li, Y.Q.; Mu, Y.G.; Chen, Y.; Liao, K. Recognition of carbon nanotube chirality by phage display. *RSC Adv.* **2012**, *2*, 1466–1476.
191. Zhang, L.Y.; Guo, C.X.; Cui, Z.M.; Guo, J.; Dong, Z.L.; Li, C.M. DNA-directed growth of Pd nanocrystals on carbon nanotubes towards efficient oxygen reduction reactions. *Chemistry* **2012**, *18*, 15693–15698.
192. Gong, J.L.; Sarkar, T.; Badhulika, S.; Mulchandani, A. Label-free chemiresistive biosensor for mercury (II) based on single-walled carbon nanotubes and structure-switching DNA. *Appl. Phys. Lett.* **2013**, *102*, doi:10.1063/1.4773569.
193. Mangalum, A.; Rahman, M.; Norton, M.L. Site-specific immobilization of single-walled carbon nanotubes onto single and one-dimensional DNA origami. *J. Am. Chem. Soc.* **2013**, *135*, 2451–2454.
194. Sha, J.J.; Hasan, T.; Milana, S.; Bertulli, C.; Bell, N.A.W.; Privitera, G.; Ni, Z.H.; Chen, Y.F.; Bonaccorso, F.; Ferrari, A.C.; *et al.* Nanotubes complexed with DNA and proteins for resistive-pulse sensing. *ACS Nano* **2013**, *7*, 8857–8869.
195. Su, H.C.; Zhang, M.L.; Bosze, W.; Lim, J.H.; Myung, N.V. Metal nanoparticles and DNA co-functionalized single-walled carbon nanotube gas sensors. *Nanotechnology* **2013**, *24*, doi:10.1088/0957-4484/24/50/505502.
196. Gong, X.; Sharma, A.K.; Strano, M.S.; Mukhopadhyay, D. Selective assembly of DNA-conjugated single-walled carbon nanotubes from the vascular secretome. *ACS Nano* **2014**, *8*, 9126–9136.
197. Guo, C.X.; Chitre, A.A.; Lu, X.M. DNA-assisted assembly of carbon nanotubes and MnO2 nanospheres as electrodes for high-performance asymmetric supercapacitors. *Phys. Chem. Chem. Phys.* **2014**, *16*, 4672–4678.
198. Tardani, F.; Pucci, C.; La Mesa, C. Confining ss-DNA/carbon nanotube complexes in ordered droplets. *Soft Matter* **2014**, *10*, 1024–1031.
199. Tardani, F.; Sennato, S. Phase behavior of DNA-stabilized carbon nanotubes dispersions: Association with oppositely-charged additives. *J. Phys. Chem. C* **2014**, *118*, 9268–9274.
200. Williams, R.M.; Nayeem, S.; Dolash, B.D.; Sooter, L.J. The effect of DNA-dispersed single-walled carbon nanotubes on the polymerase chain reaction. *PLoS One* **2014**, *9*, doi:10.1371/journal.pone.0094117.

201. Gowtham, S.; Scheicher, R.H.; Ahuja, R.; Pandey, R.; Karna, S.P. Physisorption of nucleobases on graphene: Density-functional calculations. *Phys. Rev. B* **2007**, *76*, doi:10.1103/PhysRevB.76.033401.
202. Varghese, N.; Mogera, U.; Govindaraj, A.; Das, A.; Maiti, P.K.; Sood, A.K.; Rao, C.N.R. Binding of DNA nucleobases and nucleosides with graphene. *ChemPhysChem* **2009**, *10*, 206–210.
203. He, S.J.; Song, B.; Li, D.; Zhu, C.F.; Qi, W.P.; Wen, Y.Q.; Wang, L.H.; Song, S.P.; Fang, H.P.; Fan, C.H. A graphene nanoprobe for rapid, sensitive, and multicolor fluorescent DNA analysis. *Adv. Funct. Mater.* **2010**, *20*, 453–459.
204. Liu, F.; Choi, J.Y.; Seo, T.S. DNA mediated water-dispersible graphene fabrication and gold nanoparticle-graphene hybrid. *Chem. Commun.* **2010**, *46*, 2844–2846.
205. Liu, J.B.; Li, Y.L.; Li, Y.M.; Li, J.H.; Deng, Z.X. Noncovalent DNA decorations of graphene oxide and reduced graphene oxide toward water-soluble metal-carbon hybrid nanostructures via self-assembly. *J. Mater. Chem.* **2010**, *20*, 900–906.
206. Lv, W.; Guo, M.; Liang, M.H.; Jin, F.M.; Cui, L.; Zhi, L.J.; Yang, Q.H. Graphene-DNA hybrids: Self-assembly and electrochemical detection performance. *J. Mater. Chem.* **2010**, *20*, 6668–6673.
207. Zhang, Q.A.; Qiao, Y.; Hao, F.; Zhang, L.; Wu, S.Y.; Li, Y.; Li, J.H.; Song, X.M. Fabrication of a biocompatible and conductive platform based on a single-stranded DNA/graphene nanocomposite for direct electrochemistry and electrocatalysis. *Chemistry* **2010**, *16*, 8133–8139.
208. Tang, L.H.; Wang, Y.; Liu, Y.; Li, J.H. DNA-directed self-assembly of graphene oxide with applications to ultrasensitive oligonucleotide assay. *ACS Nano* **2011**, *5*, 3817–3822.
209. Wei, L.L.; Borowiec, J.; Zhu, L.H.; Zhang, J.D. Electrochemical sensor for monitoring the photodegradation of catechol based on DNA-modified graphene oxide. *Microchim. Acta* **2011**, *173*, 439–443.
210. Wu, M.; Kempaiah, R.; Huang, P.J.J.; Maheshwari, V.; Liu, J.W. Adsorption and desorption of DNA on graphene oxide studied by fluorescently labeled oligonucleotides. *Langmuir* **2011**, *27*, 2731–2738.
211. Zhao, X.C. Self-assembly of DNA segments on graphene and carbon nanotube arrays in aqueous solution: A molecular simulation study. *J. Phys. Chem. C* **2011**, *115*, 6181–6189.
212. Kabelac, M.; Kroutil, O.; Predota, M.; Lankas, F.; Sip, M. Influence of a charged graphene surface on the orientation and conformation of covalently attached oligonucleotides: A molecular dynamics study. *Phys. Chem. Chem. Phys.* **2012**, *14*, 4217–4229.
213. Le, D.; Kara, A.; Schroder, E.; Hyldgaard, P.; Rahman, T.S. Physisorption of nucleobases on graphene: A comparative van der waals study. *J. Phys. Condes. Matter* **2012**, *24*, doi:10.1088/0953-8984/24/42/424210.
214. Premkumar, T.; Geckeler, K.E. Graphene-DNA hybrid materials: Assembly, applications, and prospects. *Prog. Polym. Sci.* **2012**, *37*, 515–529.
215. Qu, K.G.; Wu, L.; Ren, J.S.; Qu, X.G. Natural DNA-modified graphene/pd nanoparticles as highly active catalyst for formic acid electro-oxidation and for the Suzuki reaction. *ACS Appl. Mater. Interfaces* **2012**, *4*, 5001–5009.
216. Manna, A.K.; Pati, S.K. Theoretical understanding of single-stranded DNA assisted dispersion of graphene. *J. Mat. Chem. B* **2013**, *1*, 91–100.

217. Zhang, X.; Gao, F.; Cai, X.L.; Zheng, M.X.; Gao, F.; Jiang, S.L.; Wang, Q.X. Application of graphene-pyrenebutyric acid nanocomposite as probe oligonucleotide immobilization platform in a DNA biosensor. *Mater. Sci. Eng. C* **2013**, *33*, 3851–3857.
218. Joseph, D.; Seo, S.; Williams, D.R.; Geckeer, K.E. Double-stranded DNA-graphene hybrid: Preparation and anti-proliferative activity. *ACS Appl. Mater. Interfaces* **2014**, *6*, 3347–3356.
219. Shekar, S.C.; Swathi, R.S. Stability of nucleobases and base pairs adsorbed on graphyne and graphdiyne. *J. Phys. Chem. C* **2014**, *118*, 4516–4528.
220. Belyanskaya, L.; Weigel, S.; Hirsch, C.; Tobler, U.; Krug, H.F.; Wick, P. Effects of carbon nanotubes on primary neurons and glial cells. *Neurotoxicology* **2009**, *30*, 702–711.
221. Deng, X.Y.; Wu, F.; Liu, Z.; Luo, M.; Li, L.; Ni, Q.S.; Jiao, Z.; Wu, M.H.; Liu, Y.F. The splenic toxicity of water soluble multi-walled carbon nanotubes in mice. *Carbon* **2009**, *47*, 1421–1428.
222. Folkmann, J.K.; Risom, L.; Jacobsen, N.R.; Wallin, H.; Loft, S.; Moller, P. Oxidatively damaged DNA in rats exposed by oral gavage to C60 fullerenes and single-walled carbon nanotubes. *Environ. Health Perspect.* **2009**, *117*, 703–708.
223. Antonelli, A.; Serafini, S.; Menotta, M.; Sfara, C.; Pierige, F.; Giorgi, L.; Ambrosi, G.; Rossi, L.; Magnani, M. Improved cellular uptake of functionalized single-walled carbon nanotubes. *Nanotechnology* **2010**, *21*, doi:10.1088/0957-4484/21/42/425101.
224. Asakura, M.; Sasaki, T.; Sugiyama, T.; Takaya, M.; Koda, S.; Nagano, K.; Arito, H.; Fukushima, S. Genotoxicity and cytotoxicity of multi-wall carbon nanotubes in cultured chinese hamster lung cells in comparison with chrysotile a fibers. *J. Occup. Health* **2010**, *52*, 155–166.
225. Cveticanin, J.; Joksic, G.; Leskovac, A.; Petrovic, S.; Sobot, A.V.; Neskovic, O. Using carbon nanotubes to induce micronuclei and double strand breaks of the DNA in human cells. *Nanotechnology* **2010**, *21*, doi:10.1088/0957-4484/21/1/015102.
226. Lima, A.; Melo, A.M.D.; Pires, E.V.; Ferreira, R.C.D.; Sant'Ana, A.E.G.; Goulart, M.O.F.; de Abreu, F.C. Electroanalytical studies of sulfentrazone in protic medium, its degradation by the electro-fenton process, and toxicity assessment using ss-DNA. *Chemosphere* **2010**, *81*, 884–889.
227. Luszczyn, J.; Plonska-Brzezinska, M.E.; Palkar, A.; Dubis, A.T.; Simionescu, A.; Simionescu, D.T.; Kalska-Szostko, B.; Winkler, K.; Echegoyen, L. Small noncytotoxic carbon nano-onions: First covalent functionalization with biomolecules. *Chemistry* **2010**, *16*, 4870–4880.
228. Raffa, V.; Ciofani, G.; Vittorio, O.; Riggio, C.; Cuschieri, A. Physicochemical properties affecting cellular uptake of carbon nanotubes. *Nanomedicine* **2010**, *5*, 89–97.
229. Sabuncu, A.C.; Kalluri, B.S.; Qian, S.Z.; Stacey, M.W.; Beskok, A. Dispersion state and toxicity of mwCNTs in cell culture medium with different T80 concentrations. *Colloid Surf. B* **2010**, *78*, 36–43.
230. Shen, C.X.; Zhang, Q.F.; Li, J.A.; Bi, F.C.; Yao, N. Induction of programmed cell death in arabidopsis and rice by single-wall carbon nanotubes. *Am. J. Bot.* **2010**, *97*, 1602–1609.
231. Nam, C.W.; Kang, S.J.; Kang, Y.K.; Kwak, M.K. Cell growth inhibition and apoptosis by sds-solubilized single-walled carbon nanotubes in normal rat kidney epithelial cells. *Arch. Pharm. Res.* **2011**, *34*, 661–669.
232. Dong, P.X.; Wan, B.; Guo, L.H. In vitro toxicity of acid-functionalized single-walled carbon nanotubes: Effects on murine macrophages and gene expression profiling. *Nanotoxicology* **2012**, *6*, 288–303.

233. Handy, R.D.; van den Brink, N.; Chappell, M.; Muhling, M.; Behra, R.; Dusinska, M.; Simpson, P.; Ahtiainen, J.; Jha, A.N.; Seiter, J.; *et al.* Practical considerations for conducting ecotoxicity test methods with manufactured nanomaterials: What have we learnt so far? *Ecotoxicology* **2012**, *21*, 933–972.
234. Kim, J.S.; Sung, J.H.; Song, K.S.; Lee, J.H.; Kim, S.M.; Lee, G.H.; Ahn, K.H.; Lee, J.S.; Shin, J.H.; Park, J.D.; *et al.* Persistent DNA damage measured by comet assay of sprague dawley rat lung cells after five days of inhalation exposure and 1 month post-exposure to dispersed multi-wall carbon nanotubes (MWCNTs) generated by new MWCNT aerosol generation system. *Toxicol. Sci.* **2012**, *128*, 439–448.
235. Mogurampelly, S.; Panigrahi, S.; Bhattacharyya, D.; Sood, A.K.; Maiti, P.K. Unraveling siRNA unzipping kinetics with graphene. *J. Chem. Phys.* **2012**, *137*, doi:10.1063/1.4742189.
236. Van Berlo, D.; Clift, M.; Albrecht, C.; Schins, R. Carbon nanotubes: An insight into the mechanisms of their potential genotoxicity. *Swiss Med. Wkly.* **2012**, *142*, doi:10.4414/smw.2012.13698.
237. Awasthi, K.K.; John, P.J.; Awasthi, A.; Awasthi, K. Multi walled carbon nano tubes induced hepatotoxicity in Swiss albino mice. *Micron* **2013**, *44*, 359–364.
238. Mao, H.L.; Kawazoe, N.; Chen, G.P. Uptake and intracellular distribution of collagen-functionalized single-walled carbon nanotubes. *Biomaterials* **2013**, *34*, 2472–2479.
239. Zhang, Y.L.; Deng, J.J.; Zhang, Y.X.; Guo, F.; Li, C.G.; Zou, Z.; Xi, W.; Tang, J.; Sun, Y.; Yang, P.; *et al.* Functionalized single-walled carbon nanotubes cause reversible acute lung injury and induce fibrosis in mice. *J. Mol. Med.* **2013**, *91*, 117–128.
240. Chen, W.; Xiong, Q.; Ren, Q.X.; Guo, Y.K.; Li, G. Can amino-functionalized carbon nanotubes carry functional nerve growth factor? *Neural Regen. Res.* **2014**, *9*, 285–292.
241. Clichici, S.; Biris, A.R.; Catoi, C.; Filip, A.; Tabaran, F. Short-term splenic impact of single-strand DNA functionalized multi-walled carbon nanotubes intraperitoneally injected in rats. *J. Appl. Toxicol.* **2014**, *34*, 332–344.
242. Hu, C.W.; Zhang, L.J.; Wang, W.L.; Cui, Y.B.; Li, M. Evaluation of the combined toxicity of multi-walled carbon nanotubes and sodium pentachlorophenate on the earthworm *Eisenia fetida* using avoidance bioassay and comet assay. *Soil Biol. Biochem.* **2014**, *70*, 123–130.
243. Liu, Z.B.; Liu, Y.F.; Peng, D.M. Hydroxylation of multi-walled carbon nanotubes reduces their cytotoxicity by limiting the activation of mitochondrial mediated apoptotic pathway. *J. Mater. Sci. Mater. Med.* **2014**, *25*, 1033–1044.
244. McShan, D.; Yu, H.T. DNA damage in human skin keratinocytes caused by multiwalled carbon nanotubes with carboxylate functionalization. *Toxicol. Ind. Health* **2014**, *30*, 489–498.
245. Tavares, A.M.; Louro, H.; Antunes, S.; Quarre, S.; Simar, S.; de Temmerman, P.J.; Verleysen, E.; Mast, J.; Jensen, K.A.; Norppa, H.; *et al.* Genotoxicity evaluation of nanosized titanium dioxide, synthetic amorphous silica and multi-walled carbon nanotubes in human lymphocytes. *Toxicol. Vitro* **2014**, *28*, 60–69.

6

In Situ Self Assembly of Nanocomposites: Competition of Chaotic Advection and Interfacial Effects as Observed by X-Ray Diffreaction

Dilru R. Ratnaweera [1], Chaitra Mahesha [2], David A. Zumbrunnen [2] and Dvora Perahia [1,*]

[1] Department of Chemistry, Clemson University, Clemson, SC 29634-0973, USA;
E-Mail: dratnaw@clemson.edu

[2] Department of Mechanical Engineering, Clemson University, Clemson, SC 29634-0921, USA;
E-Mails: cmahesh@clemson.edu (C.M.); zdavid@clemson.edu (D.A.Z.)

* Author to whom correspondence should be addressed; E-Mail: dperahi@clemson.edu

Academic Editor: Thibaud Coradin

Abstract: The effects of chaotic advection on the *in situ* assembly of a hierarchal nanocomposite of Poly Amide 6, (nylon 6 or PA6) and platelet shape nanoparticles (NPs) were studied. The assemblies were formed by chaotic advection, where melts of pristine PA6 and a mixture of PA6 with NPs were segregated into discrete layers and extruded into film in a continuous process. The process assembles the nanocomposite into alternating pristine-polymer and oriented NP/polymer layers. The structure of these hierarchal assemblies was probed by X-rays as a processing parameter, N, was varied. This parameter provides a measure of the extent of *in situ* structuring by chaotic advection. We found that all assemblies are semi-crystalline at room temperature. Increasing N impacts the ratio of α to γ crystalline forms. The effects of the chaotic advection vary with the concentration of the NPs. For nanocomposites with lower NP concentrations the amount of the γ crystalline form increased with N. However, at higher NP concentrations, interfacial effects of the NP play a significant role in determining the structure, where the NPs oriented along the melt flow direction and the polymer chains oriented perpendicular to the NP surfaces.

Keywords: assembly; nanocomposites; chaotic advection; nylon; smart blending; X-ray diffraction

1. Introduction

Controlled assembly of structured layers of polymers that contain different types of nanoparticles (NPs) offers a means to tailor multifunctional materials. NPs play an important role in modifying the properties of polymers, where incorporation of small amounts of particles significantly changes the physical characteristics of polymers [1–6]. For example, a fivefold increase in tensile strength was observed [7] when mica NPs were added to nylon (Poly Amide 6, or PA6). Significant effects were also notable on optical transparency and nonlinearity of optical response of polymers [7]. Additional effects of NPs, such as controlled permeability [8], higher thermal stability [9] and controlled electrical conductivity [10], will benefit from well-dispersed NPs. While the dispersion and orientation of the NPs affect the composite's properties [11], effective dispersion of NPs in a polymeric matrix remains a challenge due to the tendency of NPs to aggregate [12,13]. Here we probe a new hierarchal assembly that takes advantage of inherent structuring that occurs under non-linear flow in "Smart Blending", an extrusion technique that utilizes chaotic advection to controllably structure soft materials that are processable in a viscous liquid state [14–16].

Current and potential technological benefits of polymer nanocomposites lead to numerous efforts, including chemical modifications of the NPs and matrix polymers to drive assembly into desired structures [2,17]. The interaction energies between NPs and polymers, the size of NPs and their relative dimension with respect to the polymer rigid segment as well as the shape of the NP affect the assembly of the particles and the properties of the resulting nanocomposite [18–21]. Directing the assembly of NPs via chemical modifications, coupled with large-scale extrusion that can further structure the resulting composite, provide a powerful method to move processes on a lab scale to an actual technologically viable process. One such a process is a blending technique often referred to as "Smart Blending", where chaotic advection is introduced into melts of two polymers, resulting in localization of the polymers into layers of controllable thickness and number [22–26]. These layers breakup due to a balance between shear and interfacial tension may lead to a wide variety of different blend morphologies. We have shown that using Smart Blending of two polymer-melts, where one contains NPs, results in hierarchical nanocomposites that consist of alternating layers of a polymer and polymer-NP layers [27].

Here we probe by X-ray diffraction the assembly of molecules within layers formed by chaotic advection. Morphologies observed for of a blended PA6/PA6-montmorillonite nanocomposite, as depicted by transmission electron microscopy (TEM) are shown in Figure 1. At low and medium level of the processing parameter N, for NP of 2, a clear segregation into layers, PA6 and PA6-NPs rich layers is observed, as shown in Figure 1a [27]. With increasing N, the micron-size layers become thinner and hardly any segregation is observed between PA6 and PA6-NPs. Increasing the loading of the NP results in a composite with well oriented NPs, with lower segregation of the pristine and NP layers as shown in Figure 1b. The current study probed this non-traditional assembly of nano-particles and polymers on the nm length scale.

The current study presents for the first time the effects of chaotic advection on the structure of hierarchal nanocomposites; *i.e.*, the effects of non-linear flows on the assembly, using X-ray diffraction. Chaotic advection is a process in which flow tracers move chaotically in response to flow fields that can be simple. Consequently, flow domains stretch and fold, resulting in exponentially fast reductions in domain size [22–26] forming a layered morphology. The propagation of these flows is characterized by the Lyapunov exponent by λ, where the separation of trajectories at time t is given by $l_t = l_0 e^{\lambda\tau}$ where l_0 is the separation at $t = 0$. This ideal mathematical description is mitigated by finite size effects in actual blending processes. Aref [22] demonstrated that blinking vortex flows introduced by two point vortices that are alternately activated, introduce chaotic advection. This fluid mechanics study [22] led to the development of Smart Blending instruments leading to formation of new polymer blend morphologies where the extent of structuring can be regulated by selecting the duration of chaotic advection [23–27]. A schematic representation of the agitating rods which induce flow similar to Aref's vortices are shown in Figure 2. The protocol for rotating the rods is also shown, where one set of motions for both rods constituted a one blending period ($N = 1$). N corresponds to a blending parameter and represents a sequence of rod-rotation.

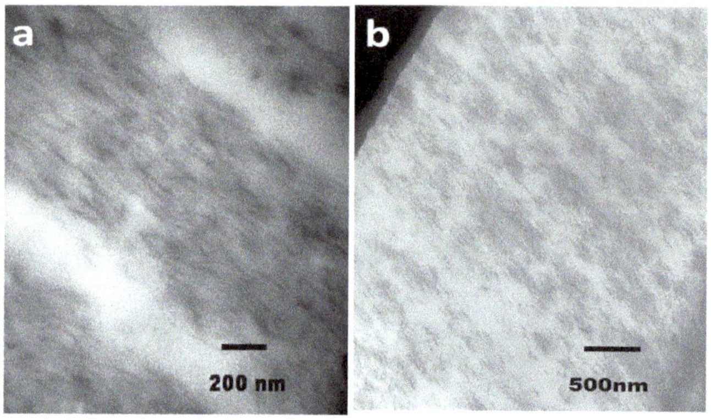

Figure 1. Transmission electron microscopy (TEM) image of (**a**) 2 vol % nanoparticles (NPs) montmorillonite nanocomposite formed by chaotic advection at $N = 9$ [27]; and (**b**) 5.6 vol % at $N = 7$, extruded as 150 micron thick films. Dark regions correspond to the NP rich areas and light regions correspond to the matrix polymer.

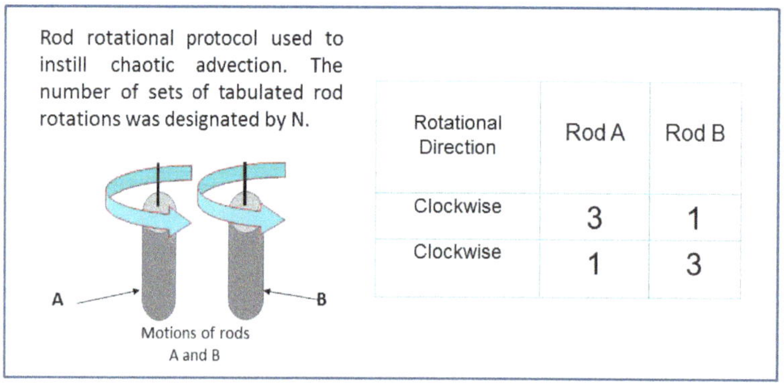

Figure 2. Schematic presentation of the agitator in chaotic advection and the time evolution of the slow. The agitation scheme used in the current study for $N = 1$ is presented.

In the experiments, several sets of such rotations were prescribed such that $N = 5$, for example, constituted five sets of rod rotations. The resulting structured melts were then extruded via a film die onto a chill roll to form films that were then studied by TEM and X-ray diffraction. Details of the smart blender are given in [25]. For polymers, quenching below their glass transition temperature retains the structures formed in the liquid phase.

Here, nanocomposites were formed in this study using two initially separate melt streams, one of nylon-6 and the other of a mixture of nylon-6/NPs. The effects of chaotic advection on the structure of the nanocomposite on a 0.1–2.5 nm length scale, where the packing of the polymer molecules is detected, were probed. PA6/montmorillonite NPs, a well-studied nanocomposite, was used as a model system. PA6 is a semi-crystalline polymer that has two major crystalline forms: α and γ. The α form consists of hydrogen bonds in between the fully extended anti parallel PA6 chains and the γ form has hydrogen bonds in between parallel pleated chains [28–30]. In the γ form, the plane of the amide group is roughly perpendicular to the plane of $(CH_2)_5$ groups where in the α form those planes are parallel. The α form has a monoclinic unit cell and the γ form closely resembles a hexagonal structure [31]. The α form is thermodynamically stable and the γ form is meta-stable. Previous studies have shown that adding NPs to PA6 enhances the formation of the γ form regardless of the blending technique [2,32]. Other factors that affect the type of crystalline phase formed include cooling rates [33], where enhancement of the γ crystalline form was observed when quenching by liquid N_2 [34].

The structure at different *N* was followed by X-ray. The blend composition and the processing conditions as described in Table 1 for a given *N* and a total loading of nanoparticles. The study attempted to compare as similar as possible blends subject to experimental variability in a multistage process.

Table 1. Samples studied at the indicated *N* and the total loading of nanoparticles.

N (vol % = 0.0)	*N* (vol % = 2.0)	*N* (vol % = 2.8)	*N* (vol % = 3.5)	*N* (vol % = 5.6)
00	06	08	00	00
30	12	10	08	07
	14	12	10	09
	20	20	12	20
	30	25	16	
			20	
			22	

The results are compared with those of the pure polymer under similar conditions. The study has shown that chaotic advection not only structured the films, but also changes the ratio of the crystalline forms of the polymer depending on the concentration of the NPs, where at higher concentrations of NPs, surface effects become significant.

2. Results and Discussion

The structuring and assembly processes take place in the melts that are then quenched, and the resulting structure is studied by X-ray diffraction. Further details of sample preparation are given in the experimental section. The alignment of the micrometer layers as well as those of the NPs was probed along three different directions, *X*, *Y* and *Z*, with respect to the final extrusion directions as demonstrated in Figure 3.

In order to set the background for the NP assembly and its impact, we first probed the pristine polymer, comparing extruded PA6 in comparison with a film formed by a melt exposed to chaotic. X-ray patterns of a pristine PA6 films that were extruded without chaotic advection and otherwise identical films that were subject to *in situ* structuring by chaotic advection are shown in Figure. The pattern of the pristine polymer without chaotic advection (Figure 4a) consists of two broad peaks superimposed by a smaller one in the center. These peaks at 4.39 Å and 3.75 Å correspond to the α crystalline form and the center at 4.16 Å corresponds to the γ crystalline form. These results are consistent with previously reported data [35,36]. The α form dominates the crystalline structure of granular PA6 and exhibit two diffraction lines: 4.39 Å refers to α1 and originates from the (200) diffraction plane and 3.75 Å (α2) corresponds to the (002) plane. The peak of the γ form corresponds to the (001) diffraction plane. These diffraction lines correspond to chain-chain correlations in the crystalline domains. These values for the positions of the α and γ peaks were used to determine peak assignments in deconvolution of the nanocomposite patterns. The patterns were deconvoluted assuming a Gaussian line shape to resolve the relative amounts of α and γ as well as to determine the ratios of crystalline to amorphous fractions [33]. The results of the de-convolution are shown as solid lines in Figure 4.

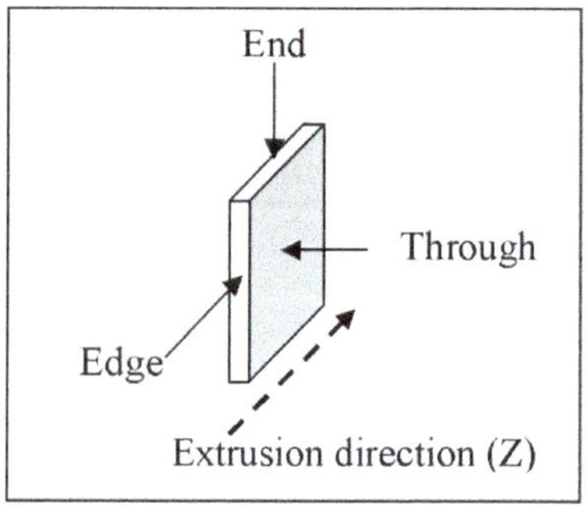

Figure 3. Directions of incident X-ray beams with respect to the extrusion direction. *X* and *Y* define the plane of the film and *Z* corresponds to the extrusion directions.

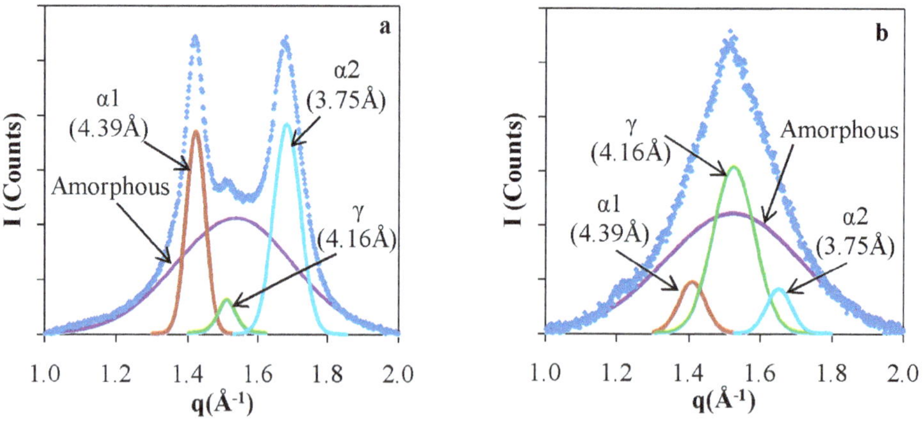

Figure 4. X-ray patterns and analysis of Poly Amide 6 (PA6) films (**a**) Pristine; and (**b**) Chaotically blended at $N = 30$.

The X-ray pattern of the chaotically blended PA6, shown in Figure 4b, consists predominantly of a peak that corresponds to the γ form, accompanied by two less intense α peaks. The ratios of α to γ changes from 27.4 to 0.8 and crystalline/amorphous change from 0.7 to 0.6 upon extruding the PA6 granules to make films at $N = 30$. All chaotically blended samples of the pristine PA6 had predominantly the γ crystalline form of the PA6. We attribute the formation of the meta-stable form to local shear effects that impact the alignment of the polymer chains [15].

For the nanocomposite films, extruded films from the chaotic blender consist of alternate layers of NP rich and NP free PA6 domains as shown in the TEM image in Figure 1. The X-ray data consist of average of scattering contributions from both NP-free polymer layers and layers containing the platelets. These layers are micron thick and are sufficiently thick to neglect interfacial effects arising from the PA6–PA6/NP boundaries. The effects of the NPs are extracted by comparing the results to those of PA6 exposed to chaotic advection. The differences are attributed to effects of the NPs.

To form the nanocomposite films, NPs were first added and well mixed with PA6 polymer to produce a melt of desired NP concentration. Then, melts of pristine PA6 and the NP containing melt were combined in the chaotic blender and were extruded after an *in situ* structuring, where the degree of structuring is represented by the processing parameter N. We found that the flow impacts both the assembly of the NPs (*i.e.*, platelet orientation and confinement within PA6 layers) and the crystal structure of the polymer matrix. When NPs are imbedded in the polymer, the crystallinity of the γ form is enhanced, as shown in the X-ray patterns for extruded PA6 with and without NPs at $N = 0$ in Figure 5. For $N = 0$, the polymer melts were not subjected to chaotic advection and were simply extruded. The peak in the pure PA6 is significantly broader than that of the films containing NPs. The X-ray line width is inversely proportional to the size of the coherently scattering domains and ordering/crystalline domain size in the system. The full peak widths at half maxima δ (FWHM) are given in Table 2. Increasing NP concentration from 0 to 5.6 vol % decreases the line widths from 0.29 $Å^{-1}$ to 0.07 $Å^{-1}$ at N = 0. These values correspond to π/δ, the size of the coherently scattering domains, of ~20 Å to ~90 Å. With increasing NP concentrations the interfacial area between the NPs and the polymers increases, enhancing the correlation between the chains.

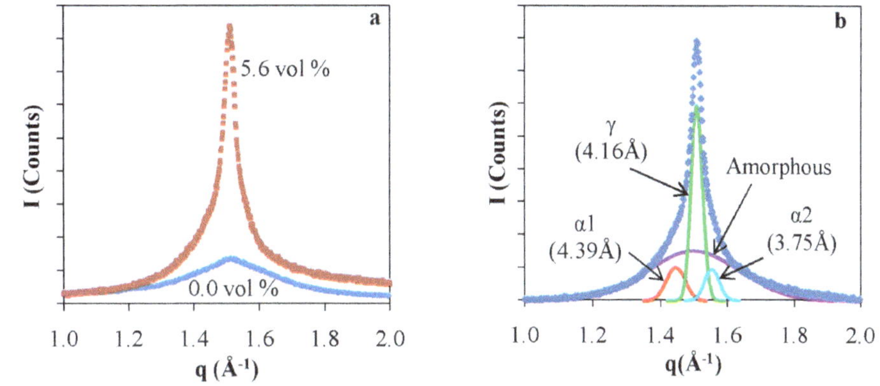

Figure 5. (**a**) Powder X-ray patterns of extruded PA6 with and without NPs at $N = 0$; and (**b**) Peak deconvolution of X-ray data for 5.6 vol % nanocomposite.

Table 2. X-ray peak widths for nanocomposite films as a function of N, the processing parameter and NP concentration in terms of volume percentage marked by vol %.

NP vol %	N	FWHM (δ) ($Å^{-1}$) (±0.001)	NP vol %	N	FWHM (δ) ($Å^{-1}$) (±0.001)	NP vol %	N	FWHM (δ) ($Å^{-1}$) (±0.001)
0.0	0	0.29	2.0	12	0.07	5.6	07	0.06
4.0	0	0.14	2.0	14	0.06	5.6	09	0.09
5.6	0	0.07	2.0	20	0.07	5.6	20	0.08
			2.0	30	0.07			

In contrast to the significant effects of N shown on the micron length scale, N, which was varied from 0 to 30, hardly affected the domain sizes over which the polymer chains scatter coherently.

The fractions of the α and γ crystalline are presented in Figure 6 as the ratio of the two crystalline forms (α/γ) as a function of N for different NP concentrations. At low NP concentrations, 2.0 and 2.8 vol %, the ratio of α/γ decreases with increasing N, where at higher concentrations this ratio remains constant. In low NP concentrations, blending at higher N values results in enhancing the γ crystalline form. This trend at low NP concentration is similar to that observed in pure PA6 films where α/γ ratio decreases from 1.8 to 0.8 as N goes from 0 to 30. Enhancing the NP concentration increases the surface area between the polymer and the NP. A large surface area results in increase cohesiveness of the polymer and NP affecting the polymer orientation.

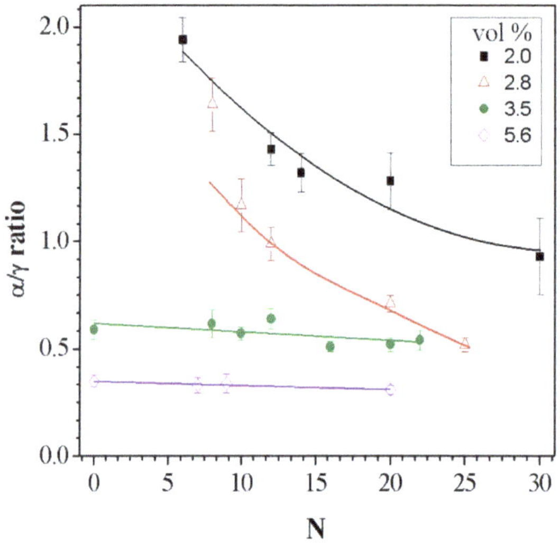

Figure 6. Variation of α/γ crystalline ratios with the processing parameter N for samples at the indicated NP compositions. The solid lines are drawn as a guideline for the eye.

Though increasing the NP content affects the melting temperature, T_m, of the polymer and the overall melt viscosity, the blending temperatures are sufficiently above T_m that *in situ* structuring occurs nearly independently for the NP concentrations studied.

The results show that while chaotic advection dominates at low NP concentrations, surface interactions of the NPs with the polymer chains control the system at higher NP concentrations. The overall degree of crystallinity is affected by both N and the concentration of NPs. The total crystalline fraction present in the nanocomposite blends, as extracted from the peak deconvolution, is shown in Figure 7. This fraction has contributions from both α and γ crystalline forms. Nanocomposites with 2.0 vol % and 2.8 vol % NPs are of lower crystallinity compared to higher NP concentrations. This is attributed to enhanced fraction of confined polymer chains within NPs, which facilitates the formation of more crystalline domains. However, no significant effects were observed as a function of N.

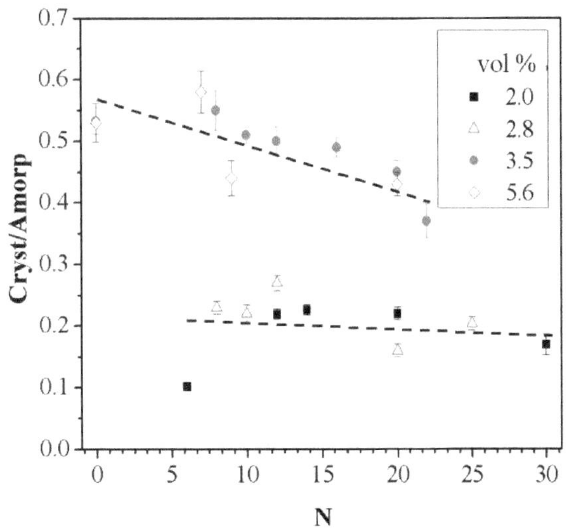

Figure 7. Total crystalline/amorphous ratio variation with respect to the processing parameter *N* at the indicated NP compositions. Lines are drawn as a visual guide.

Additional experiments were carried out using a two-dimensional detector to follow effects of *N* and the NP concentration on the orientation of the components. The X-ray patterns were measured in three directions with respect to the film plane and the direction of extrusion. The patterns corresponding to PA6 at $N = 0$ and $N = 30$ are shown in Figure 8. The intensity of the lines is affected by the amount of material in the X-ray beam. The first intense diffraction line from the center corresponds to the γ (001) diffraction of PA6, which is observed in all three directions. Two less intense arcs are observed corresponding to γ (001) diffraction in *Edge* and *End* patterns along the equator. In PA6 extruded different ways, no orientation is observed, and the intensity of this line is increased with *N*. We attribute the alignment observed to the effects of the chaotic advection, where the chains lie in the direction of the flow and the (001) plane lies perpendicular to the long axis of the polymer chain. The intensity of the γ (001) diffraction was integrated along the diffraction ring to obtain the radial distribution of the γ form. The results are shown in Figure 8b, where we observed the two peaks, which correspond to the intense arcs in the equatorial direction. The cross sections of $N = 30$ patterns in the *X* and *Y* directions further confirms the slight intensity difference of γ (001) along equatorial and meridial directions as indicated in Figure 8c. For pure PA6 there is no specific orientation as observed in injected molded samples [37–39]. Therefore, orientation of γ form observed in pure PA6 is due to the alignment of polymer chains in chaotic flow fields.

NPs enhance the packing and orientation of the polymer chains. The 2-D patterns for 5.6 vol % NP nanocomposites are presented in Figure 8a. The most center diffraction, at ~18 Å, corresponds to the inter-platelet distance between NPs. This diffraction spot is clearly observed in the *Edge* and *End* views and not in the *Through* direction. As indicated in Figure 8c, the diffraction of intercalated NPs occurs only along the equator for *Edge* and *End* directions and not from the meridial direction. Therefore the long axis of the particles orient along the extrusion direction as was previously shown by TEM. *N* hardly affects the platelets orientation. These results are consistent with previously reported studies of PA6/NP nanocomposites that were prepared by injection molding [2].

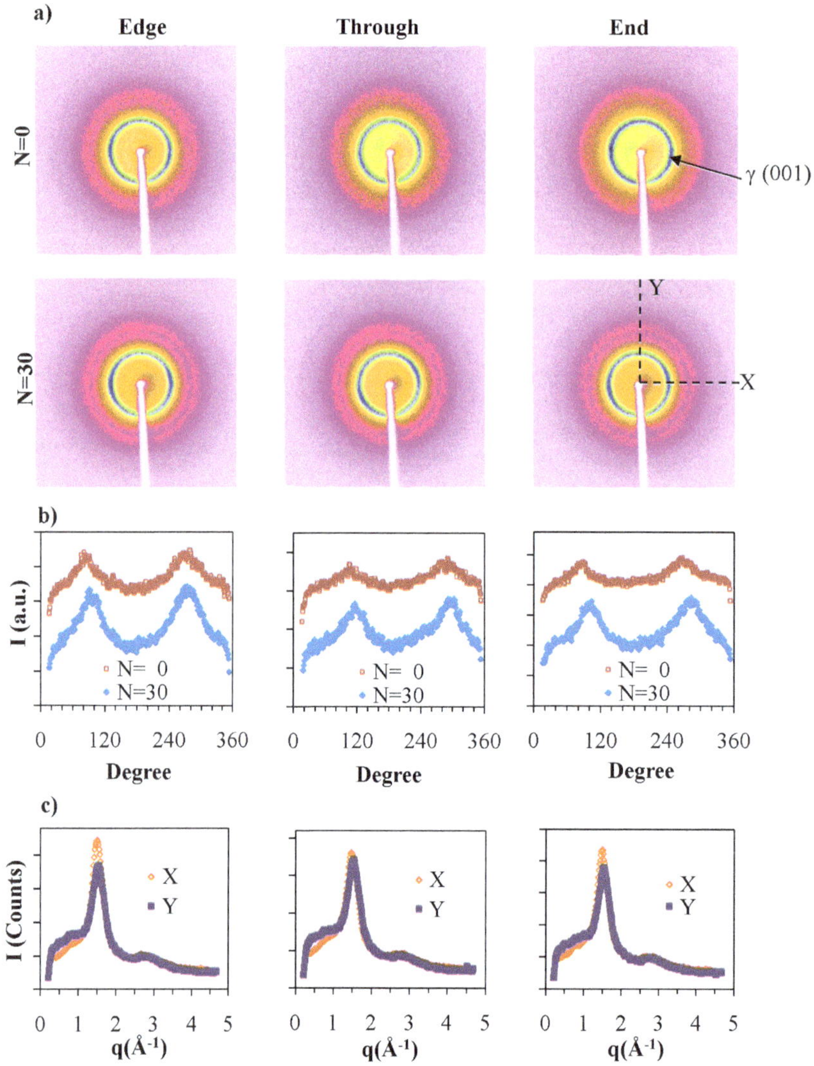

Figure 8. (**a**) Two-dimensional X-ray pattern of PA6 films at $N = 0$ and $N = 30$ from edge, through and end directions. The white regions in the middle of the images correspond to the shadow of the beam stopper; (**b**) The intensity along the γ crystalline ring marked in γ for the different directions; and (**c**) The cross sections of the $N = 30$ along the X and Y directions.

In the *Through* direction, the γ form diffraction line of PA6 is isotropic. However, in *Edge* and *End* directions, a six-fold symmetry in (001) is observed. As shown in Figure 9c, the intensities of the arcs along the equator are higher than the ones closer to the meridian. The radial distributions of γ form of PA6 for 5.6 vol % NP at $N = 0$ and $N = 20$ presented in Figure 9b, show the six-fold symmetry and relative intensities of arcs. The data are consistent with both hexagonal and monoclinic structures. The monoclinic unit cell of γ form was observed when PA6 is treated with iodine to induce the γ crystalline form [31]. The melt spun PA6 have pseudo-hexagonal symmetry [39], whereas injection molded PA6 nanocomposites exhibit a six-fold symmetry due to the presence of pseudo-orthorhombic lattice. In the current study we have observed a limited number of diffraction peaks, which do not allow us to distinguish crystalline packing that corresponds to the six-fold symmetry.

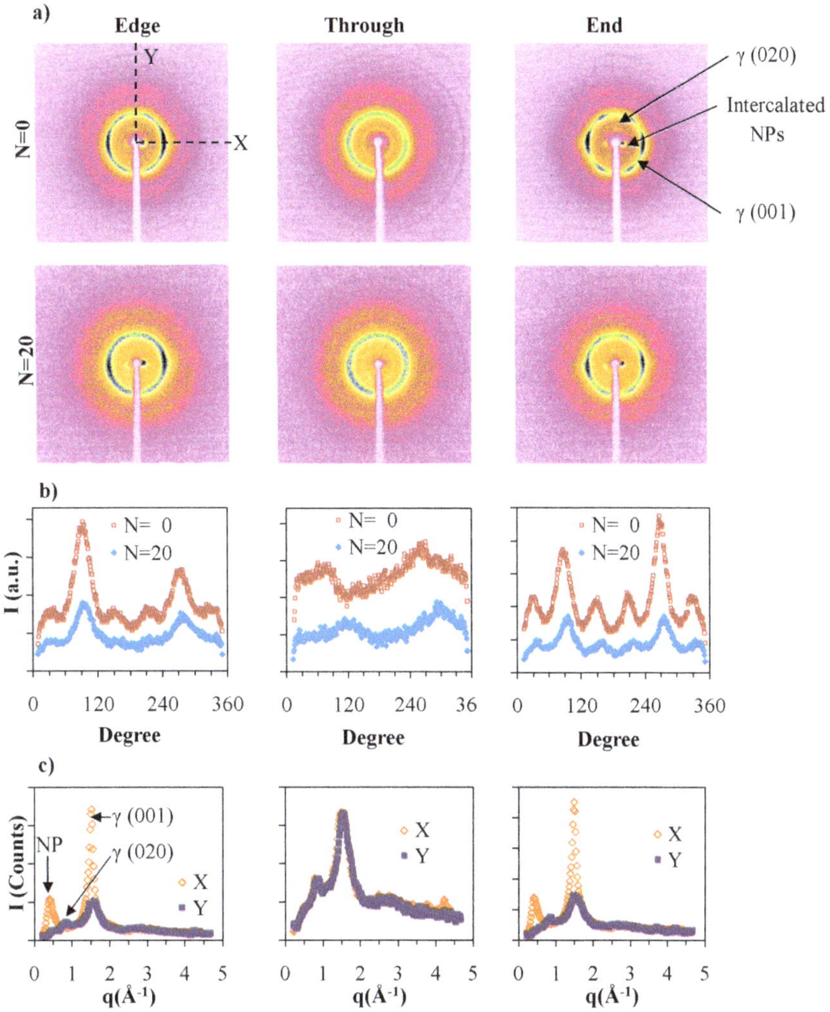

Figure 9. (**a**) Two dimensional X-ray patterns of PA6 with 5.6 vol % NP nanocomposite films at $N = 0$ and $N = 20$ from edge, through and end directions; (**b**) The intensity along the γ crystalline ring marked in γ for the different directions; and (**c**) The cross sections of the $N = 0$ along X and Y directions.

The γ (001) diffraction is in a plane perpendicular to the polymer chain axis. Since orientation of PA6 can only be observed in the *Edge* and *End* configurations, polymer chain axis lies on average perpendicular to *Edge* and *End* scanning directions. Therefore, these polymer chains are oriented perpendicular to the flow direction as well as to the surface of NPs as shown in Figure 10. Generally, polymer chains arrange parallel to the surfaces of the internal layers containing them. Such layers were themselves formed by shear acting predominantly in the Smart Blender in the direction perpendicular to the extrusion direction. This effect is consistent with studies by Kumar and co-workers [19] that observed interpenetration of the polymer matrix into the grafted layer on the NP.

The (020) diffraction is observed at 8.14 Å along the meridian of the *Edge* and *End* patterns, which further confirms the perpendicular orientation of polymer chains with respect to NP surface. The diffraction spot for the (020) plane is observed in samples with lower N values. With increasing N, the (020) diffraction is no longer visible.

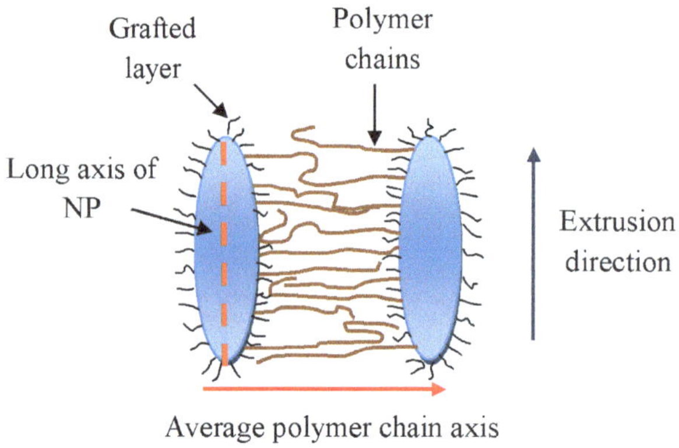

Figure 10. A schematic representation of the relative orientations of NPs and PA6 chains in nanocomposites.

3. Experimental Section

Master-batches, a mixture of the polymer with NPs of different NP percentages, were prepared from PA6 (Capron B135QP, BASF Corporation, Mount Olive, NJ, USA) and NP (Closite 30B, Southern Clay Products, Gonzales, TX, USA) using a twin-screw extruder [39]. Corresponding glass transition (T_g) and melting temperatures (T_m) of the master-batches were measured using differential scanning calorimeter (DSC) and the corresponding values are given in Table 3.

Table 3. Glass transition (T_g) and melting temperatures (T_m) of master batches with different NP volume fractions (vol %).

vol %	T_g(°C) (±0.2)	T_m(°C) (±0.2)
0 0	45.2	220.1
0 5	43.4	218.8
10	46.8	220.7

DSC measurements were performed on a Mettler Toledo 910e model (Greifensee, Switzerland) with flow rates of 10 °C/min. Samples were sealed in aluminum pans to avoid contact with air during the measurements. Nanocomposite films with NP volume fractions (vol %) of 2.0%, 2.8%, 3.5%, and 5.6% were obtained by introducing the matrix PA6 polymer and master-batches in equal proportions to the smart blender to obtain extruded films of about 150 micron thickness. The volume is reported as % of the total PA-6 in the system. Films were produced with differing extents of chaotic advection controlled through specification of N; a parameter that specifies the motion of rods that blend the polymers [22]. A list of samples measured and their volume fractions are given in Table 3.

An X-ray powder Sintag diffractometer, with Cu K_α radiation and wavelength of $\lambda = 1.54$ Å, was used to obtain powder X-ray patterns. All the samples were cut to a 1 cm^2 films. The scattering intensities were normalized to a unit thickness. Further studies were carried out on a Rigaku single crystal instrument with Mo K_α radiation $\lambda = 0.71$ Å, equipped with a two-dimensional detector. The patterns were recorded along three different directions with respect to extrusion direction as shown in Figure 3. Degrees of crystallinity and α/γ ratios were calculated from X-ray data with the PeakFit 4.2 program, which de-convolutes overlapped peaks and integrate the peak areas. Two-dimensional X-ray images were analyzed using MATLAB incorporating a home built intergraded routine to obtain radial distributions and intensity profiles along different directions.

4. Conclusions

The assembly of a hierarchical PA6/montmorillonite NP composite was investigated at different NP concentrations as the extent of *in situ* structuring by chaotic advection was varied. The segregation into two defined layers, a pristine polymer free of NPs and a NP-polymer compounded layer, is a direct result of recursive stretching and folding that characterizes chaotic motion. We found that the internal structure of the layers is a result of balance between forces induced by the flow and chaotic advection as well as interfacial forces excreted by the surface of the NPs that are present throughout the assembly process. These forces impact both the crystalline phases and the degree of orientation of the nanoparticles. At low loading of NPs, the α crystalline form transform into the γ form as layers become thinner, whereas at higher loadings no changes were observed in response to the extent of chaotic advection. At higher loadings, crystalline morphology is determined largely by the smaller inter-platelet distances. In nanocomposites, polymer chains align perpendicular to interface of the NPs. This study has demonstrated that interfacial interactions in nanocomposites could be sufficiently strong to impact the nm structure features, whereas chaotic flows impact the micron level structuring. Further studies including varying the dimensions of the NPs correlated with the degree of structuring are on the way to quantify the impact of chaotic advection on structuring of polymeric nanocomposites.

Acknowledgments

We kindly acknowledge financial support from the U.S. Army Soldier Systems Center of Natick, MA. We would like to thank Mark C. Kampf of Appvion, Inc., Appleton, WI, for helpful discussions.

Author Contributions

Chaitra Mahesha and David A. Zumbrunnen used chaotic advection and a Smart Blending extrusion system to prepare the nanocomposites and employed TEM to document nanocomposite structure.

Dilru Ratnaweera and Dvora Perahia performed the X-ray diffraction structural studies and developed the model that underlines how processing affects the assembly process.

Conflicts of Interest

The authors declare no conflict of interest.

References

1. Powell, C.E.; Beall, G.W. Physical properties of polymer/clay nanocomposites. *Curr. Opin. Solid St. M.* **2006**, *10*, 73–80.
2. Okada, A.; Usuki, A. Twenty years of polymer-clay nanocomposites. *Macromol. Mater. Eng.* **2006**, *291*, 1449–1476.
3. Xie, S.B.; Zhang, S.M.; Zhao, B.; Qin, H.L.; Wang, F.S.; Yang, M.S. Tensile fracture morphologies of nylon-6/montmorillonite nanocomposites. *Polym. Int.* **2005**, *54*, 1673–1680.

4. Weon, J.I.; Sue, H.J. Effects of clay orientation and aspect ratio on mechanical behavior of nylon-6 nanocomposite. *Polymer* **2005**, *46*, 6325–6334.
5. Kojima, Y.; Usuki, A.; Kawasumi, M.; Okada, A.; Kurauchi, T.; Kamigaito, O. One-pot synthesis of nylon 6-clay hybrid. *J. Polym. Sci. A* **1993**, *31*, 1755–1758.
6. Caseri, W.R. Nanocomposites of polymers and inorganic particles: Preparation, structure and properties. *Mater. Sci. Technol.* **2006**, *22*, 807–817.
7. Yurasova, I.V.; Antipov, O.L.; Ermolaev, N.L.; Cherkasov, V.K.; Lopatina, T.I.; Chesnokov, S.A.; Ilyina, I.G. Novel polymer nanocomposites with giant dynamical optical nonlinearity. *Phys. Solid State* **2005**, *47*, 129–132.
8. Zhu, A.P.; Cai, A.Y.; Zhang, J.; Jia, H.W.; Wang, J.Q. PMMA-grafted-silica/PVC nanocomposites: Mechanical performance and barrier properties. *J. Appl. Polym. Sci.* **2008**, *108*, 2189–2196.
9. Ma, H.Y.; Tong, L.F.; Xu, Z.B.; Fang, Z.P. Functionalizing carbon nanotubes by grafting on intumescent flame retardant: nnanocomposite synthesis, morphology, rheology, and flammability. *Adv. Funct. Mater.* **2008**, *18*, 414–421.
10. Liu, J.S.; Tanaka, T.; Sivula, K.; Alivisatos, A.P.; Frechet, J.M.J. Employing end-functional polythiophene to control the morphology of nanocrystal−polymer composites in hybrid solar cells. *J. Am. Chem. Soc.* **2004**, *126*, 6550–6551.
11. Vaia, R.A.; Maguire, J.F. Polymer nanocomposites with prescribed morphology: Going beyond nanoparticle-filled polymers. *Chem. Mater.* **2007**, *19*, 2736–2751.
12. Mackay, M.E.; Tuteja, A.; Duxbury, P.M.; Hawker, C.J.; van Horn, B.; Guan, Z.B.; Chen, G.H.; Krishnan, R.S. General strategies for nanoparticle dispersion. *Science* **2006**, *311*, 1740–1743.
13. Wang, D.Y.; Zhu, J.; Yao, Q.; Wilkie, C.A. A comparison of various methods for the preparation of polystyrene and poly(methyl methacrylate) clay nanocomposites. *Chem. Mater.* **2002**, *14*, 3837–3843.
14. Goldel A.; Marmur, A.; Kasaliwal, G.R.; Potschke, P.; Heinrich, G. Shape-dependent localization of carbon nanotubes and carbon black in an immiscible polymer blend during melt mixing. *Macromolecules* **2011**, *44*, 6094–6102.
15. Krishnamoorti, R.; Yurekli, K. Rheology of polymer layered silicate nanocomposites. *Curr. Opin. Colloid Interface Sci.* **2001**, *6*, 464–470.
16. Filippone, G.; Acierno, D. Altering the onset of cocontinuity in nanocomposite immiscible blends by acting on the melt-compounding procedure. *J. Appl. Polym. Sci.* **2011**, *122*, 3712–3719.
17. Krishnamoorti, R. Strategies for dispersing nanoparticles in polymers. *MRS Bull.* **2007**, *32*, 341–347.
18. Kumar, S.K.; Krishnamoorti, R. Nanocomposites: Structure, phase behavior, and properties. *Annu. Rev. Chem. Biomol. Eng.* **2010**, *1*, 37–58.
19. Harton, S.E.; Kumar, S.K. Mean-field theoretical analysis of brush-coated nanoparticle dispersion in polymer matrices. *J. Polym. Sci. B* **2008**, *46*, 351–358.
20. Knauert, S.T.; Douglas, J.F.; Starr, F.W. The effect of nanoparticle shape on polymer-nanocomposite rheology and tensile strength. *J. Polym. Sci. B* **2007**, *45*, 1882–1897.
21. Starr, F.W.; Douglas, J.F.; Glotzer, S.C. Origin of particle clustering in a simulated polymer nanocomposite and its impact on rheology. *J. Chem. Phys.* **2003**, *119*, 1777–1788.

22. Aref, H. Stirring by chaotic advection. *J. Fluid Mech.* **1984**, *143*, 1–21.
23. Zumbrunnen, D.A.; Inamdar, S.; Kwon, O.; Verma, P. Chaotic advection as a means to develop nanoscale structures in viscous melts. *Nano Lett.* **2002**, *2*, 1143–1148.
24. Kwon, O.; Zumbrunnen, D.A. Production of barrier films by chaotic mixing of plastics. *Polym. Eng. Sci.* **2003**, *43*, 1443–1459.
25. Dhoble, A.; Kulshreshtha, B.; Ramaswami, S.; Zumbrunnen, D.A. Mechanical properties of PP-LDPE blends with novel morphologies produced with a continuous chaotic advection blender. *Polymer* **2005**, *46*, 2244–2256.
26. Kwon, O.; Zumbrunnen, D.A. Progressive morphology development to produce multilayer films and interpenetrating blends by chaotic mixing. *J. Appl. Polym. Sci.* **2001**, *82*, 1569–1579.
27. Mahesha, C.; Zumbrunnen, D.A.; Ratnaweera, D.; Perahia, D.; Kampf, M.C. Structure-Property Investigation of Novel Multi-Layer Nanoclay Composites Formed Controllably with a Continuous Chaotic Blender. In Proceedings of the Annual Technical Conference, Cincinnati, OH, USA, 6–11 June 2007; pp. 1749–1753.
28. Liu, X.H.; Wu, Q.J. Phase transition in nylon 6/clay nanocomposites on annealing. *Polymer* **2002**, *43*, 1933–1936.
29. Mathias, L.J.; Davis, R.D.; Jarrett, W.L. Observation of α and γ crystal forms and amorphous regions of nylon 6−clay nanocomposites using solid-state 15N nuclear magnetic resonance. *Macromolecules* **1999**, *32*, 7958–7960.
30. Li, Y.Y.; Goddard, W.A. Nylon 6 crystal structures, folds, and lamellae from theory. *Macromolecules* **2002**, *35*, 8440–8455.
31. Arimoto, H.; Ishibash, M.; Hirai, M.; Chatani, Y. Crystal structure of the γ-form of nylon 6. *J. Polym. Sci. A* **1965**, *3*, 317.
32. Yoon, K.H.; Polk, M.B.; Min, B.G.; Schiraldi, D.A. Structure and property study of nylon-6/clay nanocomposite fiber. *Polym. Int.* **2004**, *53*, 2072–2078.
33. Fornes, T.D.; Paul, D.R. Crystallization behavior of nylon 6 nanocomposites. *Polymer* **2003**, *44*, 3945–3961.
34. Liu, X.H.; Breen, C. High-temperature crystalline phases in nylon 6/clay nanocomposites. *Macromol. Rapid Commun.* **2005**, *26*, 1081–1086.
35. Espinosa, A.Z.; Rodriguez, F.J.M.; Stribeck, N.; Camarillo, A.A.; Diaz, S.V.; Hsiao, B.S.; Chu, B. Complex isothermal crystallization and melting behavior of nylon 6 nanoclay hybrids. *Macromolecules* **2005**, *38*, 4246–4253.
36. Auriemma, F.; Petraccone, V.; Parravicini, L.; Corradini, P. Mesomorphic form (β) of nylon 6. *Macromolecules* **1997**, *30*, 7554–7559.
37. Bankar, V.G.; Spruiell, J.E.; White, J.L. Melt spinning of nylon 6: Structure development and mechanical properties of as-spun filaments. *J. Appl. Polym. Sci.* **1977**, *21*, 2341–2358.
38. Kojima, Y.; Usuki, A.; Kawasumi, M.; Okada, A.; Kurauchi, T.; Kamigaito, O.; Kaji, K. Novel preferred orientation in injection-molded nylon 6-clay hybrid. *J. Polym. Sci. B* **1995**, *33*, 1039–1045.

39. Carneiro, O.S.; Covas, J.A.; Vergnes, B. Experimental and theoretical study of twin-screw extrusion of polypropylene. *J. Appl. Polym. Sci.* **2000**, *78*, 1419–1430.

7

New Insights into Understanding Irreversible and Reversible Lithium Storage within SiOC and SiCN Ceramics

Magdalena Graczyk-Zajac [1,*], Lukas Mirko Reinold [1], Jan Kaspar [1], Pradeep Vallachira Warriam Sasikumar [1], Gian-Domenico Soraru [2] and Ralf Riedel [1]

[1] Institut für Materialwissenschaft, Technische Universität Darmstadt, Jovanka-Bontschits-Straße 2, 64287 Darmstadt, Germany; E-Mails: reinold@materials.tu-darmstadt.de (L.M.R.); jankaspar2@gmail.com (J.K.); pradeep@materials.tu-darmstadt.de (P.V.W.S.); riedel@materials.tu-darmstadt.de (R.R.)

[2] Dipartimento di Ingegneria Industriale, Università di Trento, Via Sommarive 9, 38123 Trento, Italy; E-Mail: soraru@ing.unitn.it

* Author to whom correspondence should be addressed; E-Mail: graczyk@materials.tu-darmstadt.de

Academic Editor: Emanuel Ionescu

Abstract: Within this work we define structural properties of the silicon carbonitride (SiCN) and silicon oxycarbide (SiOC) ceramics which determine the reversible and irreversible lithium storage capacities, long cycling stability and define the major differences in the lithium storage in SiCN and SiOC. For both ceramics, we correlate the first cycle lithiation or delithiation capacity and cycling stability with the amount of SiCN/SiOC matrix or free carbon phase, respectively. The first cycle lithiation and delithiation capacities of SiOC materials do not depend on the amount of free carbon, while for SiCN the capacity increases with the amount of carbon to reach a threshold value at ~50% of carbon phase. Replacing oxygen with nitrogen renders the mixed bond Si-tetrahedra unable to sequester lithium. Lithium is more attracted by oxygen in the SiOC network due to the more ionic character of Si-O bonds. This brings about very high initial lithiation capacities, even at low carbon content. If oxygen is replaced by nitrogen, the ceramic network becomes less attractive for lithium ions due to the more covalent character of Si-N bonds and lower electron density on the nitrogen atom. This explains the significant difference in electrochemical behavior which is observed for carbon-poor SiCN and SiOC materials.

Keywords: polymer-derived ceramics; lithium-ion battery; SiOC; SiCN; carbon content

1. Introduction

Due to increasing energy consumption and environmental aspects, there is a growing interest for new energy related materials. Lithium-ion batteries are the most promising candidates and have received much attention in recent years because of their high energy density. Nevertheless, there is still a need of new electrode materials to meet growing safety, stability and high rate capability requirements. Currently, mostly graphitic materials are used as the anode material in lithium ion batteries due to their low price and high reversibility despite their relatively low capacity (372 $mAh \cdot g^{-1}$), instability during long-time cycling and inadequacy for high power applications. Lithium plating was identified as one of the mechanisms which ends the life of a battery more rapidly due to the formation and growth of lithium dendrites. Metallic lithium is typically absent in a lithium-ion (Li-ion) cell under normal conditions of operation. However, under high current charge rates and/or low temperatures, lithium metal will deposit on the carbon anode in preference to lithium intercalation. To prevent lithium plating at the edges and to avoid anode polarization to a potential close to lithium reduction, the anode electrode has been designed with larger geometry and capacity excess [1–6]. The decomposition of the electrolyte and subsequent formation of a film surface layer (solid electrolyte interphase, SEI) causes an increase in the impedance and the consumption of recyclable lithium ions. Moreover, the graphite exfoliation being a consequence of the insertion of solvated Li^+ in case of insufficient SEI formation brings about accelerated material fatigue. The main reasons of the graphite fatigue are: (i) losses of insertion host and (ii) failure in electronic contact within the electrode material and material with current collector. The addition of various stabilizers, robust electrolyte systems, and temperature treatment are some of the methods that have been adopted to mitigate these aging effects on the electrode [7–18].

The studies summarized within this work focus on developing new anodes for lithium ion batteries and understanding the mechanisms of lithium storage. In particular, we investigate silicon-based polymer derived ceramics (PDCs), namely silicon oxycarbide (SiOC) and silicon carbonitride (SiCN), with respect to their electrochemical properties and cycling stability. Silicon-based PDCs are prepared by controlled pyrolysis, in an inert or reactive atmosphere, of preceramic polymers containing Si, H, C, O, N. At lower pyrolysis temperatures (~up to 1100 °C for SiOC and ~up to 1300 °C for SiCN, [19–21]), the product after thermal treatment is amorphous. The polymer-derived ceramics route enables designing the chemical and physical properties of the final ceramic material by chemical modification of the preceramic polymer as well as by tuning up the processing route covering the polymer to ceramic transformation [22].

Concerning silicon oxycarbide materials, a set of electrochemical data was published in the middle of the 1990s by the group of Dahn [23–28]. However, the above publications mostly address the electrochemical performance of the materials with respect to their elemental composition, without considering the mechanism of lithium storage and lithium transport in these materials. Due to the increased commercial availability of preceramic polymers and the expected industrial application of polymer-derived SiOC based anodes, there is presently a growing interest in ceramic-based electrode materials for Li-ion batteries. As a consequence, questions related to the lithium storage mechanisms within SiOC- and SiCN-based materials have recently been the focus of our research. Ahn *et al.* claim that the mixed bond configuration (tetrahedrally coordinated silicon from SiC_4 via SiC_3O, SiC_2O_2 and $SiCO_3$ to SiO_4) in SiOC ceramics acts as a major lithiation site [29–33]. Other research groups claim that the Li insertion into carbon-rich SiOC compounds occurs in the form of an adsorption and surface storage within the free carbon phase, similar to the storage of Li-ions in disordered carbons. Host sites are considered the edges of graphene sheets, interstitial and defect sites, micro-pores, graphite nano-crystallites and interfacial adsorption at carbon–crystallite interfaces. The Si–O–C glassy phase, on the contrary, is attributed a minor role in the reversible storage process [34–42]. The recent investigation of Fukui *et al.* by means of ^{7}Li-NMR (nuclear magnetic resonance) measurements demonstrates that the free carbon phase within these materials is the major hosting site for Li ions [34]. The Li-ion diffusion coefficient (D_{Li+}) determined by three different methods, namely potentiostatic intermittent titration technique (PITT), galvanostatic intermittent titration technique (GITT) and electrochemical impedance spectroscopy (EIS) revealed D_{Li+} of about 10^{-9}–10^{-11} $cm^2 \cdot s^{-1}$, similar to that reported for disordered carbons (10^{-10}–10^{-11} $cm^2 \cdot s^{-1}$) and on average faster than for graphite (10^{-9}–10^{-13} $cm^2 \cdot s^{-1}$). This agreement, again, emphasizes the Li-ion uptake in the free carbon of the SiOC microstructure; hence it determines the storing kinetics. Interestingly, the analyzed diffusion coefficient has been found to have less potential compared to disordered carbon and graphite [43].

Though Dahn *et al.* filed a patent related to the use of silazane-derived SiCN ceramics in 1997 which show reversible discharge capacities up to 560 $mAh \cdot g^{-1}$ [44], much less research has been done since that time on the application of these materials in lithium-ion batteries in comparison to SiOC. Pure polymer-derived SiCN materials obtained from polysilylethylendiamine have been investigated by Su *et al.* [45] and Feng [46]. The work of Su *et al.* showed a first discharge cycle capacity of 456 mAh g^{-1} but the material suffered from strong fading with cycling. This problem of capacity fading was solved by Feng by applying an additional heat treatment to the polymer-derived SiCN material, however relatively low capacities of about 300 mAh g^{-1} were achieved. Promising electrochemical results with regard to the capacity and stability of SiCN derived from high-carbon containing polysilylcarbodiimides have been reported by Kaspar *et al.* [47] and Graczyk-Zajac *et al.* [48]. Within the work of Reinold *et al.* [49], the excellent performance of carbon-rich SiCN materials has been demonstrated, including the discussion of the influence of the molecular polymer structure on the resulting ceramic microstructure and in consequence on the electrochemical performance of the material. Solid state NMR studies on SiCN ceramics clearly identify carbon as the main lithium storage site [50], which is in agreement with the work of Fukui *et al.* on SiOC-based materials [34]. It is also shown that composite anode materials comprised of SiCN/graphite [51] and SiCN/silicon [52] exhibit enhanced electrochemical properties as compared to pure graphite and silicon, respectively. Composite materials based on SiCN/hard carbons [53] demonstrate excellent stability with respect to high current charging/discharging performance.

Within this review, by combining the experimental results of numerous research works, we analyze which structural properties of the silicon carbonitride (SiCN) and silicon oxycarbide (SiOC) ceramics determine the reversible and irreversible lithium storage capacities, cycling stability and define the major differences in the lithium storage between SiCN and SiOC ceramic materials. Replacing oxygen with nitrogen renders the mixed bond Si-tetrahedra unable to sequester lithium. This explains why a significant difference in electrochemical behavior of carbon-poor SiCN and SiOC materials is found. For carbon-rich ceramics, the free carbon phase plays a dominant role bringing about high cycling stability and high reversible capacities but also leads to significant first cycle irreversible capacities due to lithium capturing in some pores and voids between carbon layers.

2. Results and Discussion

As the high carbon content of the aforementioned ceramic systems seems to play an important role in the reversible storage of Li-ions, our work focused on the development of carbon-rich Si-based polymer derived SiCN and SiOC anode materials for lithium-ion batteries. The following synthesis strategies were applied: (i) using phenyl-rich pre-ceramic Si-polymers as starting materials [19,38,46,48,54]; (ii) addition of carbon or carbon precursor to Si-based polymers [51,53,55] and (iii) chemical modification of pre-ceramic Si-polymers [39,42,53,56,57]. In the last few years, among the various chemical compositions of SiCN and SiOC compounds, stoichiometries with an exceptionally high content of carbon (>50 wt%) were further considered as perspective anode material in terms of high gravimetric capacity, rate capability and reliable cycling stability. The microstructure of carbon rich SiCN and SiOC is presented and discussed elsewhere [22,58–64]. Here, we present an overview of the electrochemical properties of various SiOC and SiCN based materials with respect to the amount of free carbon phase and the microstructure. Moreover, we evaluate what kind of microstructural properties of the Si(O,N)C ceramic determine the reversible and irreversible capacities and long cycling stability and we identify the major differences in the lithium storage sites within SiCN and SiOC materials.

Figures 1a and 2a present the dependence of the first cycle delithiation (extraction) capacity and cycling stability (The cycling stability has been calculated as the ratio of the delithiation capacity after prolonged cycling (>100 cycles) to the first extraction capacity), both registered with low currents (C/10–C/20). The same current was always applied for the charge (C) and for the discharge (D) processes with C/20 = 18 $mAh \cdot g^{-1}$.) from the amount of free carbon phase within SiOC (a) and SiCN (b), while in Figures 1b and 2b the first cycle lithiation (insertion) capacity and stability are plotted as a function of the amount of SiOC (a) and SiCN (b) ceramic matrix (For SiOC materials the amount of free carbon and ceramic phase has been calculated according to the general formula consisting of a stoichiometric silicon oxycarbide network, $SiC_xO_{2(1-x)}$ and a C_{free} phase [65]). In SiCN it is assumed that oxygen is bonded to silicon as SiO_2, nitrogen and silicon from Si_3N_4 and the remaining Si is bonded to C in the form of SiC. The excess carbon is assumed to exist as free carbon. (%ceramic + %free carbon = 100%). It is obvious that for SiOC delithiation, the capacity does not depend on the amount of free carbon, while for SiCN materials the capacity increases with the amount of carbon to reach a certain threshold value at about 50% of carbon phase [49,56]. The cycling stability of low carbon materials is, however, low for both ceramics.

The data reported in Figure 1b unveil a linear dependence of the insertion capacity from the amount of silicon oxycarbide phase, which is in good agreement with the model proposed by Raj *et al.* [30]. Moreover, the extrapolation of the linear fit down to zero, which means a sample of pure free carbon, indicates a capacity of 400 $mAh \cdot g^{-1}$ and is in agreement with the theoretical capacity of graphite

(372 mAh·g^{-1}). Interestingly, the insertion capacity of pure $SiC_xO_{2(1-x)}$ phase, which can be obtained by extrapolating the linear fit up to 100% amounts 1300 mAh·g^{-1}. However, according to our experience, this capacity cannot be reached practically due to the low electrical conductivity of the materials. According to DFT modelling of the lithium insertion into SiOC materials [66], insertion of Li into amorphous silica (a-SiO_2) and SiOC containing no free carbon phase is energetically unfavorable due to a large gap between their valence and conduction band.

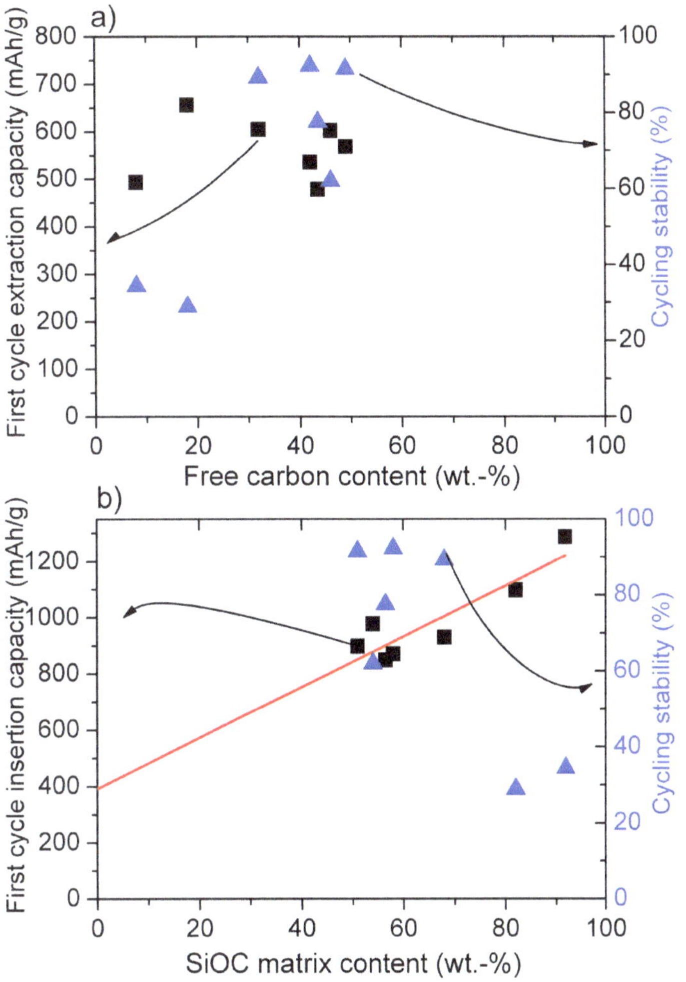

Figure 1. Dependence of the insertion and extraction capacity of SiOC-derived materials on the amount of free carbon (**a**) and SiOC matrix (**b**). Cycling stability defined as the ratio of the extraction capacity after prolonged cycling (<100 cycles) to the first extraction capacity. Experimental data for samples pyrolysed at 1000–1100 °C from [54,56,57].

In contrast, the presence of a free carbon phase provides low-laying unoccupied states in which electrons can go in, and by consequence significantly decreases the band gap. A graphical illustration of this is given in Figure 3. It shows the electronic density of states (DOS) for a-SiO_2, SiOC and a free carbon phase according to Reference [66]. A strong bonding of Li in the solid SiOC structure is provided, if the interaction of the lithium cation with the host in the form of Li-O bonds outweighs the promotion energy for the electron. Consequently, on the one hand the free carbon phase facilitates lithium bonded to oxygen sites, leading to irreversible lithium uptake; on the other hand, the segregated carbon provides a major part of the reversible lithium storage capacity [66].

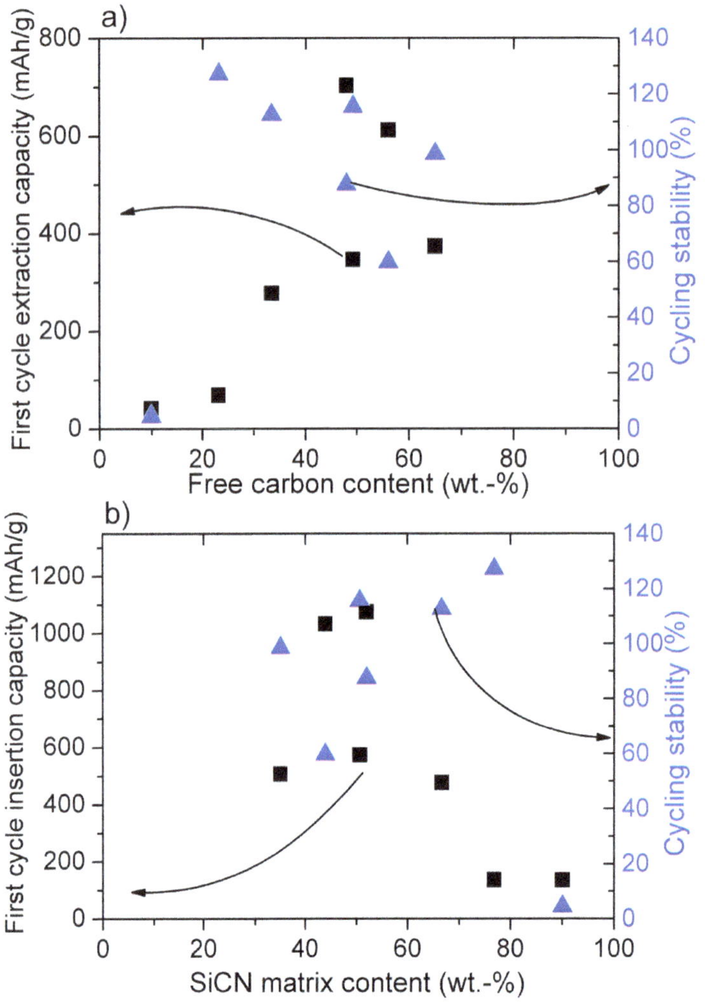

Figure 2. Dependence of the insertion and extraction capacity of SiCN-derived materials on the amount of free carbon (**a**) and SiOC matrix (**b**). Cycling stability defined as the ratio of the extraction capacity after prolonged cycling (>100 cycles) to the first extraction capacity. Experimental data for samples pyrolysed at 1000–1100 °C from [49,51,53,56].

According to Figure 2b, for silicon carbonitride there is no dependence between the first cycle insertion capacity and SiCN matrix amount. The SiCN first lithiation capacity is low for a high amount (>70%) of SiCN matrix. No modelling has been performed on lithium insertion into SiCN materials, yet. However, a qualitative experimental analysis of low carbon SiCN and SiOCN materials has been performed [33]. It has been stated that replacing oxygen with nitrogen renders the mixed bond Si-tetrahedra unable to sequester Li. In SiOC materials, the electronic structure of the mixed bonds can induce a dipole on the Li atom thereby creating a shallow energy well. Mixed bonds that are highly covalent have more localized and "stiff" electron densities, unable to induce such a dipole. Electronegativity values of oxygen and nitrogen amount 3.5 and 3.0, respectively. Therefore, Si–N bonds are more covalent than that of Si–O bonds, which confirms the above argument. This phenomenon corresponds pretty well with our experimental findings. Accordingly, lithium is strongly attracted by oxygen in the SiOC network due to the pronounced ionic character of Si–O bonds, leading to a high electron density on oxygen. This feature brings about high first lithiation capacities, even at low carbon contents. With continuing cycling, lithium is irreversibly captured within the carbon-poor

SiOC network leading to low electrochemical stability (see Figure 1a,b). Once oxygen is replaced by nitrogen, the ceramic network is much less attractive for lithium ions due to the pronounced covalent character of Si–N bonds and thus lower electron density on the nitrogen atom. These findings explain the significant difference in the electrochemical behavior of low carbon SiCN and SiOC materials. For carbon-rich SiOC and SiCN ceramics, the free carbon phase starts to play a dominant role bringing about high cycling stability and high reversible capacities, but also leading to significant first cycle irreversible capacities due to lithium capturing, as already discussed [34,50]. From the electrochemical point of view, on the one hand these results suggest that the studied ceramic anodes behave like a composite material comprised of a mixture of pure ceramic SiCN or SiOC matrix and free carbon phase. Nevertheless, it should be noted that the ceramic phase is indispensable to ensure the stability of the free carbon phase within the prolonged cycling. Thus, a certain equilibrium between the free carbon and the ceramic phase has to be maintained in order to achieve the material stability with respect to continuous lithium insertion/extraction. This equilibrium is reached at around a 1:1 weight ratio between free carbon and ceramic phase, but the exact value depends on the particular preceramic polymer and its processing to SiCN or SiOC ceramic.

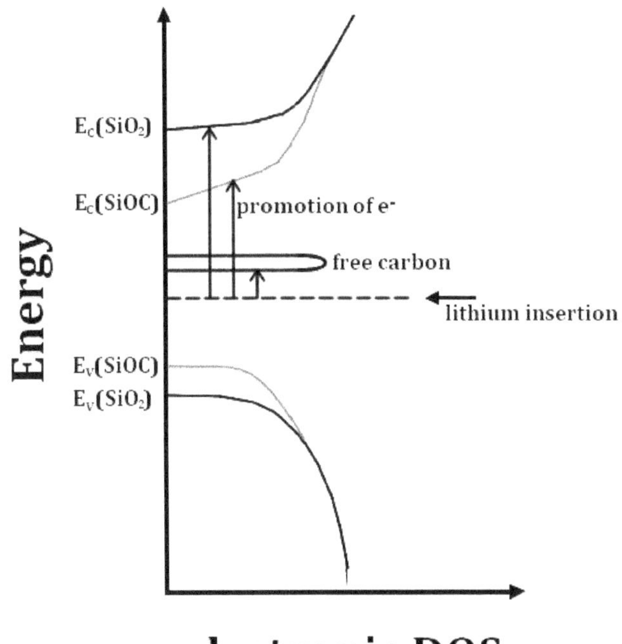

Figure 3. Schematic scheme of the electronic density of states for a-SiO2, SiOC and free carbon according to Reference [66].

3. Conclusions

In this overview on SiOC and SiCN ceramic negative electrode materials for rechargeable lithium ion batteries, we summarize the experimental results in order to find out the critical material properties influencing their electrochemical performance. For both material types, the calculated amount of SiCN/SiOC matrix or free carbon phase is correlated with the first cycle lithiation or delithiation capacity and cycling stability. It is found that the amount of free carbon phase has no significant impact on the first cycle lithiation and delithiation capacities of SiOC, whereas for SiCN the capacity increases with the amount of carbon until a threshold value is reached at about 50% of carbon phase. The cycling stability of carbon-poor ceramics is very low for SiCN and SiOC. For SiOC, a clear linear dependence of the insertion capacity on the amount of silicon oxycarbide phase is revealed, while for silicon carbonitride there is no dependence between first cycle insertion capacity and SiCN matrix amount. In contrary to the tendency observed for SiOC, the SiCN lithiation capacity is very low for high (>70%) amount of SiCN matrix.

Finally, it is stated that replacing oxygen with nitrogen renders the mixed bond Si-tetrahedra unable to sequester lithium. Lithium is more attracted by oxygen in SiOC network due to more ionic character of Si–O bonds, leading to a high electron density on oxygen. This brings about very high initial lithiation capacities, even at low carbon content. With continuing cycling, lithium is irreversibly captured within the SiOC network bringing low cycling stability. If oxygen is replaced by nitrogen, the ceramic network becomes much less attractive for lithium ions due to the more covalent character of Si-N bonds and lower electron density on the nitrogen atom. It explains why a significant difference in electrochemical behavior of carbon-poor SiCN and SiOC materials is observed. For carbon-rich ceramics, the free carbon phase leads to high cycling stability and high reversible capacity

Acknowledgments

Financial support from the German Research Foundation within SFB 595/A4 and SPP 1473/JP8 programs is greatly acknowledged. Ralf Riedel, Gian-Domenico Soraru and Pradeep Vallachira Warriam Sasikumaracknowledge the financial support from the EU through the Marie Curie ITN 7th Framework Programme, Functional Nitrides for Energy Applications (MC INT FUNEA), CT-264873. We express our gratitude to Claudia Fasel for her continuous support during this work. We also thank our present and former co-workers, Monika Wilamowska, Guanwei Liu, Andrzej Nowak and Philippe Dibandjo. Magdalena Graczyk-Zajac thanks to Helmut Ehrenberg for fruitful discussions.

Author Contributions

The authors contribution to creation of the manuscript is as follows: Magdalena Graczyk-Zajac: synthesis and characterization of SiCN/C composites, preparation and final editing of the manuscript; Lukas Mirko Reinold: synthesis and characterization of carbon-poor and carbon-rich SiCN, proof reading of the manuscript, Jan Kaspar: synthesis and characterization of carbon-poor and carbon-rich SiOC, proof reading of the manuscript, Pradeep Vallachira Warriam Sasikumar: synthesis and characterization of carbon-poor and carbon-rich SiOC, proof reading of the manuscript, Gian-Domenico Soraru: synthesis and characterization of SiOC glasses, preparation and proof reading of the manuscript, Ralf Riedel: work on SiOC, SiCN and SiCN/C materials, proof reading of the manuscript.

Conflicts of Interest

The authors declare no conflict of interest.

References

1. Bugga, R.V.; Smart, M.C. Lithium plating behavior in lithium-ion cells. *ECS Trans.* **2010**, *25*, 241–252.
2. Smart, M.C.; Ratnakumar, B.V. Effects of electrolyte composition on lithium plating in lithium-ion cells. *J. Electrochem. Soc.* **2011**, *158*, A379–A389.
3. Honbo, H.; Takei, K.; Ishii, Y.; Nishida, T. Electrochemical properties and Li deposition morphologies of surface modified graphite after grinding. *J. Power Sources* **2009**, *189*, 337–343.
4. Bhattacharyya, R.; Key, B.; Chen, H.; Best, A.S.; Hollenkamp, A.F.; Grey, C.P. *In situ* NMR observation of the formation of metallic lithium microstructures in lithium batteries. *Nat. Mater.* **2010**, *9*, 504–510.
5. Agubra, V.; Fergus, J. Lithium ion battery anode aging mechanisms. *Materials* **2013**, *6*, 1310–1325.
6. Harris, S.J.; Timmons, A.; Baker, D.R.; Monroe, C. Direct *in situ* measurements of Li transport in Li-ion battery negative electrodes. *Chem. Phys. Lett.* **2010**, *485*, 265–274.
7. Markovsky, B.; Levi, M.D.; Aurbach, D. The basic electroanalytical behavior of practical graphite-lithium intercalation electrodes. *Electrochim. Acta* **1998**, *43*, 2287–2304.
8. Aurbach, D.; Levi, M.D.; Levi, E.; Teller, H.; Markovsky, B.; Salitra, G.; Heider, U.; Heider, L. Common electroanalytical behavior of Li intercalation processes into graphite and transition metal oxides. *J. Electrochem. Soc.* **1998**, *145*, 3024–3034.
9. Aurbach, D.; Markovsky, B.; Weissman, I.; Levi, E.; Ein-Eli, Y. On the correlation between surface chemistry and performance of graphite negative electrodes for Li ion batteries. *Electrochim. Acta* **1999**, *45*, 67–86.
10. Aurbach, D.; Gnanaraj, J.S.; Levi, M.D.; Levi, E.A.; Fischer, J.E.; Claye, A. On the correlation among surface chemistry, 3D structure, morphology, electrochemical and impedance behavior of various lithiated carbon electrodes. *J. Power Sources* **2001**, *97*, 92–96.
11. Gnanaraj, J.S.; Levi, M.D.; Levi, E.; Salitra, G.; Aurbach, D.; Fischer, J.E.; Claye, A. Comparison between the electrochemical behavior of disordered carbons and graphite electrodes in connection with their structure. *J. Electrochem. Soc.* **2001**, *148*, A525–A536.
12. Aurbach, D.; Teller, H.; Koltypin, M.; Levi, E. On the behavior of different types of graphite anodes. *J. Power Sources* **2003**, *119*, 2–7.
13. Levi, M.D.; Wang, C.; Gnanaraj, J.S.; Aurbach, D. Electrochemical behavior of graphite anode at elevated temperatures in organic carbonate solutions. *J. Power Sources* **2003**, *119*, 538–542.
14. Markovsky, B.; Rodkin, A.; Cohen, Y.S.; Palchik, O.; Levi, E.; Aurbach, D.; Kim, H.J.; Schmidt, M. The study of capacity fading processes of Li-ion batteries: Major factors that play a role. *J. Power Sources* **2003**, *119*, 504–510.
15. Markervich, E.; Salitra, G.; Levi, M.D.; Aurbach, D. Capacity fading of lithiated graphite electrodes studied by a combination of electroanalytical methods, Raman spectroscopy and SEM. *J. Power Sources* **2005**, *146*, 146–150.
16. Ning, G.; Haran, B.; Popov, B.N. Capacity fade study of lithium-ion batteries cycled at high discharge rates. *J. Power Sources* **2003**, *117*, 160–169.

17. Novak, P.; Joho, F.; Lanz, M.; Rykart, B.; Panitz, J.-C.; Alliata, D.; Katz, R.; Haas, O. The complex electrochemistry of graphite electrodes in lithium-ion batteries. *J. Power Sources* **2001**, *97*, 39–46.
18. Shim, J.; Striebel, K.A. Effect of electrode density on cycle performance and irreversible capacity loss for natural graphite anode in lithium-ion batteries. *J. Power Sources* **2003**, *119*, 934–937.
19. Kaspar, J.; Graczyk-Zajac, M.; Riedel, R. Lithium insertion into carbon-rich SiOC ceramics: Influence of pyrolysis temperature on electrochemical properties. *J. Power Sources* **2013**, *244*, 450–455.
20. Haluschka, C.; Kleebe, H.J.; Franke, R.; Riedel, R. Silicon carbonitride ceramics derived from polysilazanes Part I. Investigation of compositional and structural properties. *J. Eur. Ceram. Soc.* **2000**, *20*, 1355–1364.
21. Mera, G.; Riedel, R.; Poli, F.; Müller, K. Carbon-rich sicn ceramics derived from phenyl-containing poly(silylcarbodiimides). *J. Eur. Ceram. Soc.* **2009**, *29*, 2873–2883.
22. Mera, G.; Navrotsky, A.; Sen, S.; Kleebe, H.-J.; Riedel, R. Polymer-derived SiCN and SiOC ceramics—Structure and energetics at the nanoscale. *J. Mater. Chem. A* **2013**, *1*, 3826–3836.
23. Wilson, A.M.; Reimers, J.N.; Fuller, E.W.; Dahn, J.R. Lithium insertion in pyrolyzed siloxane polymers. *Solid State Ionics* **1994**, *74*, 249–254.
24. Xing, W.; Wilson, A.M.; Zank, G.; Dahn, J.R. Pyrolysed pitch-polysilane blends for use as anode materials in lithium ion batteries. *Solid State Ionics* **1997**, *93*, 239–244.
25. Xing, W.; Wilson, A.M.; Eguchi, K.; Zank, G.; Dahn, J.R. Pyrolyzed polysiloxanes for use as anode materials in lithium-ion batteries. *J. Electrochem. Soc.* **1997**, *144*, 2410–2416.
26. Wilson, A.M.; Zank, G.; Eguchi, K.; Xing, W.; Yates, B.; Dahn, J.R. Polysiloxane pyrolysis. *Chem. Mater.* **1997**, *9*, 1601–1606.
27. Wilson, A.M.; Zank, G.; Eguchi, K.; Xing, W.; Dahn, J.R. Pyrolysed silicon-containing polymers as high capacity anodes for lithium-ion batteries. *J. Power Sources* **1997**, *68*, 195–200.
28. Wilson, A.M.; Xing, W.; Zank, G.; Yates, B.; Dahn, J.R. Pyrolysed pitch-polysilane blends for use as anode materials in lithium ion batteries II: The effect of oxygen. *Solid State Ionics* **1997**, *100*, 259–266.
29. Saha, A.; Raj, R.; Williamson, D.L. A model for the nanodomains in polymer-derived SiOC. *J. Am. Chem. Soc.* **2006**, *89*, 2188–2195.
30. Sanchez-Jimenez, P.E.; Raj, R. Lithium insertion in polymer-derived silicon oxycarbide ceramics. *J. Am. Ceram. Soc.* **2010**, *93*, 1127–1135.
31. Ahn, D.; Raj, R. Thermodynamic measurements pertaining to the hysteretic intercalation of lithium in polymer-derived silicon oxycarbide. *J. Power Sources* **2010**, *195*, 3900–3906.
32. Shen, J.; Raj, R. Silicon-oxycarbide based thin film anodes for lithium ion batteries. *J. Power Sources* **2011**, *196*, 5945–5950.
33. Ahn, D.; Raj, R. Cyclic stability and c-rate performance of amorphous silicon and carbon based anodes for electrochemical storage of lithium. *J. Power Sources* **2011**, *196*, 2179–2186.
34. Fukui, H.; Hisashi, O.; Hino, T.; Kanamura, K. A Si–O–C composite anode: High capability and proposed mechanism of lithium storage associated with microstructural characteristics. *Appl. Mater. Interfaces* **2010**, *4*, 998–1008.

35. Fukui, H.; Nakata, N.; Dokko, K.; Takemura, B.; Ohsuka, H.; Hino, T.; Kanamura, K. Lithiation and delithiation of silicon oxycarbide single particles with a unique microstructure. *ACS Appl. Mater. Interf.* **2011**, *3*, 2318–2322.
36. Fukui, H.; Ohsuka, H.; Hino, T.; Kanamura, K. Influence of polystyrene/phenyl substituents in precursors on microstructures of Si–O–C composite anodes for lithium-ion batteries. *J. Power Sources* **2011**, *196*, 371–378.
37. Fukui, H.; Ohsuka, H.; Hino, T.; Kanamura, K. Polysilane/acenaphthylene blends toward Si–O–C composite anodes for rechargeable lithium-ion batteries. *J. Electrochem. Soc.* **2011**, *158*, A550–A555.
38. Graczyk-Zajac, M.; Toma, L.; Fasel, C.; Riedel, R. Carbon-rich SiOC anodes for lithium-ion batteries: Part I. Influence of material UV-pre-treatment on high power properties. *Solid State Ionics* **2012**, *225*, 522–526.
39. Dibandjo, P.; Graczyk-Zajac, M.; Riedel, R.; Pradeep, V.S.; Soraru, G.D.A. Lithium insertion into dense and porous carbon-rich polymer-derived SiOC ceramics. *J. Eur. Ceram. Soc.* **2012**, *32*, 2495–2503.
40. Fukui, H.; Eguchi, K.; Ohsuka, H.; Hino, T.; Kanamura, K. Structures and lithium storage performance of SiOC composite materials depending on pyrolysis temperatures. *J. Power Sources* **2013**, *243*, 152–158.
41. Fukui, H.; Ohsuka, H.; Hino, T.; Kanamura, K. Silicon oxycarbides in hard-carbon microstructures and their electrochemical lithium storage. *J. Electrochem. Soc.* **2013**, *160*, A1276–A1281.
42. Pradeep, V.S.; Graczyk-Zajac, M.; Wilamowska, M.; Riedel, R.; Soraru, G.D. Influence of pyrolysis atmosphere on the lithium storage properties of carbon-rich polymer derived SiOC ceramic anodes. *Solid State Ionics* **2013**, *262*, 22–24.
43. Kaspar, J.; Graczyk-Zajac, M.; Riedel, R. Determination of the chemical diffusion coefficient of Li-ions in carbon-rich silicon oxycarbide anodes by electro-analytical methods. *Electrochim. Acta* **2014**, *115*, 665–670.
44. Dahn, J.R.; Wilson, A.M.; Xing, W.; Zank, G.A. Electrodes for Lithium Ion Batteries Using Polysilazanes Ceramic with Lithium. U.S. Patent 5631106(A), 20 May 1997.
45. Su, D.; Li, Y.-L.; Feng, Y.; Jin, J. Electrochemical properties of polymer-derived SiCN materials as the anode in lithium ion batteries. *J. Am. Ceram. Soc.* **2009**, *92*, 2962–2968.
46. Feng, Y. Electrochemical properties of heat-treated polymer-derived sicn anode for lithium ion batteries. *Electrochim. Acta* **2010**, *55*, 5860–5866.
47. Kaspar, J.; Mera, G.; Nowak, A.P.; Graczyk-Zajac, M.; Riedel, R. Electrochemical study of lithium insertion into carbon-rich polymer-derived silicon carbonitride ceramics. *Electrochim. Acta* **2010**, *56*, 174–182.
48. Graczyk-Zajac, M.; Mera, G.; Kaspar, J.; Riedel, R. Electrochemical studies of carbon-rich polymer-derived sicn ceramics as anode materials for lithium-ion batteries. *J. Eur. Ceram. Soc.* **2010**, *30*, 3235–3243.
49. Reinold, L.M.; Graczyk-Zajac, M.; Gao, Y.; Mera, G.; Riedel, R. Carbon-rich sicn ceramics as high capacity/high stability anode material for lithium-ion batteries. *J. Power Sources* **2013**, *236*, 224–229.

50. Baek, S.-H.; Reinold, L.M.; Graczyk-Zajac, M.; Riedel, R.; Hammerath, F.; Büchner, B.; Grafe, H.-J. Lithium dynamics in carbon-rich polymer-derived SiCN ceramics probed by nuclear magnetic resonance. *J. Power Sources* **2014**, *253*, 342–348.
51. Graczyk-Zajac, M.; Fasel, C.; Riedel, R. Polymer-derived-SiCN ceramic/graphite composite as anode material with enhanced rate capability for lithium ion batteries. *J. Power Sources* **2011**, *196*, 6412–6418.
52. Reinold, L.M.; Graczyk-Zajac, M.; Fasel, C.; Riedel, R. Prevention of solid electrolyte interphase damaging on silicon by using polymer derived SiCN ceramics. *ECS Trans.* **2011**, *35*, 37–44.
53. Wilamowska, M.; Graczyk-Zajac, M.; Riedel, R. Composite materials based on polymer-derived SiCN ceramic and disordered hard carbons as anodes for lithium-ion batteries. *J. Power Sources* **2013**, *24*, 80–86.
54. Kaspar, J.; Graczyk-Zajac, M.; Riedel, R. Carbon-rich SiOC anodes for lithium-ion batteries: Part II. Role of thermal cross-linking. *Solid State Ionics* **2012**, *225*, 527–531.
55. Kolb, R.; Fasel, C.; Liebau-Kunzmann, V.; Riedel, R. SiCN/C-ceramic composite as anode material for lithium ion batteries. *J. Eur. Ceram. Soc.* **2006**, *26*, 3903–3908.
56. Liu, G.; Kaspar, J.; Reinold, L.M.; Graczyk-Zajac, M.; Riedel, R. Electrochemical performance of DVB-modified sioc and sicn polymer-derived negative electrodes for lithium-ion batteries. *Electrochim. Acta* **2013**, *106*, 101–108.
57. Pradeep, V.S.; Graczyk-Zajac, M.; Riedel, R.; Soraru, G.D. New insights in to the lithium storage mechanism in polymer derived SiOC anode materials. *Electrochim. Acta* **2014**, *119*, 78–85.
58. Mera, G.; Tamayo, A.; Nguyen, H.; Sen, S.; Riedel, R. Nanodomain structure of carbon-rich silicon carbonitride polymer-derived ceramics. *J. Am. Ceram. Soc.* **2010**, *93*, 1169–1175.
59. Morcos, R.; Mera, G.; Navrotsky, A.; Varga, T.; Riedel, R.; Poli, F.; Müller, K. Enthalpy of formation of carbon-rich polymer-derived amorphous SiCN ceramics. *J. Am. Ceram. Soc.* **2008**, *91*, 3349–3354.
60. Widgeon, S.; Mera, G.; Gao, Y.; Stoyanov, E.; Sen, S.; Navrotsky, A.; Riedel, R. Nanostructure and energetics of carbon-rich SiCN ceramics derived from polysilylcarbodiimides: Role of the nanodomain interfaces. *Chem. Mater.* **2012**, *24*, 1181–1191.
61. Widgeon, S.J.; Sen, S.; Mera, G.; Ionescu, E.; Riedel, R.; Navrotsky, A. ^{29}Si and ^{13}C solid-state NMR spectroscopic study of nanometer-scale structure and mass fractal characteristics of amorphous polymer derived silicon oxycarbide ceramics. *Chem. Mater.* **2010**, *22*, 6221–6228.
62. Blum, Y.D.; MacQueen, D.B.; Kleebe, H.-J. Synthesis and characterization of carbon-enriched silicon oxycarbides. *J. Eur. Ceram. Soc.* **2005**, *25*, 143–149.
63. Gregori, G.; Kleebe, H.-J.; Blum, Y.D.; Babonneau, F. Evolution of C-rich sioc ceramics: Part II. Characterization by high lateral resolution techniques electron energy-loss spectroscopy high-resolution TEM and energy-filtered TEM. *Int. J. Mat. Res.* **2006**, *97*, 710–720.
64. Kleebe, H.-J.; Gregori, G.; Blum, Y.D.; Babonneau, F. Evolution of C-rich sioc ceramics. Part I. Characterization by integral spectroscopic techniques solid-state NMR and Raman spectroscopy. *Int. J. Mat. Res.* **2006**, *97*, 699–709.
65. Soraru, G.D.; Modena, S.; Guadagnino, E.; Colombo, P.; Egan, J.; Pantano, C. Chemical durability of silicon oxycarbide glasses. *J. Am. Ceram. Soc.* **2002**, *85*, 1529–1536.

66. Kroll, P. Tracing reversible and irreversible Li insertion in SiCO ceramics with modeling and ab-initio simulations. *MRS Proc.* **2011**, *1313*, 1–6.

8

Frontiers in Nucleic Acid Nanotechnology

Stephen F. Ralph

School of Chemistry, University of Wollongong, Northfields Avenue, Wollongong 2522, NSW, Australia; E-Mail: sralph@uow.edu.au

Academic Editor: Thomas Nann

This Special Issue of *Nanomaterials* highlights innovative work from around the world focused on harnessing the physical, chemical and topological properties of nucleic acids. It contains a mixture of articles examining composite materials that combine nucleic acids with carbon nanotubes (CNTs) or metallic nanoparticles, as well as theoretical investigations, and studies directed towards applications in a variety of areas. The latter include nanophotonics, new methods of DNA sequencing, and the development of new antibacterial agents or drug delivery vehicles. Highlights of the publications appearing in this Special Issue are discussed in the following paragraphs.

Kazuo Umemura [1] has provided a timely and comprehensive review article focused on nanobiotechnological applications of materials produced by hybridization of nucleic acids with CNTs. Since such materials were first reported a little over 10 years ago, there have been over 200 studies published in this area. The prevalence of investigations directed towards finding methods for enhancing the solubility of CNTs, or enabling their separation and purification is highlighted, before a selection of articles are described which exemplify some of the work currently being undertaken to develop new applications for hybrid DNA/CNT materials. Amongst these studies the preparation of new materials for sensing of a range of analytes, such as other DNA molecules, neurotransmitters, glucose, hormones, various gases and even bacterial cells, is a common theme. Other studies referenced in the review describe the potential of DNA/CNT hybrids for destroying cancer cells via thermal ablation, and highlight that the preparation of hybrid materials consisting of DNA and graphene is a rapidly emerging area.

The development of a new method for preparing hydrogels composed of multi-walled carbon nanotubes (MWNTs) and two types of DNA is the topic of the study undertaken by Zinchenko and co-workers [2]. A feature of this investigation was the use of relatively low molecular weight DNA extracted from salmon milt to initially produce dispersions of MWNTs. Since salmon milt is readily available in significant quantities as a by-product of the fishing industry, it offers environmental advantages for applications requiring large quantities of soluble nanotubes. Hydrogels were produced from the initial DNA/MWNT dispersions by taking advantage of the fragments of adsorbed DNA still exposed to the surrounding solution. These fragments were chemically crosslinked in a process using DNA molecules with a much higher molecular weight, and ethylene glycol diglycidyl ether. Using this procedure homogeneous hydrogels were obtained that exhibited similar swelling properties to that of DNA hydrogels containing no MWNTs. The hybrid DNA/MWNT hydrogels were shown to be significantly stronger than those not containing nanotubes in experiments measuring their extent of shrinkage in the presence of high concentrations of NaCl.

DNA obtained from salmon milt was also used in research led by Tomomi Takeshima [3]. In this case single stranded DNA containing approximately 60 nucleotides was used as the stabilizing agent in a procedure for preparing silver nanoparticles ranging in size between 2 and 20 nm. Zeta potential measurements confirmed that the surfaces of the nanoparticles were covered in negatively charged DNA molecules. The nanoparticles were shown to inhibit the growth of both a Gram-positive and a Gram-negative bacterium in a concentration dependent manner, most likely owing to the release of free Ag(I) ions. Of the two types of bacteria examined, the nanoparticles proved to be more effective towards the Gram-negative *E. coli*. As a demonstration of the potential application of their discovery, the researchers immobilized the Ag/DNA nanoparticles onto the surfaces of cotton fabrics, which had been previously modified using a cationic polymer in order to facilitate electrostatic binding of the nanoparticles. The resulting materials were shown to exhibit significant bacteriostatic and bactericidal properties in standardized tests.

One of the most exciting areas of nanotechnology research centers on the development of carrier systems for facilitating the delivery of therapeutic agents selectively to specific cells or molecular targets. As part of their efforts in this area, Kocabey and co-workers [4] have described the preparation of modified DNA nanotubes containing siRNA, folic acid or fluorescent dyes, as well as the results of experiments designed to explore the stability of these systems, and their level of uptake by HeLa cells. Initial experiments focused on DNA nanotubes measuring approximately 27 nm long, and with a diameter of ~6 nm, that were prepared using single-stranded DNA consisting of 42 nucleotides. Unfortunately, these nanotubes were incapable of entering the cytoplasm of HeLa cells and delivering their payload in order to effect gene silencing and altered gene expression. This prompted a series of insightful investigations into factors that can dramatically affect the stability of DNA-based drug delivery systems. It was shown that increasing the size of the DNA nanotubes, by doubling the number of nucleotides present in the initial DNA tiles, significantly increases their lifetime, even when in contact with buffers containing relatively low concentrations of stabilizing divalent cations. The authors conclude their article by providing a warning concerning interpretation of the results of drug-delivery experiments when the stability of the transport vehicle has not been fully investigated.

Cervantes-Salguero and co-workers [5] report the preparation of flexible DNA building blocks using the DNA origami method, as well as their self-assembly into clusters of different sizes and shapes, including triangles, squares and pentagrams, on the surfaces of pieces of mica. In a series of elegant experiments followed using atomic force microscopy (AFM), the researchers show how addition of sodium ions facilitates the disassembly of the clusters, and their subsequent reconfiguration into different shapes via diffusion of the individual blocks on the surface of the mica. AFM was also used to examine the effect of varying the length of flexible regions within the DNA building blocks on the final distribution of clusters, thereby providing insights into how this approach could be modified to yield clusters of other sizes and shapes in future.

Titration of solutions containing non-covalent DNA/MWNT complexes, with didodecyl-dimethylammonium bromide (DDAB), resulted initially in the formation of sticky precipitates, which re-dissolved when excess cationic surfactant was added. Tardani and La Mesa [6] have provided a comprehensive report on their investigations of these mixtures, which were conducted using a range of techniques. These included optical spectroscopy and microscopy, dynamic light scattering, and both ionic conductivity and zeta potential measurements. From the results obtained a (pseudo)-binary phase diagram was obtained, and the properties of the systems rationalized by considering a model in which the DNA/MWNT complexes were considered to be rigid, rod-like polyelectrolytes. Significant variations in behavior were observed compared to what is seen with the analogous surfactant dodecyltrimethylammonium bromide (DTAB), which only contains one hydrophobic chain. For example, precipitate re-dissolution is far less extensive when DTAB is added, consistent with it interacting less effectively with DNA/MWNT complexes. These insights have important implications for researchers interested in understanding how other biomolecules might bind to DNA complexes.

The ability to use DNA as a template for the controlled assembly of clusters of sliver atoms is the topic of a detailed review by Gwinn and co-workers [7]. Drawing on their own work and that of other researchers, the authors discuss methods for isolation and purification of DNA/Ag composites, as well as a number of methods, including electrospray ionization mass spectrometry and inductively coupled plasma atomic emission spectrometry, for analyzing the composition of these materials. It is revealed that DNA/Ag materials can be produced using a variety of nucleic acid structures, including the ubiquitous Watson-Crick double helical form, and different hairpin structures. The authors also explain that the observation of a single absorption band in the visible or near IR region supports the conclusion that the clusters have a rod shape, as opposed to globular. Furthermore the reason why certain fluorescence colors are more readily observed in DNA/Ag materials is accounted for by proposing that clusters containing specific ("magic") numbers of silver atoms exhibit enhanced stability compared to those with fewer or greater numbers of atoms. A brief description of the chiral properties of DNA/Ag materials is provided, along with a discussion of as yet unanswered questions concerning their preparation and properties.

Li and co-workers [8] review the state of the art with respect to our understanding of the effects of external forces on DNA structure. Advances in this area have been made possible through the advent of a number of new nano-manipulation and single-molecule techniques. The latter, in particular, have enabled researchers to explore how single stranded and double stranded DNA molecules with different base sequences respond to applied mechanical forces. Significant progress has now also been made in an effort to understand the effects of numerous factors, such as the presence of simple cations, on the strength of DNA-DNA interactions. The importance of solvent molecules in mediating such interactions is also addressed, and a limited discussion of investigations performed using DNA structures other than B-form DNA also presented. Developing a more sophisticated understanding of how DNA responds to external forces is critical if we wish to take advantage of its unique structural and chemical properties in nanotechnology, and in order to fully comprehend how DNA interacts with other complex multi-protein machines as it performs its biological functions inside cells.

The search for more rapid and reliable methods of DNA sequencing is an important objective as researchers seek to develop new biotechnological devices and products. One method of DNA sequencing that has attracted considerable interest in recent years involves the use of nanopores composed of the protein alpha-hemolysin. Manara and co-workers [9] describe the results of molecular dynamics simulations to examine how effectively a nucleotide would be captured if it were generated at different distances from the mouth of the nanopore, through the action of an exonuclease on an oligonucleotide. The author's simulations demonstrate that capture is most effective when the nucleotide is released close to the nanopore, and that the initial orientation of the nucleotide and its position relative to the center of the mouth of the protein are less important factors. Several amino acids that play a key role in preventing access of the nucleotide to the nanopore are also highlighted.

The above articles showcase the breadth of nanotechnological research currently being undertaken throughout the world that seeks to take advantage of the many unique features of DNA. We hope that you enjoy reading this Special Issue, and look forward to your contributions to this journal in this fascinating area in the years ahead.

Conflicts of Interest

The authors declare no conflict of interest.

References

1. Umemura, K. Hybrids of nucleic acids and carbon nanotubes for nanobiotechnology. *Nanomaterials* **2015**, *5*, 321–350.
2. Zinchenko, A.; Taki, Y.; Sergeyev, V.; Murata, S. DNA-assisted solubilization of carbon nanotubes and construction of DNA-MWCNT cross-linked hybrid hydrogels. *Nanomaterials* **2015**, *5*, 270–283.
3. Takeshima, T.; Tada, Y.; Sakaguchi, N.; Watari, F.; Fugetsu, B. DNA/Ag nanoparticles as antibacterial agents against gram-negative bacteria. *Nanomaterials* **2015**, *5*, 284–297.
4. Kocabey, S.; Meinl, H.; MacPherson, I.S.; Cassinelli, V.; Manetto, A.; Rothenfusser, S.; Liedl, T.; Lichtenegger, F.S. Cellular uptake of tile-assembled DNA nanotubes. *Nanomaterials* **2015**, *5*, 47–60.
5. Cervantes-Salguero, K.; Hamada, S.; Nomura, S.M.; Murata, S. Polymorphic ring-shaped molecular clusters made of shape-variable building blocks. *Nanomaterials* **2015**, *5*, 208–217.
6. Tardani, F.; La Mesa, C. Titration of DNA/carbon nanotube complexes with double-chained oppositely charged surfactants. *Nanomaterials* **2015**, *5*, 722–736.
7. Gwinn, E.; Schultz, D.; Copp, S.M.; Swasey, S. DNA-protected silver clusters for nanophotonics. *Nanomaterials* **2015**, *5*, 180–207.
8. Li, J.; Wijeratne, S.S.; Qiu, X.; Kiang, C.-H. DNA under force: Mechanics, electrostatics and hydration. *Nanomaterials* **2015**, *5*, 246–267.

9. Manara, R.M.A.; Tomasio, S.; Khalid, S. The nucleotide capture region of alpha hemolysin: Insights into nanopore design for DNA sequencing from molecular dynamics simulations. *Nanomaterials* **2015**, *5*, 144–153.

9

Separator Membrane from Crosslinked Poly(Vinyl Alcohol) and Poly(Methyl Vinyl Ether-alt-Maleic Anhydride)

Charu Vashisth Rohatgi, Naba K. Dutta and Namita Roy Choudhury *

Ian Wark Institute, University of South Australia, Mawson Lakes 5095, Australia;
E-Mails: charuvashisthrohatgi@gmail.com (C.V.R.); naba.dutta@unisa.edu.au (N.D.)

* Author to whom correspondence should be addressed; E-Mail: namita.choudhury@unisa.edu.au

Academic Editor: Thomas Nann

Abstract: In this work, we report separator membranes from crosslinking of two polymers, such as poly vinyl alcohol (PVA) with an ionic polymer poly(methyl vinyl ether-alt-maleic anhydride) (PMVE-MA). Such interpolymer-networked systems were extensively used for biomedical and desalination applications but they were not examined for their potential use as membranes or separators for batteries. Therefore, the chemical interactions between these two polymers and the influence of such crosslinking on physicochemical properties of the membrane are systematically investigated through rheology and by critical gel point study. The hydrogen bonding and the chemical interaction between PMVE-MA and PVA resulted in highly cross-linked membranes. Effect of the molecular weight of PVA on the membrane properties was also examined. The developed membranes were extensively characterized by studying their physicochemical properties (water uptake, swelling ratio, and conductivity), thermal and electrochemical properties using differential scanning calorimetry (DSC), dynamic mechanical analysis (DMA), thermo-gravimetric analysis (TGA) and electrochemical impedance spectroscopy (EIS). The DSC study shows the presence of a single T_g in the membranes indicating compatibility of the two polymers in flexible and transparent films. The membranes show good stability and ion conductivity suitable for separator applications.

Keywords: poly vinyl alcohol; separator membrane; ionic polymer; interpolymer-networked systems; battery separator; hydrogen bonding; electrochemical impedance spectroscopy; thermal analysis; dynamic mechanical analysis

---

1. Introduction

A separator is a membrane, whose main function is to keep the positive and negative electrodes apart to prevent electrical short circuits and, at the same time, allow rapid transport of ionic charge carriers by intrinsic ionic conductor [1,2]. It should minimize any processes that adversely affect the electrochemical energy efficiency of the batteries. A variety of separators have been used in batteries over the years. Separators have been manufactured from cellulosic papers and cellophane to nonwoven fabrics, foams, ion exchange membranes, and microporous flat sheet membranes made from polymeric materials. A separator has significant role in all battery systems. The thin membrane performs both as a medium to transport ions during electrochemical reactions and also prevents/isolates the positive and negative electrode from short circuits. They should be chemically and electrochemically inert and should exhibit high mechanical integrity against the electrolyte and electrode materials during cell operation. Current and primary demands on the separator membrane are ion conductivity, excellent mechanical strength, dimensional stability, chemical stability towards oxidation, reduction and hydrolysis, and low cost [2]. For example, to fabricate the separator for a lithium-based battery, the following criteria should be satisfied: (a) electronic insulator; (b) minimal electrolyte resistance ($<2\ \Omega\cdot cm^2$); (c) dimensional stability (shrinkage: <5%); (d) mechanical strength enough to allow the assembly process; (e) chemical stability against degradation by electrolyte or electrode reactants (should be stable for long cycle life); (f) effective prevention of the migration of particles or soluble species between the electrodes; (g) good wettability in electrolyte solution (should be wet completely in liquid electrolytes); (h) uniform thickness and pore distribution (thickness: <25 μm; pore size: <1 μm; porosity: ~40%); increased ionic conductivity, good wettability, and excellent interfacial adhesion between the electrodes and the separators [2]. As batteries have become more sophisticated, the separator function has also become more demanding and complex. Very little work is directed toward developing and characterizing new separators for flow batteries, which can be made readily and scalable. Thus, in this work, we aim to develop a facile route for the fabrication of separators for redox flow batteries, e.g., zinc-bromine.

Amongst a variety of different approaches to synthesize new separator membranes, acid-base polymer blends can offer a potential route to designing improved separator materials due to the interaction between polymers. In multi-component polymer systems, the interfacial interaction between component phases can be effectively utilized to enhance homogeneous processing and achieve unique performance. Such interactions between the component phases can be brought about by either chemical and/or specific interaction between the components or by adding block copolymer of component phases. Amongst various types of specific interactions, hydrogen bonding is an effective route to complexation of two polymers in their blends to result in unique properties. Despite a number of different strategies having been used for membrane development, research on inter polymer ion exchange membranes, prepared from compatible casting solutions of two water soluble polymers is scanty.

A variety of approaches, including thermal cross-linking [3–6], electrospinning [7–11], chemical cross-linking [12–14] are being applied to synthesize new membranes. Although chemical cross-linking is a versatile method to tune polymer microstructure and physical properties; however, to avoid the use of toxic chemicals, environmentally benign routes are gaining importance in membrane fabrication. Among them the acid-base polymer blends have become one of the most facile approach in the design of improved membrane materials due to the interaction (ionically cross-linked) between the polymers.

A series of different polymer blends, such as sulfonated poly(ether ether ketone) (SPEEK) blended with poly(ether imide) (PEI), poly(amide imide) (PAI), and poly(benzimidazole) (PBI), have been reported in the literature [15–17] for such a use. In most of the previous studies [3,4], these membranes were used for the desalination of aqueous salt solutions, for various potential in biomedical applications and also for their oxygen barrier properties [18]. It has been observed that the association of a polyacid with polybase can significantly reduce the swelling and the methanol permeability of membranes [19,20] due to stabilization of interpolymer complexes through hydrogen bonding [21]. In addition, unique properties of these complexes include increased impact and tensile strength, which are useful in a wide range of applications.

In this work, poly(vinyl alcohol) (PVA) has been selected as base polymer matrix in view of its film-forming characteristics, hydrophilic properties, and high density of reactive chemical functions that are favorable for cross-linking using thermal treatments [5]. It also exhibits interaction ability with complimentary polymers [22,23]. Several cross-linking methods have been published for different uses. As a rule, all multifunctional compounds capable of reacting with hydroxyl groups can be used to obtain three-dimensional networks in PVA [24]. Poly(methyl vinyl ether-alt-maleic anhydride) (PMVE-MA) was selected due to its very high hydrophilicity and also because it can act as a cross-linking agent for PVA due to the presence of maleic acid groups in its repeating units.

Here we report the preparation of cross-linked membranes, based on PVA and PMVE-MA, and investigate the cross linking process in detail. We also establish the influences of the interaction on the membrane properties including dimensional stability, water uptake, ionic conductivity with increasing PMVE-MA ratios in the blends.

2. Experimental Section

2.1. Materials

Poly vinyl alcohol (PVA) (98% hydrolyzed, average Mw = 13,000–26,000 $g\cdot mol^{-1}$), (99% hydrolyzed, average Mw = 89,000–98,000 $g\cdot mol^{-1}$) and (99% hydrolyzed, average Mw = 146,000–186,000 $g\cdot mol^{-1}$) and Poly(methyl vinyl ether-alt-maleic anhydride) (PMVE-MA) average Mw = 1,080,000 $g\cdot mol^{-1}$) were procured from Aldrich and were used as received. Hydrochloric acid (98%) from Aldrich was used as esterification catalyst. High-purity Milli-Q water was used throughout the study (resistivity ~18.2 $M\cdot\Omega\cdot cm$).

2.2. Blend Membranes

Three different molecular weight PVA samples have been chosen for crosslinking study. The cross-linked PVA/PMVE-MA membranes were prepared using the following two-step procedure. First, aqueous 10 wt% PVA (Table 1) solution was prepared by dissolving a pre-weighed amount polymer at 90 °C for 6 h. Aqueous 10 wt% PMVE-MA solution was prepared by dissolving pre-weighted quantities of dry PMVE-MA in Milli-Q-water at room temperature. Finally, PVA solution was mixed together with PMVE-MA solution at a fixed ratio (Table 1) for 5 h at room temperature until a homogeneous solution was formed. The mixed solution was poured onto a petri dish, and dried at room temperature to prepare the film. The cast polymer solutions were allowed to dry in a vacuum oven at 30 °C for 24 h. The dried membranes were peeled off the Petri dish, and then heated in an oven at 120 °C for 2 h to induce the cross-linking reaction. Insolubilization of the polymer blend films was accomplished via formation of interchain ester crosslinks during heat treatment. The samples were then characterized using spectroscopic, thermal and rheological studies. For all the spectroscopic and thermal studies, we have used blend samples of PVA with a molecular weight 89 K.

Table 1. Water uptake and dimensional change of cross-linked PVA/PMVE-MA blend membranes for different molecular weight PVA with various blend ratios. Values are means ± SD.

Sample No.	Effect of molecular weight of PVA					
	Mw-13,000–23,000		Mw-89,000–98,000		Mw-124,000–186,000	
$S_{PMVE\text{-}MA:PVA}$	Water uptake (%)	Dimensional change (%)	Water uptake (%)	Dimensional change (%)	Water uptake (%)	Dimensional change (%)
$S_{10:90}$	36	35	42	47	49	50
$S_{20:80}$	19	23	24	25	32	21
$S_{30:70}$	22	26	27	30	39	42
$S_{40:60}$	34	29	37	32	44	41
$S_{50:50}$	40	39	45	42	51	58

3. Characterization

3.1. Water Uptake and Dimensional Stability

The weight change of the samples was monitored to investigate the water uptake by the membranes. In a typical experiment 10 mm diameter discs were punched out from each sample and the weight of the pre-dried samples was measured (Wd_1). The samples were then immersed in de-ionized water at 25 °C for 24 h, and then surface-attached water on the membranes was removed using a Kim wipe. Then, the wet sample weight (W_{wet}) was determined as quickly as possible. The weight of dry sample was measured after completely drying the hydrated membranes at 60 °C (Wd_2) for 24 h. The weight change of the membranes was calculated as follow:

$$\text{Weight change (\%)} = [(Wd_1 - Wd_2)/Wd_1] \times 100 \quad (1)$$

The total water uptake (%) was determined by

$$\text{Total water uptake (\%)} = [(W_{wet} - W_{dry})/W_{dry}] \times 100 \quad (2)$$

$$\text{Dimensional change (\%)} = [(D_{\text{wet}} - D_{\text{dry}})/D_{\text{dry}}] \times 100 \quad (3)$$

where W_{wet}, D_{wet} and W_{dry}, D_{dry} are the wet and dried sample weight and dimensions after immersing in water, respectively. The water uptake measurement was done in triplicate and the average values are reported.

3.2. Spectroscopic Analysis

In order to examine the nature of interaction photo-acoustic fourier transform infrared spectroscopy (PA-FTIR) study was performed on a Nicolet Magna-IR Spectrometer 750 (Thermo Fisher Scientific, Waltham, MA, USA) equipped with an MTEC model 300 photoacoustic cell under helium purge with carbon black as the reference. IR spectra were recorded with a wave number resolution of 8 cm^{-1} and 526 scans were performed in the range of 450–4000 cm^{-1}. Samples in the form of thin films were placed into the sample holder and placed in the Photo acoustic cell. Carbon black was employed as a background reference.

3.3. Cross-Linking of PVA/PMVE-MA by Rheological Measurements

Cross linking kinetics and gel point measurement of PVA/PMVE-MA blend systems were examined using dynamic rheological measurement using a cone-plate geometry (40 mm 2° steel cone) at different temperature range from 30 °C to 45 °C using AR 1000-N Rheometer (TA Instruments, New Castle, DE, USA) and at a single frequency of 1 Hz.

3.4. Thermal Analyses

Thermogravimetric analysis (TGA) was done on a TA 2950 Thermal Analyzer (TA Instruments). Samples were dried in an oven at 80 °C prior to analysis and the experiment was performed on approximately 5 mg of the sample under nitrogen with a temperature ramp of 10 °C·min^{-1}. Differential scanning calorimetric measurements were done using (DSC) Q2000 (TA Instruments). Each sample was heated from 30 °C to 250 °C at a heating rate of 10 °C·min^{-1} under nitrogen. Then, it was cooled down to 30 °C at the same rate. A second heating was carried out in the same conditions as the first one. The glass transition temperature (T_g) was determined from the second heating scan. The thermal stability of the samples was evaluated in a TA Instrument's thermogravimetric analyzer at 30–600 °C and 10 °C·min^{-1} under nitrogen/oxygen atmosphere at a flow rate of 50 mL·min^{-1}. The degradation temperatures were determined. Dynamic-mechanical analyses (DMA) were performed using a TA Q800 (TA Instruments). It was employed to determine the thermal transitions of the membranes. Scans were carried out at a heating rate of 2 °C·min^{-1} and a frequency of 1 Hz. The loss modulus E' (viscous component) and the storage modulus E" (elastic component) of the membranes were obtained from the analysis. The T_gs were obtained from the peak of Tan δ (E'/E") *versus* temperature data.

3.5. Conductivity Measurement

The ionic conductivity measurements were carried out at ambient temperature after equilibrating the membrane in de-ionized water for one day. A normal two-point probe technique at relative humidity 95% was used where the membrane sample was sandwiched between the platinum electrodes. Each sample was cut in 4 × 1 cm^2 prior to mounting on the cell. The impedance spectral measurement of the membrane was carried out using a Solartron Frequency Response Analyzer 1260 A connected to an Electrochemical Interface 1287 (Solartron Metrology, Cambridge, UK). The impedance analyzer worked in galvanostatic mode with AC current amplitude of 0.1 mA over frequency range from 100 kHz to 0.1 Hz. The ionic conductivity (σ) was obtained as follows:

$$\sigma = \frac{L}{R \cdot S} \tag{4}$$

where σ is the ionic conductivity (in S/cm); L is the distance between the electrodes used to measure the potential (cm); R is the impedance of the membrane (in Ω); and S is the surface area for ion to penetrate the membrane (in cm^2).

4. Results and Discussion

4.1. Interactions Study by PA-FTIR

FTIR is a powerful tool to investigate specific interaction between polymers. Chemical investigation of the interfacial interaction was performed using PA-FTIR, for all the composition, to gain insight into the degree of inter-polymer complexation that can take place due to hydrogen bonding between carbonyl (C=O) groups of the PMVE-MA and the hydroxyl (–OH) groups of the PVA. Figure 1 represents the FTIR spectra for pristine PVA, PMVE-MA, and PMVE-MA/PVA blend membranes. The spectrum of PMVE-MA shows the presence of a band at 1107 cm^{-1} due to C–O–C group. All the blends show band in the region of 1650–1800 cm^{-1} arising from −C=O group. In pristine PVA, intra-molecular and inter-molecular hydrogen bonds are expected to occur among PVA chains due to high hydrophilic force [25]. The −OH group of pristine PVA (3345 cm^{-1}) progressively increases in height after incorporation of PMVE-MA in the blend membranes. This change is caused due to strong interaction between hydroxyl groups on PVA and carboxyl groups on PMVE-MA, and weakening of the hydrogen bonds between PVA chains. Hence, intermolecular interactions are preferred with increasing PMVE-MA ratio, leading to a more effective mixing of both constituents. This observation indicates that the cross linking reaction occurred between PVA and PMVE-MA by creating new covalent bonds and weakening hydrogen bonds between PVA chains. The band in the 1800–1700 cm^{-1} region arises from C=O group vibrations [26]. As seen from Figure 1 (highlighted portion), the bands for −COOH groups show shifting and increase in the intensity after cross-linking when PMVE-MA was initially mixed with PVA, and resulted in the formation of a peak around 1718 cm^{-1} characteristic of COOH group vibration with a gradual increase in the shoulder of the peak which is quite evident from the graph as the PMVE-MA content is increased. The PVA/PMVE-MA blend films show a new absorption band around 1615 cm^{-1} for (C=O) stretching, whereas around 1237 cm^{-1} for (C–O) stretch mode of ester group in sample $S_{10:90}$. The absorption peaks for sample $S_{30:70}$ was around 1610 cm^{-1}, 1235 cm^{-1}, around 1608 cm^{-1}, 1196 cm^{-1} for $S_{40:60}$ and around 1603 cm^{-1}, 1204 cm^{-1} for $S_{50:50}$, are assigned as peaks for (C=O) and (C–O) ester group stretching, respectively, and showed shift towards lower wavenumber as we increase the PMVE-MA content [26] due to the formation of ester bonds (C–O–C) between alcohol group of PVA and carboxyl group of PMVE-MA.

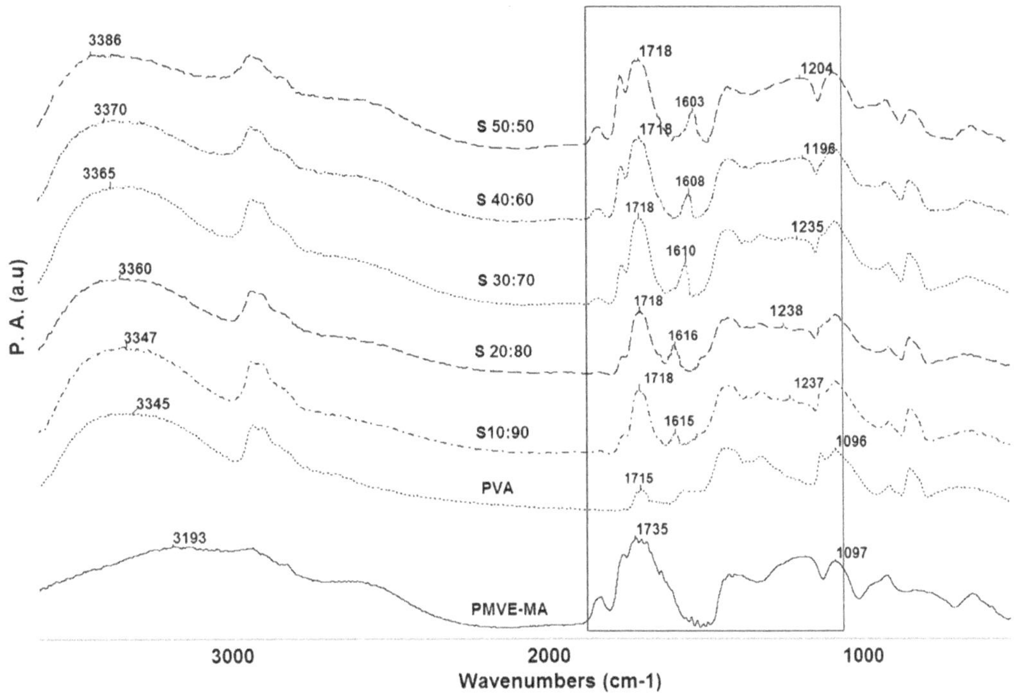

Figure 1. Fourier transform infrared spectroscopy (FTIR) spectra of poly vinyl alcohol (PVA), poly(methyl vinyl ether-alt-maleic anhydride) (PMVE-MA) and cross-linked PMVE-MA/PVA blends. The highlighted area represents the ester bonds formed during cross linking.

4.2. Kinetics of Cross-Linking of PVA/PMVE-MA Solutions

As the strength of interpolymer complexation interactions are strongly determined by the stoichiometric ratios of the component polymers; therefore, optimization of blend ratio is critical. One of the rapid ways of determining the optimum blend ratio and the strength of interactions is through evaluation of blend viscosities. Figure 2 compares the viscosity of each blend with different molecular weight PVA and compares it to the calculated (rule of additivity) value in order to highlight deviations. As observed from the graph, an increase in PMVE-MA content in the blend led to a decrease in blend viscosity due to the dissociation of the carboxylic acid groups. When PVA with Mw-13 K and 89 K was used, a positive deviation in viscosity from linear mixing values for all blend compositions was observed, with a maximum positive deviation of about 0.05 Pa·s and 0.5 Pa·s respectively at 40 wt% PMVE-MA indicating favorable interaction. However, at 10 wt% PMVE-MA resulted in negative deviation from calculated values for PVA Mw-89 K due to decrease in the viscosity of solution. However, further addition of PMVE-MA content leads to positive deviation again. An identical trend was observed with PVA Mw-124 K with the only difference being that the maximum positive deviation was seen ~30 wt% PMVE-MA and, unlike the other two molecular weight PVAs, further addition of PMVE-MA content led to decrease in viscosity until a negative deviation occurred around 60–100 wt% PMVE-MA (positive negative deviation behavior-PNDB). The PNDB phenomenon can be related to concentration-dependent interaction and explained by partial miscibility at low concentration and a concentration-dependent change of the flow mechanism in the resultant blend. The reduction in viscosity and broadening of the peak with a decrease in viscosity can be attributed to higher intramolecular hydrogen-bonding within PMVE-MA, and the inter-polymer interaction is reduced, leading to a reduction in viscosity. The broadening of the peak is probably due to a larger number of possible hydrogen-bonding interactions that can occur at similar energy levels [27].

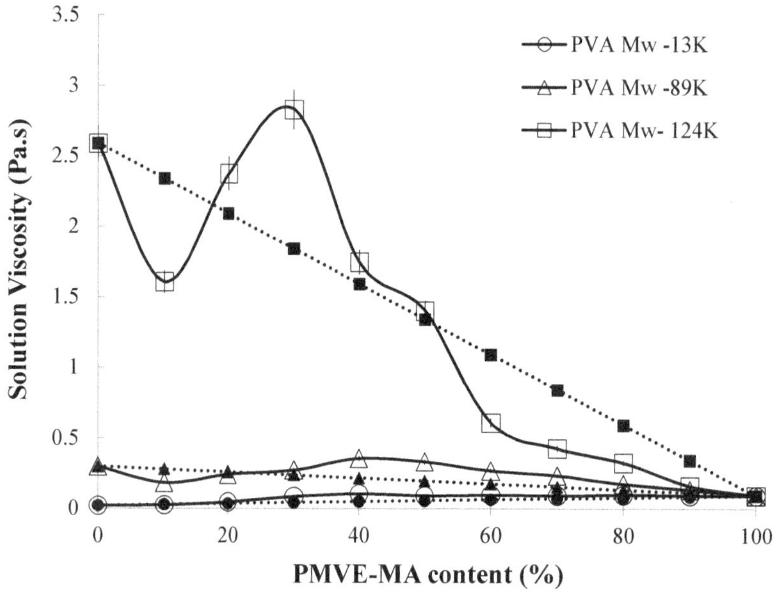

Figure 2. Viscosity of PVA/PMVE-MA blends for different molecular weight PVA as a function of PMVE-MA content by weight, compared to the linear additive viscosity values (dotted lines). Values are means ± SD.

Various rheological methods have been widely used to determine the gel point of polymer gels. One of the methods involves the use of the peak maximum in tangent angle of mechanical loss $(\tan\delta)_{max}$. However, the commonly used method is the crossover point to determine the gel point, in part, due to the ease of measurement. Here we have used the later approach for determining the occurrence of the gelation point during crosslinking of PVA/PMVE-MA. The crossover point of G'-t and G"-t curves was studied, *i.e.*, when the system shows the same level of elastic and viscous behavior, meaning that the same quantity of energy is stored and dissipated. In some cases, the intersection point appeared even before the reaction commenced and it is impossible to apply the crossover point of G' and G", so the peak maximum in tanδ, *i.e.*, where there is maximum difference between the elastic and viscous behavior, is taken as the gel time [28].

In this system, upon addition of PMVE-MA to PVA solution, the anhydride group first gets hydrolyzed and then reacts with PVA. Figure 3 shows the effect of mixing PVA with PMVE-MA on the elastic modulus (G') and viscous modulus (G") of a solution blend. For PVA/PMVE-MA solution in sample $S_{20:80}$ (Figure 3) and at certain temperature the loss modulus G", is above the elastic modulus G', which demonstrates the relative fluidity of the system, whereas, at the cross over point; G' approaches G" and almost overlaps. The crossing over point of G' and G" is the gel time of the sample and after cross over point the G' becomes higher than G" and is also parallel to G', hence, formation of gel due to the cross linking between PVA and PMVE-MA.

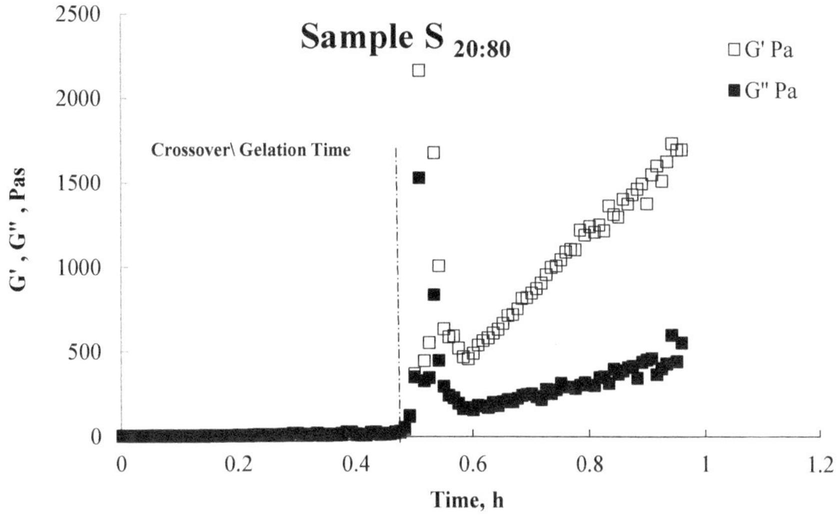

Figure 3. Elastic modulus (G') and viscous modulus (G") of sample $S_{20:80}$ (representative for all ratios) as a function of frequency.

This trend was observed to be identical for all the samples with different gel times depending on the composition and temperature. Figure 4 represents the effect of temperature and PMVE-MA concentration on the gel time of the blends. Sample $S_{40:60}$ at room temperature took 0.38 h to gel, whereas it took less than 20 min to gel at a high temperature. This trend was identical with all the blends at high temperature regardless of the PMVE-MA content; however, at room temperature, a difference in gelling time has been observed. The blends at both far ends, *i.e.*, high PVA or high PMVE-MA ratio took more time to form gel as compared to blends with PMVE-MA/PVA ratio in between.

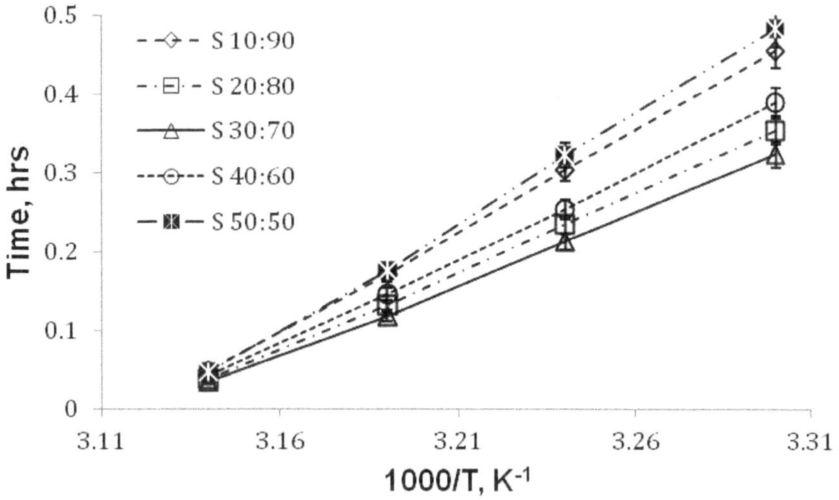

Figure 4. Gelation time *versus* (1000/T), where temperature *T* is expressed in Kelvin. Comparisons of all PMVE-MA/PVA blend solutions. Values are means ± SD.

The expected cross-linking between the two polymers primarily occurs from the chemical reactions and hydrogen bonding between the hydroxyl groups of PVA with carboxyl groups and ether groups from PMVE-MA. Scheme 1 represents the various proposed reactions occurring during mixing of PVA and PMVE-MA. The anhydride ring of PMVE-MA undergoes hydrolysis in aqueous solution to form free carboxylic acid groups, which further reacts with the PVA. PVA has been reported to crosslink by esterification with di/polyacids, radiation induced chemical reactions. In the present study, the cross-linking reaction has been carried out at a higher temperature for all the samples. The expected cross-linking between the polymers comes from the chemical reactions between hydroxyl groups in PVA with either maleic anhydride groups directly (a) or carboxyl groups (b) from PMVE-MA and also hydrogen bonding between the ether group and hydroxyl groups.

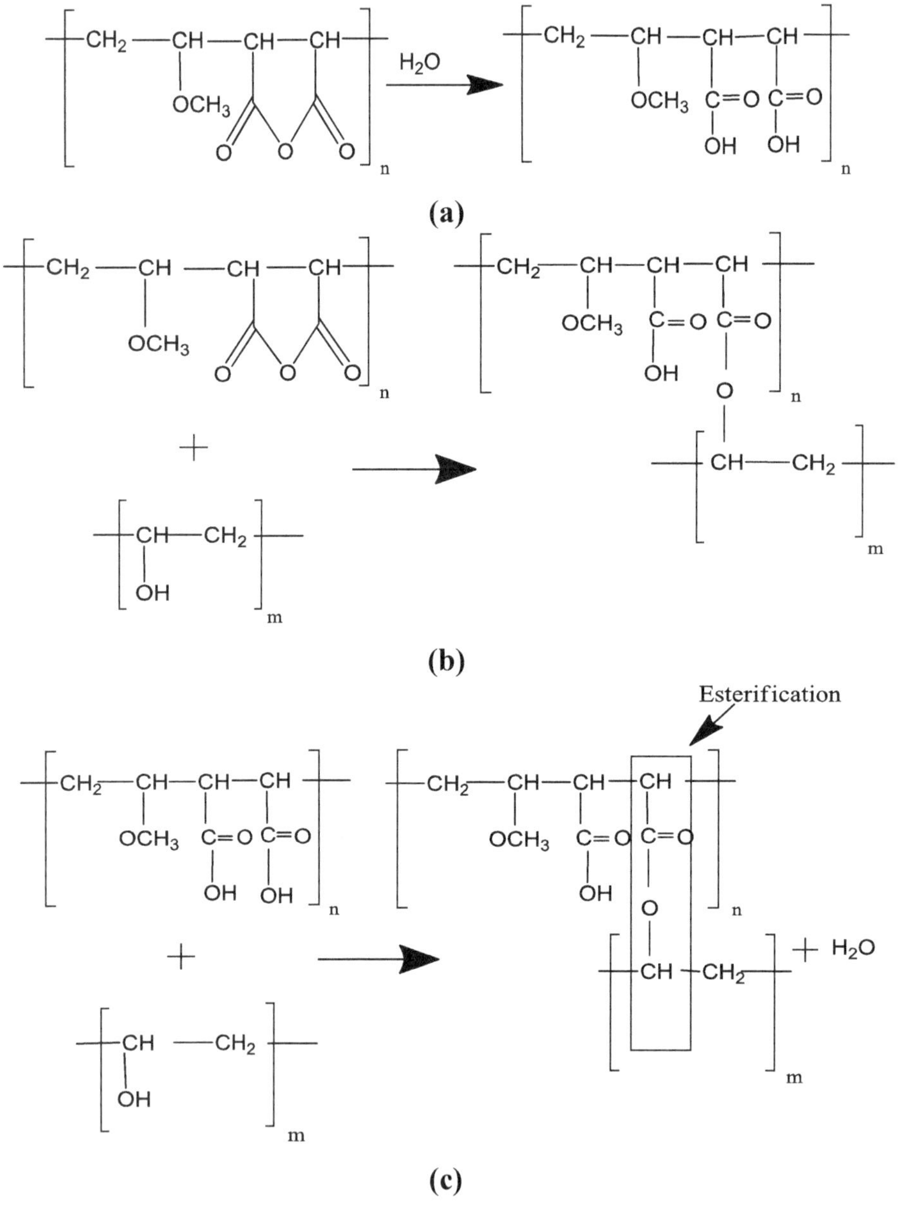

Scheme 1. Reaction during mixing of PVA and PMVE-MA (**a**) PMVE-MA with water; (**b**) PVA with partially hydrolyzed PMVE-MA; (**c**) esterification reaction.

4.3. Membrane Characterization by Water Uptake and Dimensional Stability

Table 1 gives the comparison of water uptake and dimensional stability of cross-linked PVA/PMVE-MA blend membranes for different molecular weight PVAs. The presence of water in the membrane facilitates channel for ion transport and increases the mobility of ionic charge carrier. From the table it can be observed that the data reveals a parabolic trend, *i.e.*, the minimum is indicated by the blend ratio $S_{30:70}$ for all molecular weight PVAs. It can be explained by the fact that incorporation of 20%–40% PMVE-MA in the blend is leading to optimum water uptake, whereas below and above this ratio it causes excessive swelling, indicating that 60%–80% PVA is needed in order to attain maximum interaction between PVA and PMVE-MA, which leads to controlled water uptake and dimensional stability. Then, at low concentrations of PMVE-MA; PVA intermolecular interactions are preferred and only when the PVA concentration is sufficiently high does the intermolecular interaction becomes prominent [29]. As can be observed from Table 1, the water uptake behavior of the films was affected by the ratio of PVA to PMVE-MA content in the blends. When the ratio of PMVE-MA is high compared to PVA it leads to highly water swollen membranes. For each molecular weight a similar trend was observed, which reveals better interaction of low molecular weight PVA compared to high molecular weight PVA. This can be due to PMVE-MA moiety being much larger than the PVA and introduction of the PMVE-MA moieties increases the interchain space within the PVA matrix.

Therefore, more free volume is available for the sorbed water molecules. In addition, dimensionally stable membranes resulted at optimum ratios of PVA/PMVE-MA in the blends regardless of the PVA molecular weight as more −COOH groups are available for the cross linking with −OH groups of PVA and increased hydrogen bonding results in reducing the mobility of the polymer chains, hence better stability also evident from the FTIR data. However, it was also observed that if a higher ratio of PMVE-MA was used than that of PVA; it increases the brittleness of the membrane, even in the water swollen membranes [4]. From Table 1 it is clear that PVA with Mw-13 K has most efficiently cross-linked with PMVE-MA followed by PVA Mw-89 K and PVA Mw-124 K, *i.e.*, the smaller the molecular weight of PVA the more stable it is. This observation can be explained in that with smaller molecular volume, a relatively easily accessible hydroxyl groups per unit mass favoring the complexation. Therefore, the low molecular-weight PVA may have diffused into, and interacted more effectively with, the PMVE-MA chains resulting in restricting their mobility and reducing the swelling of the polymer matrix. Furthermore, it is clear that lower molecular weight PVA has increased miscibility with the PMVE-MA polymer molecules, which may have resulted in a further increase in cross-linking efficiency as observed with other polyalcohol [30,31].

4.4. Thermal Properties

Figure 5 shows the TGA thermograms of the PVA/PMVE-MA interpolymer membranes with various PMVE-MA ratios. TGA curves of the PVA/PMVE-MA polymer films revealed three main weight loss regions. The first region, at a temperature of 100–200 °C ($T_{p,1}$ at 124 °C), degradation of PVA polymer chains; the weight loss of the membrane was about 2.0–3.0 wt%. The second transitional region, at around 210–300 °C ($T_{p,2}$ at 240 °C), appeared to be due to the degradation of the side-chain of the PMVE-MA polymer membrane; the total weight loss corresponding to this stage was about 11–12 wt%. The peak of the third stage at 295 °C ($T_{p,3}$ at 295 °C) occurred due to the cleavage of ester groups of PVA/PMVE-MA composite polymer membrane with a total weight loss at about 90 wt% at 600 °C. The degradation peaks of the cross-linked PVA/PMVE-MA composite polymer samples were less intense and shifted towards higher temperatures. The weight loss results of the samples are recorded at various temperatures in Table 2. It could, therefore, be concluded that the thermal stability was improved due to addition of PMVE-MA and the chemical crosslinking reaction between the −OH groups on the PVA and the −COOH group on the PMVE-MA.

PVA, PMVE-MA, and PVA/PMVE-MA samples were also analyzed by differential scanning calorimetry. The DSC measurements were carried out by a heating-cooling-heating cycle. The purpose of the first heating cycle was to remove any thermal history of the PVA composite polymer membrane. The DSC curves for the pure PVA polymer and the PVA/PMVE-MA membranes with various PMVE-MA compositions are shown in Figure 6.

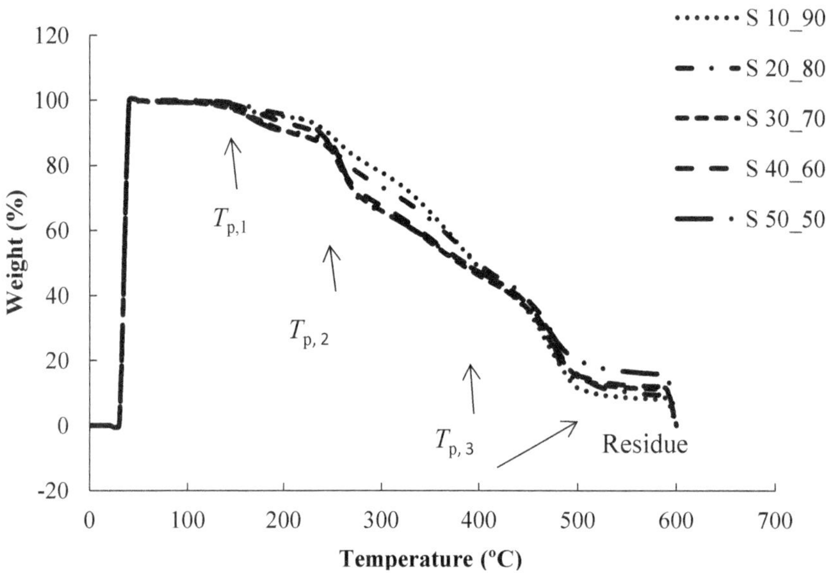

Figure 5. TGA thermograms of the PVA/PMVE-MA interpolymer membranes.

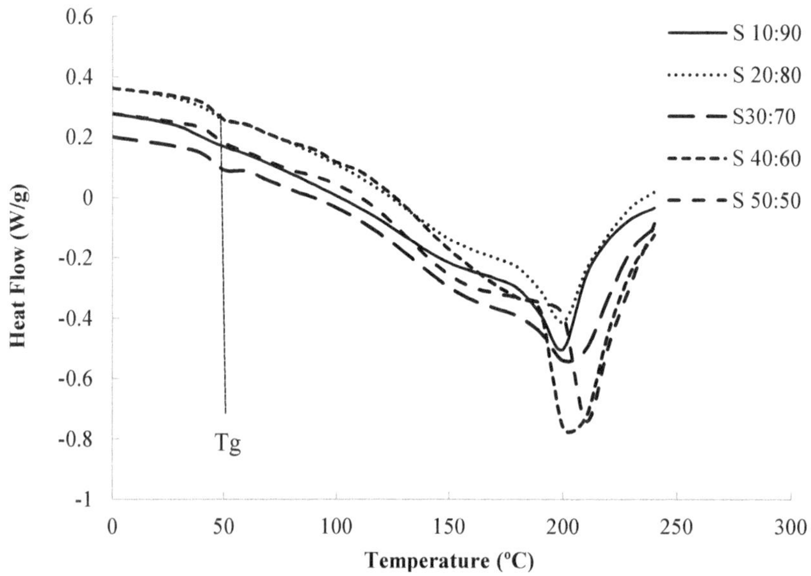

Figure 6. DSC thermograms of PMVE-MA/PVA blends.

Table 2. TGA results of weight loss of the PVA/PMVE-MA blend polymer membranes at various temperatures.

$S_{PMVE-MA:PVA}$ Blends	Weight Loss (%) at different temperature				
	200 °C	300 °C	400 °C	500 °C	600 °C
$S_{10:90}$	4.31	16.57	52.70	88.93	92.10
$S_{20:80}$	5.23	21.98	51.76	86.65	90.23
$S_{30:70}$	6.18	25.29	52.28	84.26	88.24
$S_{40:60}$	7.64	30.07	49.38	78.60	84.26
$S_{50:50}$	9.92	30.69	55.81	85.51	88.83

The T_g of a polymer represents the amorphous regions of a polymer and is, therefore, useful to reflect changes that occur in these regions. From Table 3, it was observed that only one T_g is found in the dry samples, indicating miscibility over entire composition. It had been reported that the T_m and T_g of the pure PVA polymer with 98%–99% hydrolysis degree are at 226 °C and 81.59 °C, respectively [32]. It was found that the glass transition temperature (T_g) and the melting temperature T_m, decreased from 81.59 to 40.46 °C, and from 226 °C to 199 °C, respectively, in sample $S_{10:90}$, and increased again with increasing PMVE-MA content in the samples. A change of T_g and T_m of PVA/PMVE-MA composite polymer membrane indicated a change between a semi-crystalline phase and an amorphous phase and the decrease in T_g and T_m is due to plasticization effect at this lower ratio ($S_{10:90}$).

Table 3. DSC results of PVA and PMVE-MA blends.

Temperature (°C)	PVA	PMVE-MA	Blend membranes				
			$S_{10:90}$	$S_{20:80}$	$S_{30:70}$	$S_{40:60}$	$S_{50:50}$
T_g	81	77	40	42	44	47	49
T_m	226	163	199	199	204	204	209
ΔH	57	406	154	141	189	164	216

4.5. Conductivity

To use a polymer membrane separator it is necessary to provide a facilitated channel for ion transport by providing ionic exchange sites, such as carboxylic acid. During the thermal treatment the hydrogen bonds between –OH of PVA and –COOH of PMVE-MA are formed due to the proximity of the polymer chains [15]. This physical interaction between the functional groups results in the formation of hydrophilic ionic channels or micro-domains by the arrangement of hydrophilic polymeric groups. Figure 7 represents the effect on the conductivity of the membrane as the PMVE-MA ratio is increased. Since the conductivity under the conditions used (room temperature and 100% humidity) are largely ascribed to the movement of free water within the material, a systematic trend in conductivity is expected with blend interactions and water uptake behavior. When the ratio of PMVE-MA in the system is increased, a gradual increase in the conductivity of the membrane was observed. This trend is identical with different molecular weights of PVA used. This increase is due to the presence of several ion acceptor site contained in each PMVE-MA unit compared to the one ion donor of the PVA repeat unit. As the ionic conductivity is dependent on the state of hydration of the membrane, as in a water-swollen membrane, the ions are solvated and enabled to transport through the membrane, which can be easily seen from Figure 7. The crosslinked samples with high molecular weight PVA show two times higher (~4.5 mS/cm) conductivity values as compared to highly cross-linked blends with low molecular weight PVA due to excellent strength of interaction and reasonable amount of water sorption.

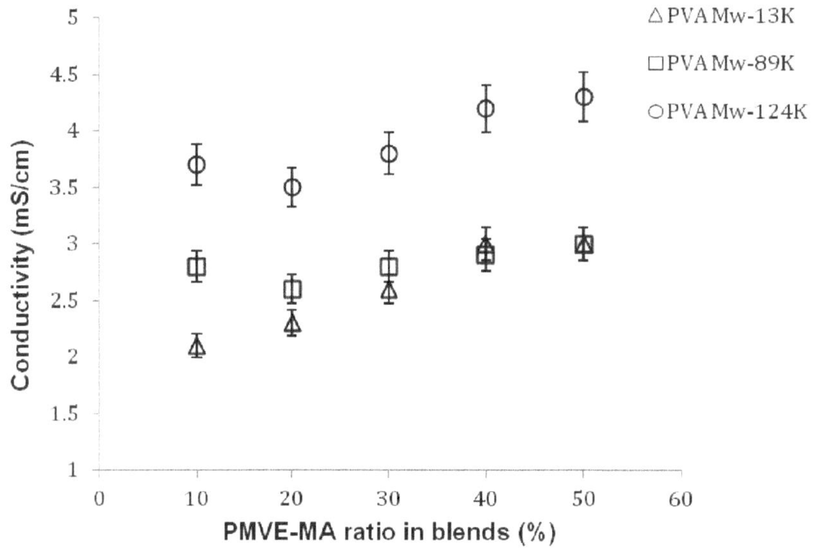

Figure 7. Effect of PMVE-MA ratio and PVA molecular weight on the conductivity of the blend membranes. Values are means ± SD.

5. Conclusions

This study confirmss that both PVA and PMVE-MA are able to form excellent membranes from homogenous and miscible solutions as a result of inter-polymer complexation that takes place between the carboxylic acid groups of PMVE-MA and the hydroxyl groups of PVA. The carboxylic acid (*i.e.*, hydrolyzed PMVE-MA) in this work served two purposes: to provide effective cross-linking and, at the same time, to impart a fixed charge character to the cross-linked membranes. The charge character results from the presence of carboxylate groups of the hydrolyzed but unesterified PMVE-MA. This was first evidenced by insolubility and dimensional stability of the blend films in water. FTIR results confirm the presence of these hydrogen bonding interactions by the appearance of two new bands at ~1600 and ~1200 cm^{-1} characteristic of the ester groups arising from the C=O and C–O groups, respectively. The appearance of a single T_g for each of the blends by DSC proves that miscible blends were formed, resulting from the hydrogen bonding. The chemical interaction between the functional groups results in the formation of hydrophilic ionic channels or micro domains by the arrangement of hydrophilic polymeric groups leading to an increase in ionic conductivity in membranes. The optimized membranes from PVA and PMVE-MA show good ion conductivity making them suitable for separator applications.

Acknowledgments

The authors would like to thank the Australian Research Council-ARC Linkage Grant for funding this research.

Author Contributions

Namita Roy Choudhury conceived the idea and the plan, Naba K. Dutta designed the rheology experiment, Charu Vasisth Rohatgi did the experimental work, and Namita Roy Choudhury and Naba K. Dutta finalized the manuscript.

Conflicts of Interest

The authors declare that there is no competing interest in this work.

References

1. Linden, D.; Reddy, T.B. *Handbook of Batteries*, 3rd ed.; McGraw-Hill: New York, NY, USA, 2002.
2. Arora, P.; Zhang, Z. Battery separators. *Chem. Rev.* **2004**, *104*, 4419–4462.
3. Immelman, E.; Sanderson, R.D.; Jacobs, E.P.; van Reenen, A.J. Poly(vinyl alcohol) gel sub-layers for reverse osmosis membranes II. Insolubilization by crosslinking with poly(methyl vinyl ether-alt-maleic anhydride). *Desalination* **1993**, *94*, 37–54.
4. Guo, X.; Deng, F.; Li, L.; Prud'homme, R.K. Synthesis of biocompatible polymeric hydrogels with tunable adhesion to both hydrophobic and hydrophilic surfaces. *Biomacromolecules* **2008**, *9*, 1637–1642.
5. Rhim, J.-W.; Park, H.B.; Lee, C.-S.; Jun, J.-H.; Kim, D.S.; Lee, Y.M. Crosslinked poly(vinyl alcohol) membranes containing sulfonic acid group: Proton and methanol transport through membranes. *J. Membr. Sci.* **2004**, *238*, 143–151.
6. Kim, D.S.; Park, H.B.; Rhim, J.-W.; Lee, Y.M. Preparation and characterization of crosslinked PVA/SiO_2 hybrid membranes containing sulfonic acid groups for direct methanol fuel cell applications. *J. Membr. Sci.* **2004**, *240*, 37–48.
7. Lebrun, L.; Silva, E.D.; Metayer, M. Elaboration of ion-exchange membranes with semi-interpenetrating polymer networks containing poly(vinyl alcohol) as polymer matrix. *J. Appl. Polym. Sci.* **2002**, *84*, 1572–1580.
8. Li, X.; Cheruvally, G.; Kim, J.-K.; Choi, J.-W.; Ahn, J.-H.; Kim, K.-W.; Ahn, H.-J. Polymer electrolytes based on an electrospun poly(vinylidene fluoride-*co*-hexafluoropropylene) membrane for lithium batteries. *J. Power Sources* **2007**, *167*, 491–498.
9. Cho, T.H.; Sakai, T.; Tanase, S.; Kimura, K.; Kondo, Y.; Tarao, T.; Tanaki, M. Electrochemical performances of polyacrylonitrile nanofiber-based nonwoven separator for lithium-ion battery. *Electrochem. Solid State Lett.* **2007**, *10*, A159–A162.
10. Gao, K.; Hu, X.; Dai, C.; Yi, T. Crystal structures of electrospun PVDF membranes and its separator application for rechargeable lithium metal cells. *Mater. Sci. Eng.* **2006**, *B131*, 100–105.
11. Lee, S.W.; Choi, S.W.; Jo, S.M.; Chin, B.D.; Kim, D.Y.; Lee, K.Y. Electrochemical properties and cycle performance of electrospun poly(vinylidene fluoride)-based fibrous membrane electrolytes for Li-ion polymer battery. *J. Power Sources* **2006**, *163*, 41–46.
12. Kim, J.R.; Choi, S.W.; Jo, S.M.; Lee, W.S.; Kim, B.C. Characterization and properties of P(VdF-HFP)-based fibrous polymer electrolyte membrane prepared by electrospinning. *J. Electrochem. Soc.* **2005**, *152*, A295–A300.

13. Chen, J.; Asano, M.; Maekawa, Y.; Yoshida, M. Polymer electrolyte hybrid membranes prepared by radiation grafting of *p*-styryltrimethoxysilane into poly(ethylene-*co*-tetrafluoroethylene) films. *J. Membr. Sci.* **2007**, *296*, 77–82.
14. Kuo, P.L.; Chen, W.F.; Liang, W.J. Proton transportation in an organic-inorganic hybrid polymer electrolyte based on a polysiloxane/poly(allylamine) network. *J. Polym. Sci. A* **2005**, *43*, 3359–3367.
15. Smitha, B.; Sridhar, S.; Khan, A.A. Polyelectrolyte complexes of chitosan and poly(acrylic acid) as proton exchange membranes for fuel cells. *Macromolecules* **2004**, *37*, 2233–2239.
16. Kerres, J.; Ullrich, A.; Meier, F.; Haring, T. Synthesis and characterization of novel acid-base polymer blends for application in membrane fuel cells. *Solid State Ion.* **1999**, *125*, 243–249.
17. Wu, H.-L.; Ma, C.-C.M.; Li, C.-H.; Lee, T.-M.; Chen, C.-Y.; Chiang, C.-L.; Wu, C. Sulfonated poly(ether ether ketone)/poly(amide imide) polymer blends for proton conducting membrane. *J. Membr. Sci.* **2006**, *280*, 501–508.
18. Chen, J.; Asano, M.; Yamaki, T.; Yoshida, M. Effect of crosslinkers on the preparation and properties of ETFE-based radiation-grafted polymer electrolyte membranes. *J. Appl. Polym. Sci.* **2006**, *100*, 4565–4574.
19. Labuschagne, P.W.; Germishuizen, W.A.; Verryn, S.M.C.; Moolman, F.S. Improved oxygen barrier performance of poly(vinyl alcohol) films through hydrogen bond complex with poly(methyl vinyl ether-*co*-maleic acid). *Eur. Polym. J.* **2006**, *44*, 2146–2152.
20. Walker, M.; Baumgärtner, K.-M.; Kaiser, M.; Kerres, J.; Ullrich, A.; Rauchle, E. Proton-conducting polymers with reduced methanol permeation. *J. Appl. Polym. Sci.* **1999**, *74*, 67–73.
21. Kerres, J. Development of ionomer membranes for fuel cells. *J. Membr. Sci.* **2001**, *185*, 3–27.
22. Zaidi, S.M.J. Preparation and characterization of composite membranes using blends of SPEEK/PBI with boron phosphate. *Electrochim. Acta* **2005**, *50*, 4771–4777.
23. Nurkeeva, Z.S.; Mun, G.A.; Dubolazov, A.V.; Khutoryanskiy, V.V. pH effects on the complexation, miscibility and radiation-induced crosslinking in poly(acrylic acid)-poly(vinyl alcohol) blends. *Macromol. Biosci.* **2005**, *5*, 424–432.
24. Kitano, S.; Koyama, Y.; Kataoka, K.; Okano, T.; Sakurai, Y. A novel drug delivery system utilizing a glucose responsive polymer complex between poly(vinyl alcohol) and poly(*N*-vinyl-2-pyrrolidone) with a phenylboronic acid moiety. *J Control. Release* **1992**, *19*, 161–170.
25. Mansur, H.S.; Oréfice, R.L.; Mansur, A.A.P. Characterization of poly(vinyl alcohol)/poly(ethylene glycol) hydrogels and PVA-derived hybrids by small-angle X-ray scattering and FTIR spectroscopy. *Polymer* **2004**, *45*, 7193–7202.
26. Lin, C.W.; Huang, Y.F.; Kannan, A.M. Semi-interpenetrating network based on cross-linked poly(vinyl alcohol) and poly(styrene sulfonic acid-*co*-maleic anhydride) as proton exchange fuel cell membranes. *J. Power Sources* **2007**, *164*, 449–456.
27. Markovic, S.; Dunjic, B.; Zlatanic, A.; Djonlagic, J. Dynamic mechanical analysis study of the curing of phenol-formaldehyde novolac resins. *J. Appl. Polym. Sci.* **2001**, *81*, 1902–1913.
28. Mooman, F.S.; Meunier, M.; Labuschagne, P.W.; Truter, P.A. Compatibility of polyvinyl alcohol and poly(methyl vinyl ether-*co*-maleic acid) blends estimated by molecular dynamics. *Polymer* **2005**, *46*, 6192–6200.

29. Krumova, M.; Lopez, D.; Benavente, R.; Mijangos, C.; Perena, J.M. Effect of crosslinking on the mechanical and thermal properties of poly(vinyl alcohol). *Polymer* **2000**, *41*, 9265–9272.
30. Qussi, B.; Suess, W.G. The influence of different plasticizers and polymers on the mechanical and thermal properties, porosity and drug permeability of free shellac films. *Drug Dev. Ind. Pharm.* **2006**, *32*, 403–412.
31. Repka, M.A.; Gerding, T.G.; Repka, S.L.; McGinity, W. Influence of plasticizers and drugs on the physical-mechanical properties of hydroxypropylcellulose films prepared by hot melt extrusion. *Drug Dev. Ind. Pharm.* **1999**, *25*, 625–633.
32. Xu, W.; Liu, C.; Xue, X.; Su, Y.; Lv, Y.; Xing, W.; Lu, T. New proton exchange membranes based on poly(vinyl alcohol) for DMFCs. *Solid State Ion.* **2004**, *171*, 121–127.

10

Graphene Oxide Synthesis from Agro Waste

Thirunavukkarasu Somanathan [1,2,*], Karthika Prasad [1,3], Kostya (Ken) Ostrikov [3], Arumugam Saravanan [1] and Vemula Mohana Krishna [1]

[1] Department of Nanoscience, School of Basic Sciences, Vels University, Pallavaram, Chennai, Tamil Nadu 600117, India; E-Mails: karthikaprasad@live.com (K.P.); saravanan.con@gmail.com (A.S.); krishnachem07@gmail.com (V.M.K.)

[2] Center for Advanced Research and Development (CARD), Vels University, Pallavaram, Chennai, Tamil Nadu 600117, India

[3] Institute of Future Environments, Nanotechnology and Molecular Science, School of Chemistry, Physics and Mechanical Engineering, Queensland University of Technology, Brisbane, QLD 4000, Australia; E-Mail: kostya.ostrikov@qut.edu.au

* Author to whom correspondence should be addressed; E-Mail: soma_nano@yahoo.co.in

Academic Editor: Jiye Fang

Abstract: A new method of graphene oxide (GO) synthesis via single-step reforming of sugarcane bagasse agricultural waste by oxidation under muffled atmosphere conditions is reported. The strong and sharp X-ray diffraction peak at $2\theta = 11.6°$ corresponds to an interlayer distance of 0.788 nm (d_{002}) for the AB stacked GOs. High-resolution transmission electron microscopy (HRTEM) and selected-area electron diffraction (SAED) confirm the formation of the GO layer structure and the hexagonal framework. This is a promising method for fast and effective synthesis of GO from sugarcane bagasse intended for a variety of energy and environmental applications.

Keywords: graphene oxide (GO); natural precursors; value-added reforming

1. Introduction

Graphene and related materials have attracted a great deal of recent interest owing to their unique structure and physical, chemical, thermal, and other properties [1]. Graphene's two-dimensional (2D)

hexagonal lattice consists of a network of sp^2-bonded carbon atoms and represents the "thinnest material" which is stable in its free form [2]. Excellent electronic properties [3], thermal conductivity [4], and high surface area, combined with unusual mechanical properties and good dispersion performance, make graphene a promising candidate for structural modification of composite materials and several other applications [5–8].

Graphene oxide (GO) has also attracted major interdisciplinary attention because of the broad range of envisaged applications across several scientific and engineering fields including physics, chemistry, biology, and medicine [9–13]. The original method to synthesize GO is based on the addition of potassium chlorate to a slurry of graphite in fuming nitric acid [14]. This synthetic protocol can be improved by using concentrated sulfuric acid as well as fuming nitric acid and adding the chlorate in multiple aliquots over the course of the reaction [15]. In the commonly used Hummers' method [16], graphite is oxidized throughout treatment of $KMnO_4$ and $NaNO_3$ in concentrated H_2SO_4 acid.

All these procedures involve the generation of NO_2, N_2O_4, and ClO_2, toxic gases which are also explosive. In the past few years, various techniques such as mechanical exfoliation techniques, chemical vapor deposition (CVD) techniques and other chemical techniques have been developed for the synthesis of graphene and graphene oxide materials. However, many of these techniques are highly sophisticated and expensive [17]. Most of the commercially available GO samples were synthesized by Hummers' method or a modified version of it. The cost of the commercially available graphene oxide per gram is approximately $200. In comparison with the Hummers' method, our method is more environment friendly as it avoids toxic gas emission during synthesis. The synthesis technique reported here is a quite efficient asset demonstrating a new procedure for the production of graphene oxide from agro waste in simple steps. This simple and low-cost process could lead to new opportunities for cost-effective production of GO [18,19].

Here we report on the synthesis of graphene oxide by directly oxidizing sugarcane bagasse under muffled atmosphere. The GO produced by this method is called SOMA-GO (sugarcane oxidized under muffled atmosphere for graphene oxide). Furthermore, the structural characteristics of the obtained product are confirmed by X-ray diffraction, Fourier transmission infrared spectroscopy, field emission scanning electron microscopy, high-resolution transmission electron microscopy, and Raman spectroscopy.

2. Experimental Section

2.1. Synthesis of GO from Agro Waste

Sugarcane bagasse is a type of agricultural waste (agro waste). After juice was extracted, the remaining fiber was taken. The fiber was crushed and ground well in order to produce powder. This crushing and separating process was repeated several times to obtain fine powder. About 0.5 g of ground sugarcane bagasse powder mixed with 0.1 g of ferrocene was placed in a crucible and put directly into a muffle furnace at 300 °C for 10 min under atmospheric conditions. Next, the as-produced black solid was collected at room temperature. The black solid product was subjected to further analysis (Scheme 1).

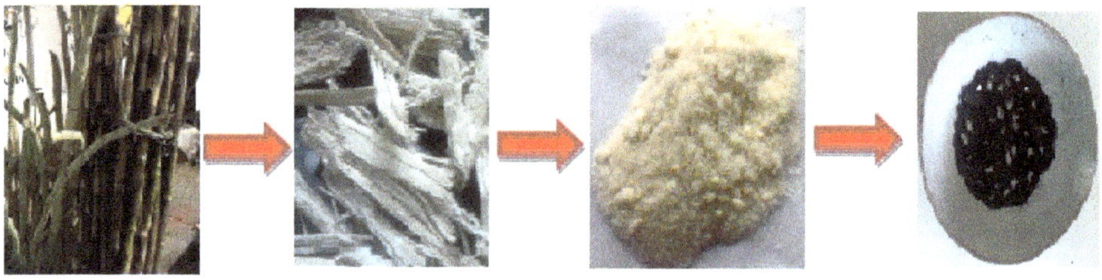

Scheme 1. Schematic representation of graphene oxide synthesis from agro waste.

2.2. Characterization of GO

The X-ray diffraction (XRD) pattern was obtained using a Bruker D8 Advance X-ray diffractometer (Berlin, Germany). This device is equipped with a Ni filter, and generates monochromated Cu-Kα radiation (λ = 0.154 nm) operated at 40 kV accelerating voltage and 30 mA current. The samples were scanned in step mode with a 2° min^{-1} scan rate. Fourier transform infrared spectroscopy (FT-IR), Avatar 370, Thermo Nicolet (Berlin, Germany) was carried out to study the bonding configuration in the synthesized graphene oxide. High resolution transmission electron microscopy (HRTEM) was performed using a JEOL JEM-3000F microscope (Tokyo, Japan) operating at 300 kV to study the microstructure of the GO structures. Field emission scanning electron microscopy was carried out using a JEOL JSM 7800F revealed the morphological features of the GO structure.

3. Results and Discussion

3.1. X-Ray Diffraction Measurements

The structural and chemical analysis of sugarcane bagasse is clearly described by previous investigators [20–22]. In this study, we have presented the XRD pattern of the GO which is shown in Figure 1. The peak at 2θ = 11.6° indicates that the agricultural sugarcane bagasse is fully oxidized into graphene oxide with the interlayer distance of 0.79 nm. The 2θ peaks of the GO obtained by the single-step oxidation of sugarcane bagasse are consistent with the results of previous studies [23,24]. This XRD pattern can be attributed to well graphitized two-dimensional structures made of GO sheets.

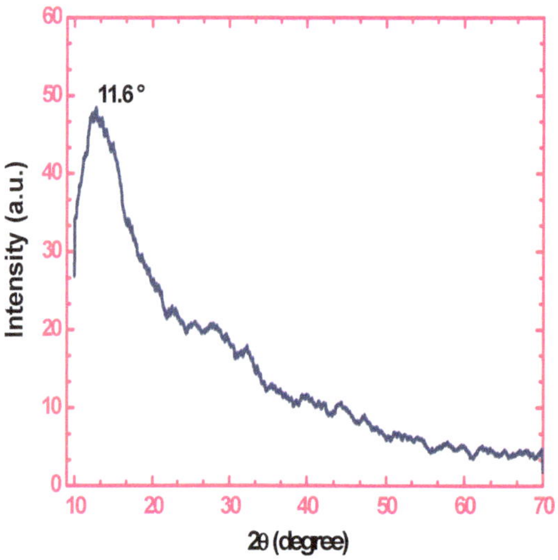

Figure 1. X-ray diffraction (XRD) pattern of graphene oxide.

3.2. FT-IR Spectroscopy

The FT-IR spectrum of GO shown in Figure 2 confirms the introduction of oxygen containing groups such as functional hydroxyl, epoxy and carboxylic groups upon oxidation of sugarcane bagasse. The strong band at 1708 cm^{-1} is attributed to stretching vibration modes of C=O in carboxylic acid and carbonyl groups. The peak at 1594 cm^{-1} is assigned to the skeletal vibrations of un-oxidized graphitic domains. The band at 1060 cm^{-1} is assigned to C–O (epoxy) groups while the band at 1210 cm^{-1} is usually attributed to C–OH stretching vibrations [25–27]. The strong peak around 3500–4000 cm^{-1} can be attributed to the O–H stretching vibrations of the C–OH groups and water [27].

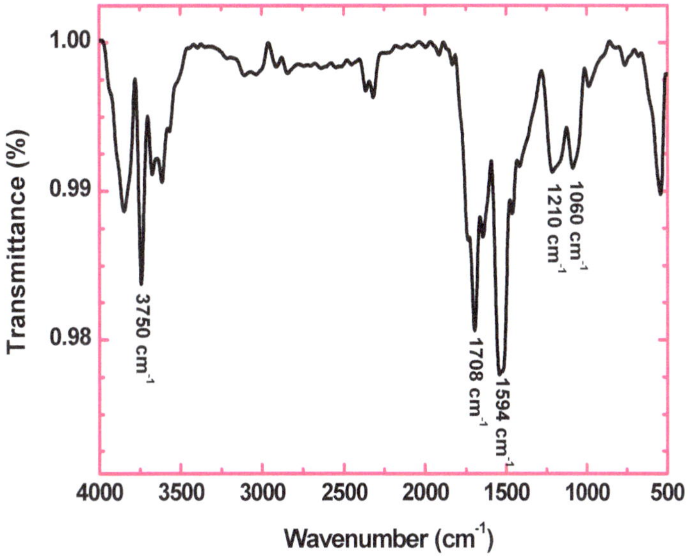

Figure 2. Fourier transform infrared spectroscopy (FT-IR) spectrum of graphene oxide.

3.3. SEM and HRTEM Analysis

The surface morphology of the graphene oxide (GO) produced by sugarcane bagasse oxidation is identified by FESEM. The SEM image of the GO shown in Figure 3a,c clearly reveals its sheet-like structure. The morphology of the synthesised GO resembles flake-like structures as shown in Figure 3b,d. Quite similar textures have been reported by other authors [28].

HRTEM micrographs of GO are shown in Figure 4. Bends wrinkles, and edges on graphene oxide nanosheets are seen in several places in Figure 4a,b. Various defects and functional groups containing sp^3 hybridized carbon atoms may also be introduced during the oxidation process. Figure 4c shows that the flakes contain a few layers of graphene oxide. The selected area of electron diffraction (SAED) pattern shown in Figure 4d indicates the presence of a polycrystalline structure corresponding to graphene oxide [29–33]. Thus, transmission electron microscopy (TEM) analysis confirms the presence of graphene oxide in the oxidized carbon samples.

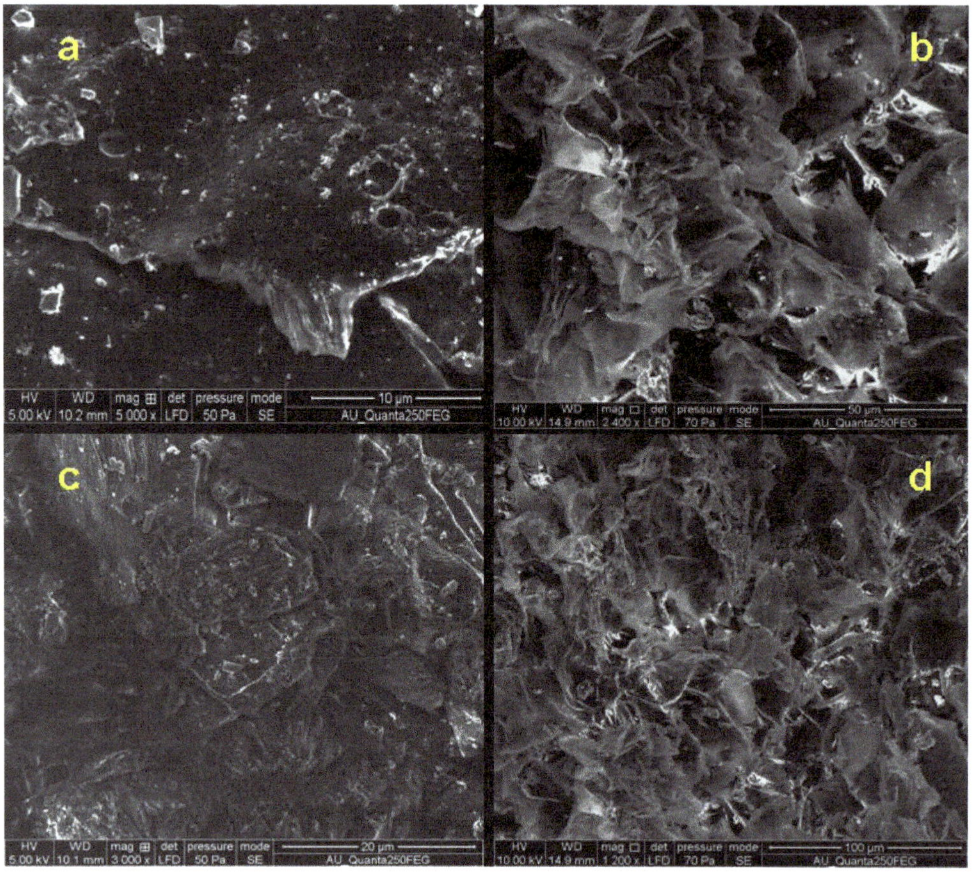

Figure 3. FESEM images of GO obtained from sugarcane bagasse.

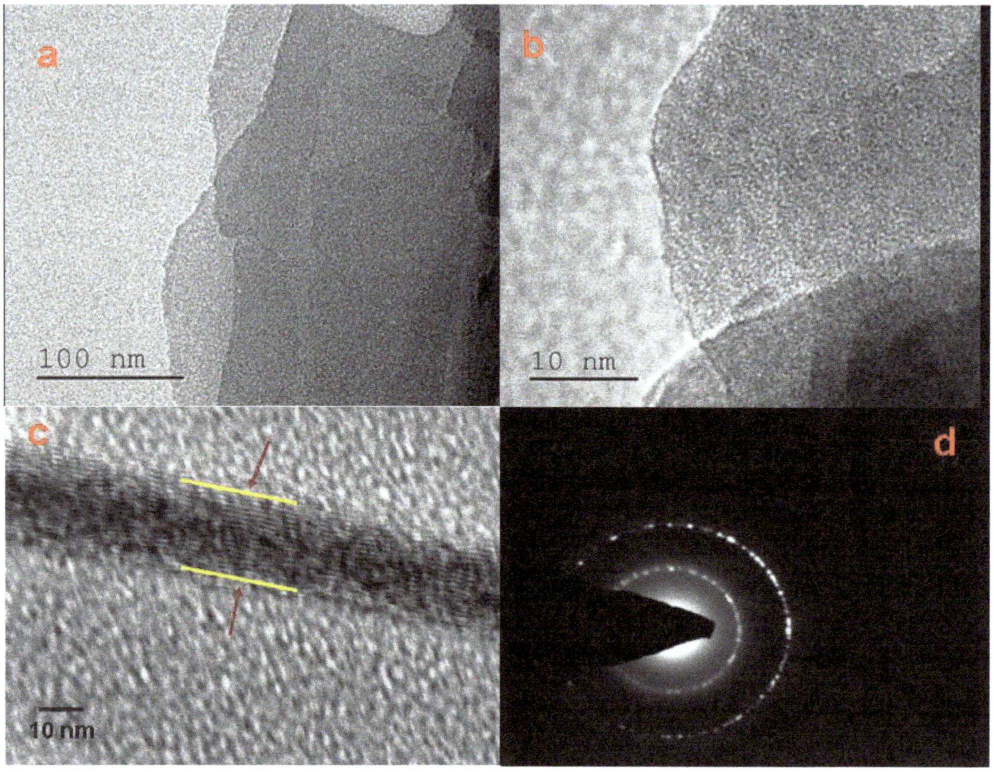

Figure 4. High-resolution transmission electron microscopy (HRTEM) images of GO obtained from sugarcane bagasse.

3.4. Raman Spectroscopy

The Raman spectrum of graphene oxide synthesized from sugarcane bagasse is shown in Figure 5. The spectrum of GO displays two prominent peaks at 1358 and 1585 cm^{-1}, which are attributed to the local defects (found at the edges of graphene sheets) and the sp^2 graphitized structure, respectively. The *G* peak formation is due to the double degenerated zone center $E_{2}g$ mode [34,35]. The functionality and reduction in particle size of sp^2 plane domains are possibly due to the extensive oxidation of sugarcane bagasse [24]. The two *D* and *G* bands at 1358 cm^{-1} and 1585 cm^{-1} [36] indicate the clear sp^2 carbon hybridization in the observed multi-layer stacks [31]. Thus, I_D/I_G peak intensity ratios are assigned to lower defects/disorders. Raman spectra show an intensity ratio of I_D/I_G at 0.76 for GO which is in line with previous investigations [24,37,38].

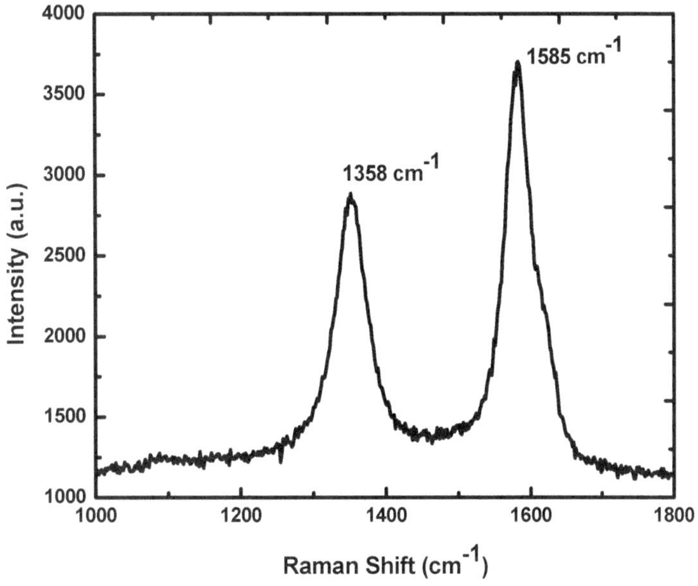

Figure 5. Raman spectrum of GO obtained from sugarcane bagasse.

4. Conclusions

A simple and rapid method is shown here to effectively convert solid sugarcane bagasse waste into value-added graphene oxide. The produced graphene oxide presents a well-graphitized structure. In comparison with the commonly used methods, our method is more environment friendly as it avoids toxic gas emission during the synthesis. This simple and low-cost synthesis process could lead to new opportunities for cost-effective production of graphene-based materials for gas sensors, energy storage, and other functional devices.

Acknowledgments

Thirunavukkarasu Somanathan thanks the Department of Science and Technology (DST) for the Fast Track Young Scientist Award and financial support (SR/FT/CS-111/2011). Karthika Prasad acknowledges support from the SEF Postgraduate Research Scholarship (Queensland University of Technology, Australia). This work was partially supported by the Australian Research Council.

Author Contributions

The manuscript was prepared using contributions from all authors. Thirunavukkarasu Somanathan designed the work, conducted the synthesis of graphene oxide from agro waste, analysed the results and led the manuscript preparation. Karthika Prasad prepared agro waste materials and conducted the synthesis and characterization experiments. Kostya (Ken) Ostrikov contributed to the analysis of the results and manuscript writing. Arumugam Saravanan and Vemula Mohana Krishna conducted FE-SEM and HRTEM observation and analyzed the results. All authors have given approval to the final version of the manuscript.

Conflicts of Interest

The authors declare no conflict of interest.

References

1. Chae, H.K.; Siberio-Perez, D.Y.; Kim, J.; Go, Y.; Eddaoudi, M.; Matzger, A.J.; O'Keeffe, M.; Yaghi, O.M. A route to high surface area, porosity and inclusion of large molecules in crystals. *Nature* **2004**, *427*, 523–527.
2. Novoselov, K.; Geim, A.; Morozov, S.; Jiang, D.; Zhang, Y.; Dubonos, S.; Grigorieva, I.; Firsov, A. Electric field effect in atomically thin carbon films. *Science* **2004**, *306*, 666–669.
3. Schwierz, F. Graphene transistors. *Nat. Nanotechnol.* **2010**, *5*, 487–496.
4. Balandin, A.A.; Ghosh, S.; Bao, W.; Calizo, I.; Teweldebrhan, D.; Miao, F.; Lau, C.N. Superior thermal conductivity of single-layer graphene. *Nano Lett.* **2008**, *8*, 902–907.
5. Song, H.; Zhang, L.; He, C.; Qu, Y.; Tian, Y.F.; Lv, Y. Graphene sheets decorated with SnO_2 nanoparticles: *In situ* synthesis and highly efficient materials for cataluminescence gas sensors. *J. Mater. Chem.* **2011**, *21*, 5972–5977.
6. Si, Y.C.; Samulski, E.T. Exfoliated graphene separated by platinum nanoparticles. *Chem. Mater.* **2008**, *20*, 6792–6797.
7. Lu, G.; Park, S.; Yu, K.; Ruoff, R.S.; Ocola, L.E.; Rosenmann, D.; Chen, J. Toward practical gas sensing with highly reduced graphene oxide: A new signal processing method to circumvent run-to-run and device-to-device variations. *ACS Nano* **2011**, *5*, 1154–1164.
8. Lee, C.; Wei, X.; Kysar, J.W.; Hone, J. Measurement of the elastic properties and intrinsic strength of monolayer grapheme. *Science* **2008**, *321*, 385–388.
9. Georgakilas, V.; Otyepka, M.; Bourlinos, A.B.; Chandra, V.; Kim, N.; Kemp, K.C.; Hobza, P.; Zboril, R.; Kim, K.S. Functionalization of graphene: Covalent and non-covalent approaches, derivatives and applications. *Chem. Rev.* **2012**, *112*, 6156–6214.
10. Novoselov, K.S.; Fal'ko, V.I.; Colombo, L.; Gellert, P.R.; Schwab, M.G.; Kim, K. A roadmap for grapheme. *Nature* **2012**, *490*, 192–200.
11. Chung, C.; Kim, Y.K.; Shin, D.; Ryoo, S.R.; Hong, B.H.; Min, D.H. Biomedical applications of graphene and graphene oxide. *Acc. Chem. Res.* **2013**, *46*, 2211–2224.

12. Mao, H.Y.; Laurent, S.; Chen, W.; Akhavan, O.; Imani, M.; Ashkarran, A.A.; Mahmoudi, M. Graphene: Promises, facts, opportunities, and challenges in nanomedicine. *Chem. Rev.* **2013**, *113*, 3407–3424.
13. Eigler, S.; Hu, Y.; Ishii, Y.; Hirsch, A. Controlled functionalization of graphene oxide with sodium azide. *Nanoscale* **2013**, *5*, 12136–12139.
14. Brodie, B.C. On the atomic weight of graphite. *Philos. Trans. R. Soc. Lond.* **1859**, *149*, 249–259.
15. Staudenmaier, L. Verfahren zur Darstellung der Graphitsäure. *Ber. Dtsch. Chem. Ges.* **1898**, *31*, 1481–1487. (In German)
16. Hummers, W.S.; Offeman, R.E. Preparation of graphitic oxide. *J. Am. Chem. Soc.* **1958**, *80*, 1339.
17. Guo, S.; Dong, S. Graphene nanosheet: Synthesis, molecular engineering, thin film, hybrids, and energy and analytical applications. *Chem. Soc. Rev.* **2011**, *40*, 2644–2672.
18. Seo, D.H.; Rider, A.E.; Kumar, S.; Randeniya, L.K.; Ostrikov, K. Vertical graphene gas- and bio-sensors via catalyst-free, reactive plasma reforming of natural honey. *Carbon* **2013**, *60*, 221–228.
19. Seo, D.H.; Han, Z.J.; Kumar, S.; Ostrikov, K. Structure-controlled, vertical grapheme-based, binder-free electrodes from plasma-reformed butter enhance super capacitor performance. *Adv. Energy Mater.* **2013**, *10*, 1316–1323.
20. Chandel, A.K.; Antunes, F.A.F.; Anjos, V.; Bell, M.J.V.; Rodrigues, L.N.; Polikarpov, I.; de Azevedo, E.R.; Bernardinelli, O.D.; Rosa, C.A.; Pagnocca, F.C.; *et al.* Multi-scale structural and chemical analysis of sugarcane bagasse in the process of sequential acid-base pretreatment and ethanol production by Scheffersomycesshehatae and Saccharomyces cerevisiae. *Biotechnol. Biofuels* **2014**, *7*, doi:10.1186/1754-6834-7-63.
21. Rezende, C.A.; de Lima, M.A.; Maziero, P.; de Azevedo, E.R.; Garcia, W.; Polikarpov, I. Chemical and morphological characterization of sugarcane bagasse submitted to a delignification process for enhanced enzymatic digestibility. *Biotechnol. Biofuels* **2011**, *11*, 4–54.
22. Velmurugan, R.; Muthukumar, K. Utilization of sugarcane bagasse for bioethanol production: Sono-assisted acid hydrolysis approach. *Bioresour. Technol.* **2011**, *102*, 7119–7123.
23. Muralidharan, M.N.; Ansari, S. Thermally reduced graphene oxide/thermoplastic polyurethane nanocomposites as photomechanical actuators. *Adv. Mater. Lett.* **2013**, *4*, 927–932.
24. Kellici, S.; Acord, J.; Ball, J.; Reehal, H.S.; Morgan, D.; Saha, B. A single rapid route for the synthesis of reduced graphene oxide with antibacterial activities. *RSC Adv.* **2014**, *4*, 14858–14861.
25. Marcano, D.C.; Kosynkin, D.V.; Berlin, J.M.; Sinitskii, A.; Sun, Z.; Slesarev, A.; Alemany, L.B.; Lu, W.; Tour, J.M. Improved synthesis of graphene oxide. *ACS Nano* **2010**, *4*, 4806–4814.
26. Xu, Y.; Bai, H.; Lu, G.; Li, C.; Shi, G. Flexible graphene films via the filtration of water-soluble noncovalent functionalized graphene sheets. *J. Am. Chem. Soc.* **2008**, *130*, 5856–5857.
27. Verma, S.; Mungse, H.P.; Kumar, N.; Choudhary, S.; Jain, S.L.; Sain, B.; Khatri, O.P. Graphene oxide: An efficient and reusable carbocatalyst for aza-Michael addition of amines to activated alkenes. *Chem. Commun.* **2011**, *47*, 12673–12675.
28. Satish, B.; Venkata Rao, K.; Shilpa Chakra, C.H.; Tejaswi, T. Synthesis and characterization of graphene oxide and its antimicrobial activity against klebseilla and staphylococus. *Int. J. Adv. Biotechnol. Res.* **2013**, *4*, 142–146.

29. Sun, Z.; Yan, Z.; Yao, J.; Beitler, E.; Zhu, Y.; Tour, J.M. Growth of graphene from solid carbon sources. *Nature* **2010**, *468*, 549–552.
30. Roy, M.; Kusurkar, T.S.; Maurya, S.K.; Meena, S.K.; Singh, S.K.; Sethy, N.; Bhargava, K.; Sharma, R.K.; Goswami, D.; Sarkar, S.; *et al*. Graphene oxide from silk cocoon: A novel magnetic fluorophore for multi-photon imaging. *3 Biotech* **2014**, *4*, 67–75.
31. Bo, Z.; Shuai, X.; Mao, S.; Yang, H.; Qian, J.; Chen, J.; Yan, J.; Cen, K. Green preparation of reduced graphene oxide for sensing and energy storage applications. *Sci. Rep.* **2014**, *4*, 1–5.
32. Mao, S.; Pu, H.; Chen, J. Graphene oxide and its reduction: Modeling and experimental progress. *RSC Adv.* **2012**, *2*, 2643–2662.
33. Mao, S.; Yu, K.; Cui, S.; Bo, Z.; Lu, G.; Chen, J. A new reducing agent to prepare single-layer, high-quality reduced graphene oxide for device applications. *Nanoscale* **2011**, *3*, 2849–2853.
34. Vidano, R.P.; Fischbach, D.B.; Willis, L.J.; Loehr, T.M. Observation of Raman band shifting with excitation wavelength for carbons and graphites. *Solid State Commun.* **1981**, *39*, 341–344.
35. Tuinstra, F.; Koenig, J.L. Raman spectrum of graphite. *J. Chem. Phys.* **1970**, *53*, 1126–1130.
36. Chun, O.W.; Chen, M.L.; Zhang, K.; Zhang, F.J. The effect of thermal and ultrasonic treatment on the formation of graphene-oxide nanosheets. *J. Korean Phys. Soc.* **2010**, *56*, 1097–1102.
37. Kaniyoor, A.; Ramaprabhu, S. A Raman spectroscopic investigation of graphite oxide derived graphene. *AIP Adv.* **2012**, *2*, doi:10.1063/1.4756995.
38. Eigler, S.; Dotzer, C.; Hirsch, A. Visualization of defect densities in reduced graphene oxide. *Carbon* **2012**, *50*, 3666–3673.

11

Modeling the Dispersibility of Single Walled Carbon Nanotubes in Organic Solvents by Quantitative Structure-Activity Relationship Approach

Hayriye Yilmaz [1,2,*], Bakhtiyor Rasulev [2,3] and Jerzy Leszczynski [2]

1 Department of Biomedical Devices and Technology, Kayseri Vocational School, Erciyes University, Kayseri 38039, Turkey

2 Interdisciplinary Center for Nanotoxicity, Department of Chemistry and Biochemistry, Jackson State University, Jackson, MS 39217, USA; E-Mails: rasulev@icnanotox.org (B.R.); jerzy@icnanotox.org (J.L.)

3 Center for Computationally Assisted Science and Technology (CCAST), North Dakota State University, Fargo, ND 58108, USA

* Author to whom correspondence should be addressed; E-Mail: hayriyey@erciyes.edu.tr or hayriyey@icnanotox.org

Academic Editor: Robert Tanguay

Abstract: The knowledge of physico-chemical properties of carbon nanotubes, including behavior in organic solvents is very important for design, manufacturing and utilizing of their counterparts with improved properties. In the present study a quantitative structure-activity/property relationship (QSAR/QSPR) approach was applied to predict the dispersibility of single walled carbon nanotubes (SWNTs) in various organic solvents. A number of additive descriptors and quantum-chemical descriptors were calculated and utilized to build QSAR models. The best predictability is shown by a 4-variable model. The model showed statistically good results ($R^2_{training}$ = 0.797, Q^2 = 0.665, R^2_{test} = 0.807), with high internal and external correlation coefficients. Presence of the X0Av descriptor and its negative term suggest that small size solvents have better SWCNTs solubility. Mass weighted descriptor ATS6m also indicates that heavier solvents (and small in size) most probably are better solvents for SWCNTs. The presence of the Dipole *Z* descriptor indicates that higher polarizability of the solvent molecule increases the solubility. The developed model and contributed descriptors can help to understand the mechanism of the dispersion process and predictorganic solvents that improve the dispersibility of SWNTs.

Keywords: dispersibility; organic solvent; single walled carbon nanotubes (SWNTs); QSAR

1. Introduction

With the rapid development of nanoscience and nanotechnology, carbon nanotubes (CNTs) have attracted a great deal of attention due to their unique and versatile properties since single-walled carbon nanotubes (SWCNTs) were discovered by Iijima in 1993 [1]. SWCNTs areknown for exhibiting unique mechanical, electrical, and thermal properties that are useful for a widerange of applications in materials. However, SWCNTs represent quite unusual systems and possess extremely low solubility or poor dispersibility in water and many known organic solvents [2]. Because of high polarizability, hydrophobic surface, and substantial van der Waals interactions, CNTs are able to aggregate, with each other, as well as with other chemical and biological systems to give mixture aggregates, especially in water [3,4]. The treating of CNTs by various active chemicals can change the surface properties and therefore the ability to aggregate. Some active chemicals can destruct the CNTs. This can be achieved by oxidizing CNTs by strong acids, such as refluxing in a mixture of sulfuric acid and nitric acid [5,6], "piranha" solution (sulfuric acid-hydrogen peroxide) [7], boiling in nitric acid [8], or treating with oxidative gases, such as ozone [4,9]. However, treatment under such harsh conditions clearly deviates from green chemistry and results in the opening of the tube tips [5], shortening of the tubes [7], and fragmentation of the sidewalls [8]. Therefore, the stability of CNTs may decrease, along with other important properties.

In recent years there have been several studies describing the preparation of stable suspensions of SWCNTs in a range of known solvents [9–16]. To get detailed structural information and dispersion values of SWCNTs in organic solvents, Hildebrand or Hansen solubility parameters have received close attention by researchers [17–20]. The main purpose of these studies was to improve the dispersion of nanotubes and to understand the dispersion process.

Quantitative structure activity relationships (QSARs) are often used to predict various physicochemical and biological properties of chemicals. Considering the difficulties in obtaining experimental data in CNTs research, theoretical approaches including Quantitative Structure-Property Relationships (QSPR) can provide useful information regarding predicted dispersibility values of carbon nanotubes directly from the structure. Discovering and developing effective new organic solvents for SWCNTs and C_{60} solubility have attracted many researchers'attention [21–26]. The first attempt to explain SWCNT dispersibility with the application of QSAR approach was carried out by Rofouei *et al.* [27], and the second study was made by Salahinejad *et al.* [28]. Our study is focused on developing an improved QSPR model that is able to predict the dispersibility of SWCNTs in various organic solvents.

2. Materials and Methods

2.1. Data Set

The QSAR modeling was applied for a set of single-walled carbon nanotubes in a pool of 29 different solvents (molecular structures are represented in Figure 1) which were selected from Bergin *et al.* [29]. The set consist of 29 organic solvents, which are randomly divided into training (22 compounds) and test (7 compounds) sets. The splitting to training and test sets was balanced across the variables. The dispersibilities of studied compounds were expressed in terms of the C (mg/mL) for SWCNTs (Table 1). All original concentrations data were converted to molar LogC(exp) variables.

Table 1. List of 29 organic solvents used in the study, including corresponding names, experimental and calculated dispersibility of single-walled carbon nanotubes (SWCNTs) in organic solvents.

No	Name	C (mg/mL)	LogC(exp) **	LogC(cal)
1	*N*-Cyclohexyl-pyrrolidinone	3.5	0.544	0.317
2 *	1,3-Dimethyltetrahydro-2(1H)-pyrimidinone	0.65	−0.187	−0.617
3	1-Butylpyrrolidin-2-one	0.279	−0.554	−0.480
4	1-Benzylpyrrolidin-2-one	0.18	−0.745	−0.566
5 *	1-Methylpyrrolidin-2-one	0.116	−0.935	−0.972
6	3-(2-Oxo-1-pyrrolidinyl)propanenitrile	0.115	−0.939	−0.510
7	*N*-Ethyl-pyrrolidinone	0.101	−0.996	−1.200
8	*N*-Octyl-pyrrolidone	0.092	−1.036	−1.241
9	*N*-Vinyl-pyrrolidinone	0.084	−1.076	−1.459
10	Dimethyl-imidazolidinone	0.083	−1.080	−1.111
11	Dimethylacetamide	0.041	−1.387	−1.534
12 *	*N*-Formyl-piperidine	0.039	−1.409	−1.126
13	*N*-Dodecyl-pyrrolidone	0.03	−1.553	−1.157
14	Dimethylformamide	0.023	−1.638	−2.374
15	Benzyl acetate	0.0192	−1.717	−2.040
16	Propionitrile	0.015	−1.824	−1.850
17	Acrylic acid	0.0138	−1.860	−2.345
18	2,2'-thiodiethanol	0.0136	−1.866	−2.100
19 *	Ethanolamine	0.0133	−1.876	−1.424
20 *	Cyclopentanone	0.0129	−1.889	−1.757
21 *	Chlorophenol	0.012	−1.921	−2.064
22	Acetone	0.011	−1.959	−1.530
23	Benzyl benzoate	0.0109	−1.963	−2.160
24	Isopropyl alcohol	0.0105	−1.979	−2.038
25 *	Cyclohexanone	0.0068	−2.168	−2.330
26	Toluene	0.005	−2.301	−2.056
27	Triethyleneglycol	0.0037	−2.432	−2.647
28	Formamide	3.00×10^{-4}	−3.523	−2.374
29	Benzyl alcohol	2.79×10^{-4}	−3.554	−3.007

Notes: * Test Set; ** Experimental data is taken from [29].

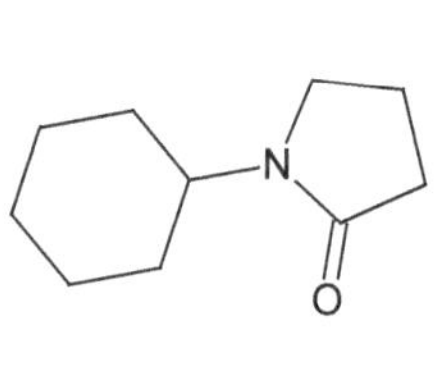
1

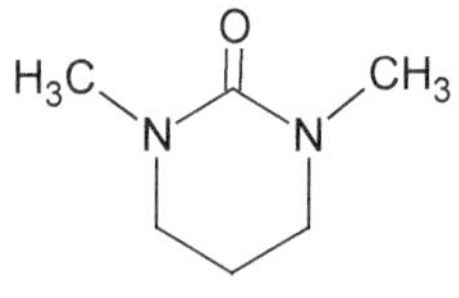
2

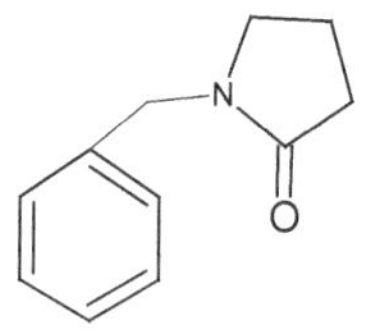
3

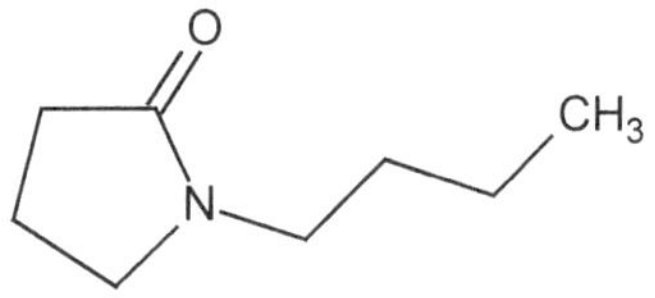
4

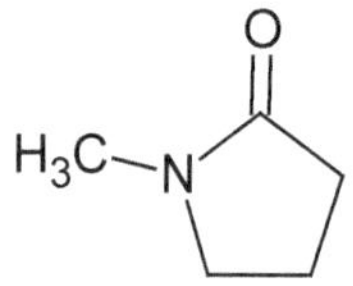
5

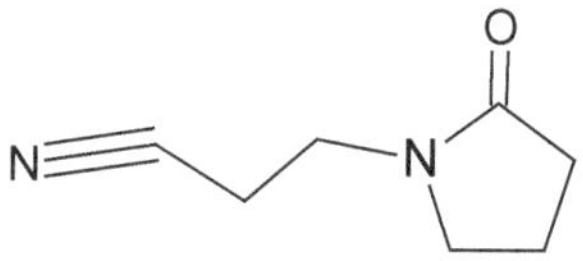
6

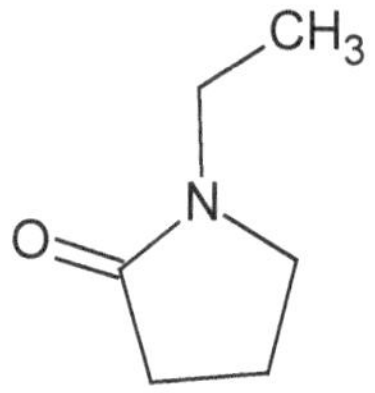
7

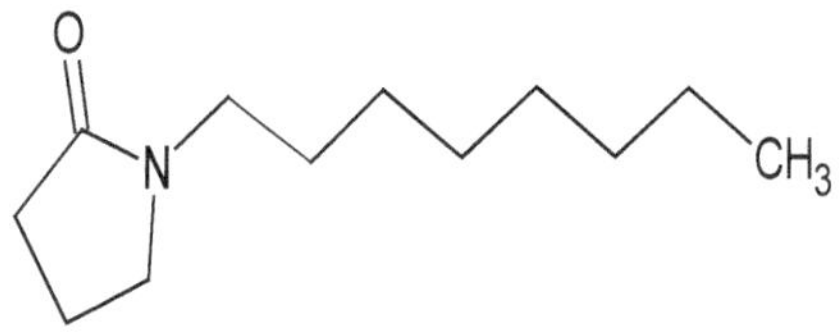
8

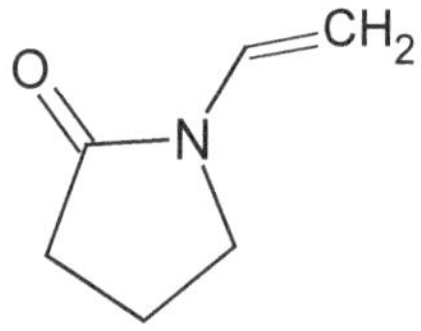
9

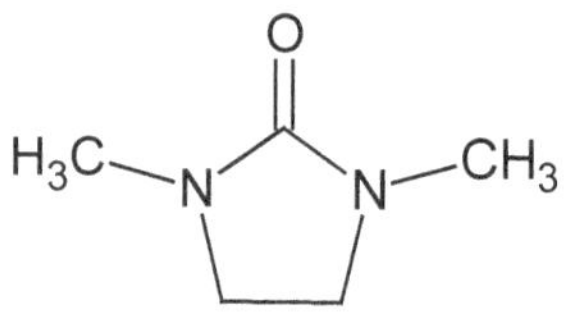
10

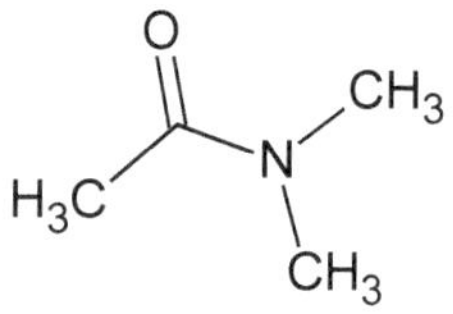
11

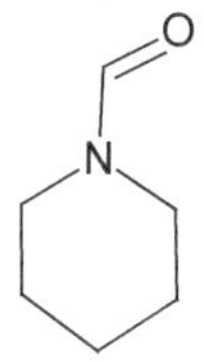
12

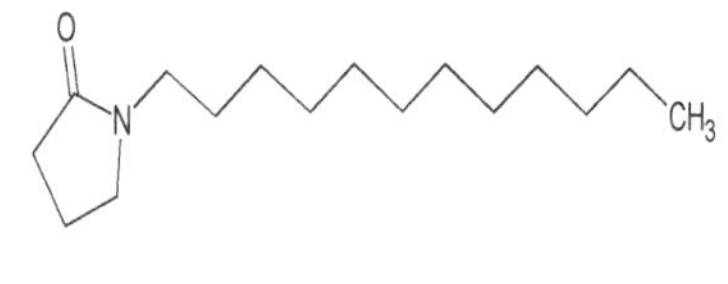
13

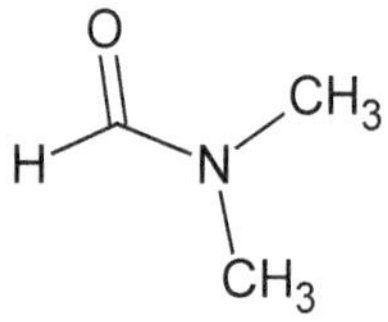
14

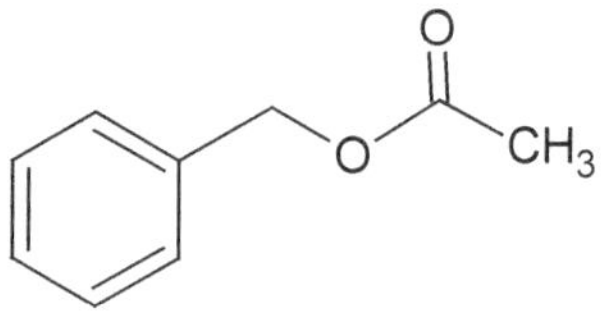
15

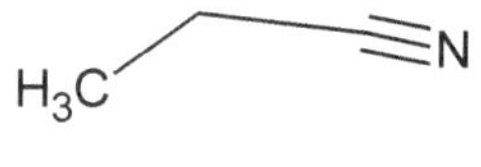
16

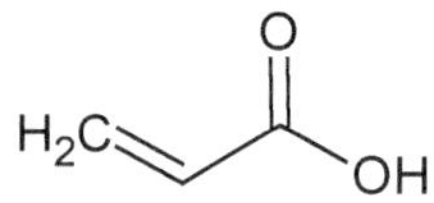
17

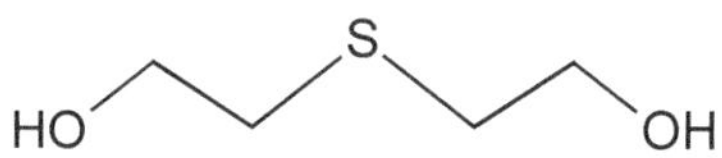
18

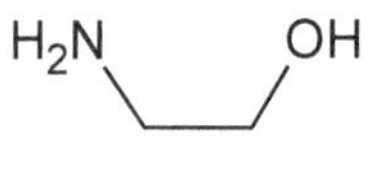
19

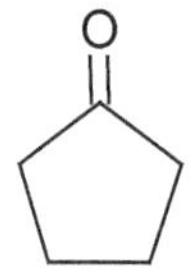
20

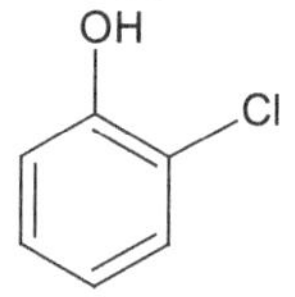
21

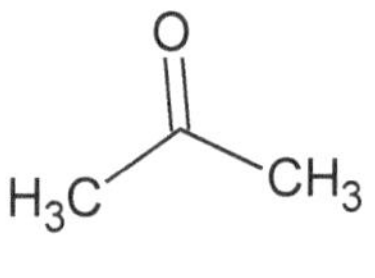
22

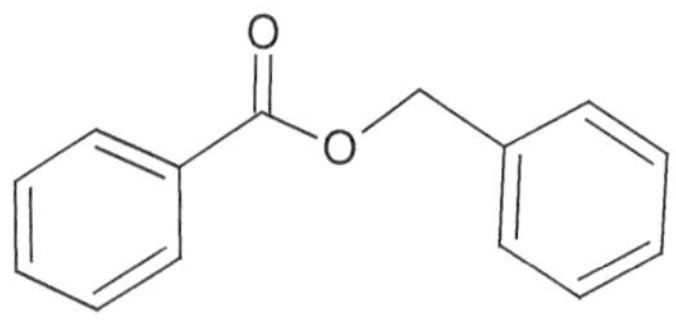
23

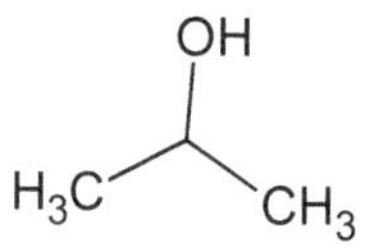
24

Figure 1. *Cont.*

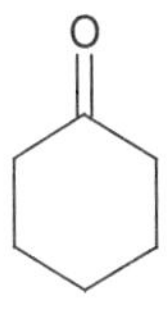

25

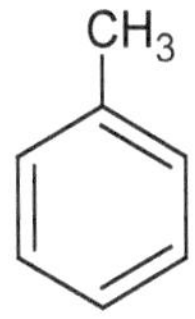

26

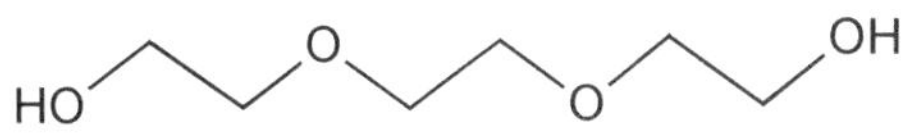

27

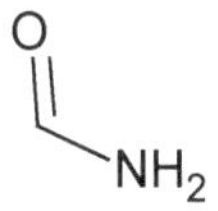

28

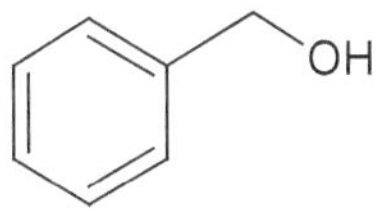

29

Figure 1. The structures of 29 organic solvents for single-walled carbon nanotube (SWCNT) derivatives.

2.2. Quantum Chemical Calculations

The initial structures of investigated organic solvents for SWCNTs were built using HyperChem 7.5 package [30]. After that step, the structures of compounds were firstly pre-optimized with the Molecular Mechanics Force Field (MM+) procedure included in the HyperChem. The semiempirical quantum chemical descriptors (including total energy, binding energy, electronic energy, nuclear energies, heats of formation, total dipole moment, *X*, *Y*, and *Z* components of dipole moment, E_{HOMO}, E_{LUMO}, surface area, volume, hydration energy, refractivity, LogP, polarizability, mass) were calculated by the RM1 method implemented in HyperChem. An initial set of 258 DRAGON software generated [31] theoretical descriptors was selected from the entire set of generated descriptors and used to describe the chemical diversity of the compounds. The software provides about 4000 various descriptors corresponding to 0D-, 1D-, 2D-, and 3D- descriptor modules. The outlined modules are comprised of 20 different classes of descriptors, namely, the constitutional, the topological, the walk and path counts, the connectivity indices, the information indices, the 2D autocorrelations, the edge adjacency indices, Burden eigenvalues, the topological charge indices, the eigenvalue based indices, the randic molecular profiles, the geometrical descriptors, the RDF descriptors, the 3D-MoRSE descriptors, the WHIM descriptors, the GETAWAY descriptors, the functional groups, the atom-centered fragments, the charge descriptors, and the molecular properties descriptors [32,33]. In addition, the density functional theory (DFT) with the hybrid meta exchange-correlation functional M06-2X/6-311G(d,p) [34] calculations were applied to obtain another set of quantum-chemically generated physico-chemical parameters of studied SWCNTs solvents—including dipole moments (total dipole moment, *X*, *Y*, and *Z* components); orbital energies, E_{HOMO}, E_{LUMO} and heats of formation. All DFT calculations were performed using the Gaussian 09 software [35].

2.3. QSAR Modeling and Statistical Analysis

The correlation between biological activity and structural properties was obtained by using the variable selection Genetic Algorithm (GA) and Multiple Linear Analysis (MLRA) methods. Preliminary models selection was performed by means of the GA-MLRA [36–38] technique as implemented in the BuildQSAR [39] program. Genetic Algorithms have been applied in recent studies as a powerful tool to address many problems in QSAR studies [36–38]. This method based on the mechanism of evaluation of species, in which the higher descriptor weights are the more preserved in the mathematic model, while the lower weight is eliminated. In this form, the best model which represents the experimental biological activity isobtained [36–38,40,41]. We selected the resulting model in the range of 1–5 variables per model by limiting the GA variable selection algorithm. The MLR technique was used to develop QSAR models since it is transparent, easy interpretable, and ideal to obtain reproducible results. Several QSAR models developed were followed by statistical analysis with evaluation by squared correlation coefficient R^2, standard error s, Fisher coefficient F, and non-collinearity of descriptors in the model. A final set of QSARs was tested by applying the "leave-one-out" technique (the process of removing a molecule from the set, then creating and validating the model against the individual molecules, which was performed for the entire training set). The mean was taken of all the Q^2 based on the predictive error sum of squares (PRESS).

The selection of robust and well predictive QSAR models on the basis of only R^2, Q^2 and R^2_{pred} might mislead the search for the ideal predictive model, so additional statistical analysis was done on the basis of a few other parameters, such as Average r_m^2, Delta r_m^2. For an acceptable QSAR model, the value of "Average r_m^2" should be >0.5 and "Delta r_m^2" should be <0.2 [42,43].

3. Results and Discussion

Our study was focused on developing a valid model that is able to predict the dispersibility of SWCNTs in various organic solvents. For this purpose we utilized a QSAR approach. Dragon software-generated additive descriptors, as well as semi-empirical and quantum mechanical descriptors were calculated and a total of 280 descriptors were used to build a QSAR model.

The correlation matrix for the most populated 2D-3D descriptors and LogC(cal) used in the present study is shown in Table 2. The sign of the correlation tells us whether the two variables are positively (more X means more Y) or negatively (more X means less Y) related. LogC(cal) and SRW09 had a good positive correlation (r=0.706) and strongly associated with dispersibility. In addition, LogC(cal), was found to be correlated to the Ram descriptor with r = 0.530 and to Dipole Z with r = 0.377, respectively. However, the correlations for any of these two descriptors considered as a single descriptor in the model were not sufficient to be considered significant in predicting dispersibility.

Table 2. Correlation matrix for physicochemical, 2D-3D descriptors and LogC(cal) used in the study.

Descriptor	SRW09	Dipole Z	piPC05	Ram	X0Av	ATS6m	LogC(cal)
SRW09	1						
Dipole Z	−0.004	1					
piPC05	−0.002	0.009	1				
Ram	0.567	0.068	0.631	1			
X0Av	0.053	−0.112	−0.275	0.075	1		
ATS6m	0.427	0.118	0.532	0.545	0.042	1	
LogC(cal)	0.706	0.377	−0.037	0.530	0.159	0.276	1

In Table 3 the performances for all developed models with 1–5 variables, for the training and test sets are listed. The 4-variable GA-MLRA based model showed the best predictive ability

($R^2_{training}$ = 0.797, Q^2 = 0.665, R^2_{test} = 0.807), with high internal and correlation coefficients. It is clearly noticeable that R^2 values in the case of the training set follow increasing order with increase of the number of variables: 1-variable model < 2-variable model < 3-variable model < 4-variable model and for the test set follow increasing order: 2-variable model < 4-variable model < 3-variable model < 1-variable model. In the results, according to Figure 2 the 4-variable model was chosen as the most predictive and robust model.

Table 3. Descriptor names and statistical values for the developed models (statistics are shown for split sets into training (22 compounds) and test (7)).

No.	Descriptors	Training Set						Test Set			
		N	R^2	Q^2	*F*	SDEP	Spress	*N*	R^2	$R_{m(avr)}^2$	ΔR^2
1	SRW09	22	0.548	0.453	24.249	0.674	0.690	7	0.830	0.773	0.112
2	SRW09, Dipole *Z*	22	0.679	0.573	20.080	0.595	0.626	7	0.786	0.717	0.137
3	Ram,piPC05, Dipole *Z*	22	0.736	0.601	16.687	0.575	0.621	7	0.813	0.751	0.122
4	ATS6m,SRW09, X0Av, Dipole *Z*	22	0.797	0.666	16.722	0.527	0.585	7	0.807	0.744	0.125
5	X3A,ATS6m, SRW09, X0Av, Dipole *Z*	22	0.817	0.632	14.366	0.552	0.633	7	0.736	0.702	0.162

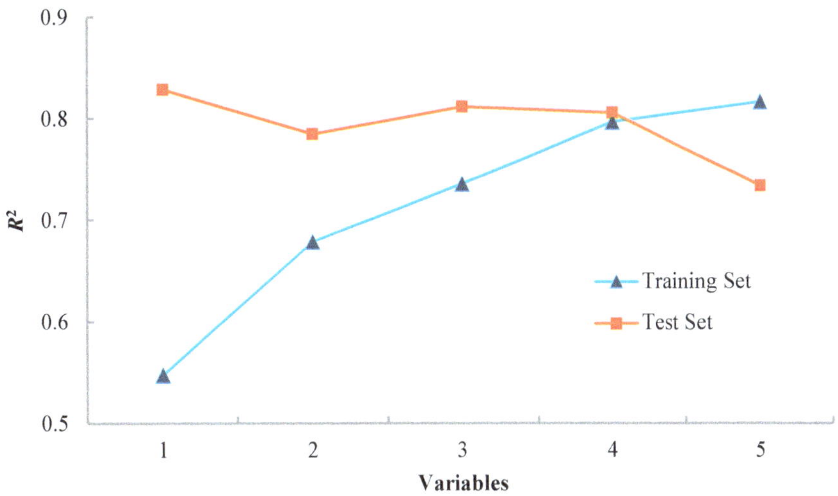

Figure 2. Comparison of the regression coefficients (R^2) for the training and test set.

Table 4 represents the descriptor values selected using GA-MLR variable selection for dispersibility of SWCNTs in organic solvents. The one, two, and four variable models include SRW09 (9th order self-returning walk count). Organic solvents like 1, 3–10, 13, and 20 which have 5-membered rings, show a high value of SRW09, with a positive impact on the SWCNTs dispersibility.

The 1-variable model is represented by the following Equation (1):

$$\text{Log(C)}_{max} = +\, 0.0020\ (\pm\ 0.0008)\ \mathbf{SRW09} - 2.1522\ (\pm\ 0.3620) \qquad (1)$$

where n = 22, R^2 = 0.548, s = 0.627, F = 24.249, p = 0.0001, Q^2 = 0.453, SPress = 0.690, SDEP = 0.674.

Table 4. The descriptor values for organic solvents.

No.	SRW09	Dipole *Z*	piPC05	Ram	X0Av	ATS6m
1	702	2.225	30	3	0.632	6.731
2	0	−0.166	10	1	0.576	0.000
3	684	1.312	78	3	0.582	7.343
4	684	0.459	14	2	0.660	6.580
5	504	0.515	4	1	0.588	0.000
6	684	1.110	18	2	0.584	6.472
7	684	−1.214	8	2	0.648	2.583
8	684	−0.743	18	2	0.673	7.406
9	684	−1.345	10	2	0.595	2.215
10	504	−0.006	4	1	0.554	0.000
11	0	0.000	0	2	0.726	0.000
12	0	2.665	12	1	0.621	3.566
13	684	−0.590	22	2	0.681	7.797
14	0	0.000	0	0	0.521	0.000
15	0	1.192	63	2	0.583	6.898
16	0	0.000	0	0	0.649	0.000
17	0	0.000	0	1	0.528	0.000
18	0	0.001	2	0	0.707	5.869
19	0	0.887	0	0	0.610	0.000
20	504	0.454	4	1	0.623	0.000
21	0	0.000	52	2	0.611	0.703
22	0	0.000	0	1	0.727	0.000
23	0	−0.763	132	3	0.550	7.631
24	0	−1.142	0	1	0.756	0.000
25	0	1.892	10	1	0.635	0.000
26	0	0.000	44	1	0.627	1.401
27	0	0.000	5	0	0.595	6.923
28	0	0.000	0	0	0.521	0.000
29	0	−0.973	50	1	0.568	3.508

The first descriptor given in the Equation (1) above is SRW09 (from the MWC class). It is among the 2D-descriptors representing self-returning walk counts of different lengths. The SRW count of any even order indicates the length and shape (branching) of the entire molecular graph. It increases when the number of atoms increases, when a molecule becomes more branched or contains even-membered rings. But on the other hand, the SRW count of the odd order represents only local surrounding of odd-membered rings [44,45]. To clarify the above, the self-returning walk of the 9th order of the molecules 1 and 4 structures is given in Figure 3.

SRW09 is the self-returning walk count of the 9th order and represents the surroundings of odd membered rings (five-membered in our case). It provides valuable insight into the relationship between structure and dispersibility action.

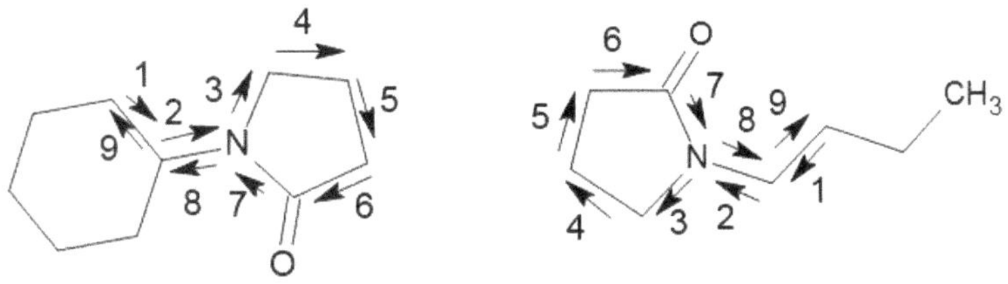

1 4

Figure 3. Self-returning walk of the 9th order of molecules 1 and 4.

The 2-variable model is represented by following Equation (2):

$$\text{Log(C)}_{\text{max}} = +\,0.0019\ (\pm\,0.0007)\ \textbf{SRW09} + 0.3809\ (\pm\,0.2865)\ \textbf{Dipole}\ \boldsymbol{Z} - 2.1228\ (\pm\,0.3149) \qquad (2)$$

where $n = 22$, $R^2 = 0.679$, $s = 0.543$, $F = 20.080$, $p < 0.0001$, $Q^2 = 0.573$, SPRESS = 0.626, SDEP = 0.595.

The second descriptor in Equation (2) is a dipole moment descriptor, which is a 3D electronic descriptor that indicates the strength and orientation behavior of a molecule in an electrostatic field. Both the magnitude and the components (*X, Y, Z*) of the dipole moment are calculated. The descriptor is estimated by utilizing partial atomic charges and atomic coordinates. The presence and sign of the dipole moment contribution indicates that the higher polarity of the attached fragment the higher will be the overall dispersibility value of the SWCNT derivative.

The 3-variable model is represented by following Equation (3):

$$\text{Log(C)}_{\text{max}} = +0.8389\ (\pm\,0.2960)\ \textbf{Ram} - 0.0209\ (\pm\,0.0122)\ \textbf{piPC05} + 0.3855\ (\pm\,0.2920)\ \textbf{Dipole}\ \boldsymbol{Z} - 2.3590\ (\pm\,0.4189) \qquad (3)$$

where $n = 22$, $R^2 = 0.736$, $s = 0.506$, $F = 16.687$, $p < 0.0001$, $Q^2 = 0.601$, SPress = 0.621, SDEP = 0.575.

Here one can see the presence of the Ram descriptor, which positively contributes to the dispersibility. The topological descriptor Ram addresses the branching in the molecule [46]. Its regression coefficient suggests in favor of more branched molecular structures for increased activity. Another descriptor represents the molecular multiple path count of order 05 (piPC05). This descriptor reflects the length of the molecule. In Equation (3) piPC05 contributes negatively to the dispersibility.

The 4-variable model is represented by the following Equation (4):

$$\text{Log(C)}_{\text{max}} = +0.0022\ (\pm\,0.0008)\ \textbf{SRW09} + 4.0933\ (\pm\,3.1813)\ \textbf{ATS6m} - 0.0832\ (\pm\,0.0857)\ \textbf{X0Av} + 0.5488\ (\pm\,0.2716)\ \textbf{Dipole}\ \boldsymbol{Z} - 4.5062\ (\pm\,1.9741) \qquad (4)$$

where $n = 22$, $R^2 = 0.797$, $s = 0.456$, $F = 16.722$, $p < 0.0001$, $Q^2 = 0.666$, SPRESS = 0.585, SDEP = 0.527.

In the model (4) one can see the presence of other descriptors, ATS6m and X0Av. Walk and path counts class descriptors, SRW09, the 2D-AUTO class descriptors, the atomic mass weighted terms ATS6m show the higher weights with positive influences. In this model, the atomic mass weighted term (ATS6m) showed the highest contribution. The activity exhibits negative linear relationship with the connectivity index descriptor (X0Av), which is an average valence connectivity index chi-0.

Analysis of these influential molecular descriptors can lead to the revealing of the mechanism of the dispersion process and thus the model is able to predictnew organic solvents that improve the dispersibility of SWCNTs. For example, X0Av descriptor and its negative term suggest that small size solvents have better influence on the solubility of SWCNTs. Also, mass weighted descriptor ATS6m indicates that a heavier solvent (at the same time having a small size) is most probably a better solvent for SWCNTs. The presence of the Dipole *Z* descriptor indicates that higher polarizability of the solvent molecule increases the solubility.

The dependence of the number of variables in the models, for training and test sets, on the R^2 values is displayed in Figure 2. The correlation graph of the best QSAR model (Equation (4)) is shown in Figure 4. The GA-MLRA based QSAR model with four variables showed better results than other models ($R^2_{training}$ = 0.797, Q^2 = 0.666, R^2_{test} = 0.807), with high internal and external correlation coefficients. A good agreement between the predictions and the experimental values confirmed the reliability of the QSAR model.

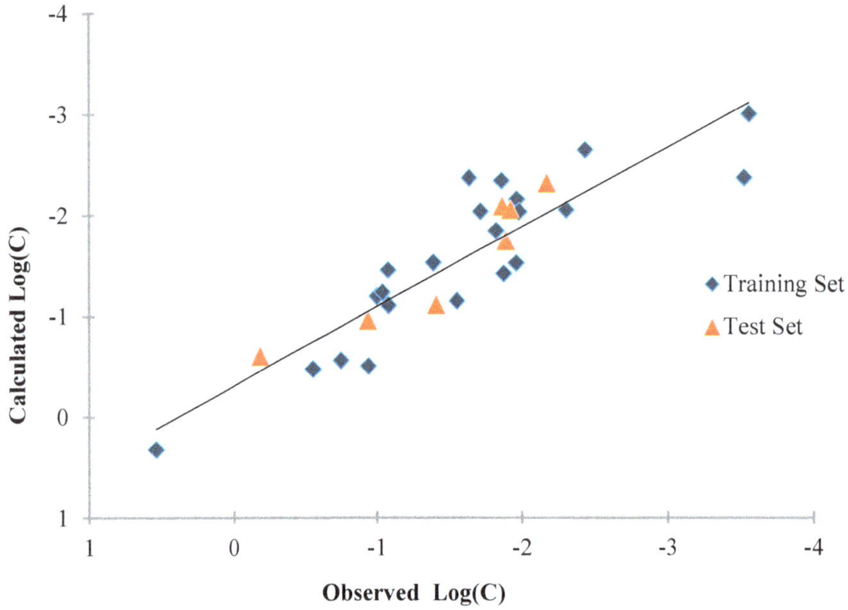

Figure 4. A plot of observed and predicted $LogC_{max}$ values for the entire set (29 compounds) calculated by the 4-variable model.

4. Conclusions

In the present work we investigated the influence of the characteristics of a series of organic solvents on the dispersibility of SWCNTs. For this purpose both the additive descriptors (DRAGON-software based) and quantum-chemical descriptors were generated, a total of 280 descriptors.

The 4-variable model retains also a good ratio of the number of descriptors and their predictive ability. TheGA-MLRA based model showed good results ($R^2_{training}$ = 0.797, Q^2 = 0.665, R^2_{test} = 0.807), with high internal and external correlation coefficients. The model (4) developed here showed the highest performance with the presence of the following four descriptors, SRW09, ATS6m, Dipole Z, and X0Av. In this model the atomic mass weighted term (ATS6m) showed the highest contribution. The other significant descriptors are graph elements weighted SRW09 (as walk and path counts) and Dipole-*Z* (as a 3D descriptor). Descriptors SRW09 and Dipole *Z* exhibit a positive influence on the dispersibility. A molecular multiple path count of the order 05 (piPC05) and an average valence connectivity index of the order 0 (X0Av) showed negative influence on the considered activity. The X0Av descriptor and its negative term suggest that small size solvents have better influence on the solubility of SWCNTs. Also, the mass weighted descriptor ATS6m indicates that a heavier solvent (at the same time having a small size) most probably is the better solvent for SWCNTs. The presence of the Dipole *Z* descriptor indicates that a higher polarizability of the solvent molecule increases the solubility. Analysis of these influential molecular descriptors can lead to details of the mechanism of the dispersion process and thus enable predictions of new organic solvents to improve the dispersibility of SWCNTs.

Acknowledgments

This work was financially supported by The Scientific and Technological Research Council of Turkey (TÜBİTAK-2219) and by NSF CREST Interdisciplinary Nanotoxicity Center NSF-CREST Grant # HRD-0833178. Authors also thank the Extreme Science and Engineering Discovery Environment (XSEDE) for the award allocations (TG-DMR110088 and CHE140005) and Mississippi Center for Supercomputer Research (Oxford, MS) for a generous allotment of computer time. B.R. gratefully acknowledges support from the North Dakota State University Center for Computationally Assisted Science and Technology and the U.S. Department of Energy through Grant No. DE-SC0001717.

Author Contributions

H.Y., B.R., and J.L. conceived the concept and designed the research. H.Y. analyzed data, wrote the manuscript, and produced all figures. B.R. advised the chemo informatics workflow and quantum-mechanical calculations. J.L. supervised and directed the project. All authors discussed results, reviewed and edited the manuscript.

Conflicts of Interest

The authors declare no conflict of interest.

References

1. Lijima, S.; Ichihashio, T. Single-shell carbon nanotubes of 1-nm diameter. *Nature* **1993**, *363*, 603–605.
2. Nakashima, N. Soluble carbon nanotubes: Fundamental and applications. *Int. J. Nanosci.* **2005**, *4*, 119–137.
3. Britz, D.A.; Khlobystov, A.N. Noncovalent interactions of molecules with single walled carbon nanotubes. *Chem. Soc. Rev.* **2006**, *35*, 637–659.
4. Girifalco, L.A.; Hodak, M.; Lee, R.S. Carbon nanotubes, buckyballs, ropes, and a universal graphitic potential. *Phys. Rev. B.* **2000**, *62*, 13104.

5. Liu, J.; Rinzler, A.G.; Dai, H.; Hafner, J.H.; Bradley, R.K.; Boul, P.J.; Lu, A.; Iverson, T.; Shelimov, K.; Huffman, C.B.; *et al*. Fullerene pipes. *Science* **1998**, *280*, 1253–1256.
6. Pavese, M.; Musso, S.; Bianco, S.; Giorcelli, M.; Pugno, N. An analysis of carbon nanotube structure wettability before and after oxidation treatment. *J. Phys. Condens. Matter* **2008**, *20*, doi:10.1088/0953-8984/20/47/474206.
7. Ziegler, K.J.; Gu, Z.; Peng, H.; Flor, E.L.; Hauge, R.H.; Smalley, R.E. Controlled oxidative cutting of single-walled carbon nanotubes. *J. Am. Chem. Soc.* **2005**, *127*, 1541–1547.
8. Worsley, K.A.; Kalinina, I.; Bekyarova, E.; Haddon, R.C. Functionalization and dissolution of nitric acid treated single-walled carbon nanotubes. *J. Am. Chem. Soc.* **2009**, *131*, 18153–18158.
9. Ausman, K.D.; Piner, R.; Lourie, O.; Ruoff, R.S.; Korobov, M. Organic solvent dispersions of single-walled carbon nanotubes: Toward solutions of pristine nanotubes. *J. Phys. Chem. B* **2000**, *104*, 8911–8915.
10. Bahr, J.L.; Mickelson, E.T.; Bronikowski, M.J.; Smalley, R.E.; Tour, J.M. Dissolution of small diameter single-wall carbon nanotubes in organic solvents. *Chem. Commun.* **2001**, *2*, 193–194.
11. Furtado, C.A.; Kim, U.J.; Gutierrez, H.R.; Pan, L.; Dickey, E.C.; Eklund, P.C. Debundling and dissolution of single walled carbon nanotubes in amide solvents. *J. Am. Chem. Soc.* **2004**, *126*, 6095–6105.
12. Landi, B.J.; Ruf, H.J.; Worman, J.J.; Raffaelle, R.P. Effects of alkyl amide solvents on the dispersion of single-wall carbon nanotubes. *J. Phys. Chem. B* **2004**, *108*, 17089–17095.
13. Maeda, Y.; Kimura, S.; Hirashima, Y.; Kanda, M.; Lian, Y.F.; Wakahara, T.; Akasaka, T.; Hasegawa, T.; Tokumoto, H.; Shimizu, T.; *et al*. Dispersion of single-walled carbon nanotube bundles in nonaqueous solution. *J. Phys. Chem. B* **2004**, *108*, 18395–18397.
14. Umek, P.; Vrbanic, D.; Remskar, M.; Mertelj, T.; Venturini, P.; Pejovnik, S.; Mihailovic, D. An effective surfactant-free isolation procedure for single-wall carbon nanotubes. *Carbon* **2002**, *40*, 2581–2585.
15. Bergin, S.D.; Nicolosi, V.; Streich, P.V.; Giordani, S.; Sun, Z.; Windle, A.H.; Ryan, P.; Peter, N.; Niraj, P.; Wang, Z.-T.T.; *et al*. Towards solutions of single-walled carbon nanotubes in common solvents. *Adv. Mater.* **2008**, *20*, 1876–1881.
16. Giordani, S.; Bergin, S.D.; Nicolosi, V.; Lebedkin, S.; Kappes, M.M.; Blau, W.J.; Coleman, J.N. Debundling of single walled nanotubes by dilution: Observation of large populations of individual nanotubes in amide solvent. *J. Phys. Chem. B* **2006**, *110*, 15708–15718.
17. Usrey, M.L.; Chaffee, A.; Jeng, E.S.; Strano, M.S. Application of polymer solubility theory to solution phase dispersion of single-walled carbon nanotubes. *J. Phys. Chem. C* **2009**, *113*, 9532–9540.
18. Maiti, A.; Wescott, J.; Kung, P. Nanotube-polymer composites: Insights from Flory-Huggins theory and mesoscale simulations. *Mol. Simul.* **2005**, *31*, 143–149.
19. Detriche, S.; Zorzini, G.; Colomer, J.F.; Fonseca, A.; Nagy, J.B. Application of the Hansen solubility parameters theory to carbon nanotubes. *J. Nanosci. Nanotechnol.* **2007**, *8*, 6082–6092.
20. Ham, H.T.; Choi, Y.S.; Chung, I.J. An explanation of dispersion states of single-walled carbon nanotubes in solvents and aqueous surfactant solutions using solubility parameters. *J. Colloid Interface Sci.* **2005**, *286*, 216–223.

21. Turabekova, M.; Rasulev, B.; Theodore, M.; Jackman, J.; Leszczynska, D.; Leszczynski, J. Immunotoxicity of nanoparticles: A computational study suggests that CNTs and C60 fullerenes might be recognized as pathogens by Toll-like receptors. *Nanoscale* **2014**, *7*, 3488–3495.
22. Rasulev, B.; Gajewicz, A.; Puzyn, T.; Leszczynska, D.; Leszczynski, J. Nano-QSAR: Advances and challenges. *RSC Nanosci. Nanotechnol.* **2012**, *25*, 220–256.
23. Petrova, T.; Rasulev, B.F.; Toropov, A.A.; Leszczynska, D.; Leszczynski, J. Improved model for fullerene C60 solubility in organic solvents based on quantum-chemical and topological descriptors. *J. Nanopart. Res.* **2011**, *13*, 3235–3247.
24. Cook, S.M.; Aker, W.G.; Rasulev, B.F.; Hwang, H.-M.; Leszczynski, J.; Jenkins, J.J.; Shockley, V. Choosing safe dispersing media for C60 fullerenes by using cytotoxicity tests on the bacterium *Escherichia coli*. *J. Hazard. Mater.* **2010**, *176*, 367–373.
25. Toropov, A.A.; Rasulev, B.F.; Leszczynska, D.; Leszczynski, J. Multiplicative SMILES-based optimal descriptors: QSPR modeling of fullerene C60 solubility in organic solvents. *Chem. Phys. Lett.* **2008**, *457*, 332–336.
26. Toropov, A.A.; Rasulev, B.F.; Leszczynska, D.; Leszczynski, J. Additive SMILES based optimal descriptors: QSPR modeling of fullerene C60 solubility in organic solvents. *Chem. Phys. Lett.* **2007**, *444*, 209–214.
27. Rofouei, M.K.; Salahinejad, M.; Ghasemi, J.B. An alignment independent 3D-QSAR modeling of dispersibility of single-walled carbon nanotubes in different organic solvents. *Fuller. Nanotub. Carbon Nanostruct.* **2014**, *22*, 605–617.
28. Salahinejad, M.; Zolfonoun, E. QSAR studies of the dispersion of SWNTs in different organic solvents. *J. Nanopart. Res.* **2013**, *15*, 2028–2039.
29. Bergin, S.D.; Sun, Z.Y.; Rickard, D.; Streich, P.V.; Hamilton, J.P.; Coleman, J.N. Multicomponent solubility parameters for single-walled carbon nanotube-solvent mixtures. *ACS Nano* **2009**, *3*, 2340–2350.
30. HyperChem (HyperCube) Homepage. Available online: http://www.hyper.com (accessed on 2 June 2009).
31. DRAGON 6.0. Available online: http://www.disat.unimib.it/chm. (accessed on 11 December 2014).
32. Todeschini, R.; Consonni, V. *DRAGON Software for the Calculation of Molecular Descriptors*; web version 3.0 for Windows; TALETE S.R.L.: Milano, Italy, 2003.
33. Todeschini, R.; Consonni, V. *Handbook of Molecular Descriptors*; Wiley-VCH: Weinheim, Germany; New York, NY, USA, 2000.
34. Zhao, Y.; Truhlar, D.G. The M06 suite of density functionals for main group thermochemistry, kinetics, noncovalent interactions, excited states, and transition elements: Two new functionals and systematic testing of four M06 functionals and twelve other functionals. *Theor. Chem. Acc.* **2008**, *120*, 215–241.
35. Frisch, M.J.; Trucks, G.W.; Schlegel, H.B.; Scuseria, G.E.; Robb, M.A.; Cheeseman, J.R.; Scalmani, G.; Barone, V.; Mennucci, B.; Petersson, G.A.; *et al. Gaussian 09, Revision B.01*; Gaussian Inc: Wallingford, CT, USA, 2009.
36. Devillers, J. *Genetic Algorithms in Molecular Modeling*; Academic Press: London, UK, 1996.

37. Rasulev, B.F.; Saidkhodzhaev, A.I.; Nazrullaev, S.S.; Akhmedkhodzhaeva, K.S.; Khushbaktova, Z.A.; Leszczynski, J. Molecular modelling and QSAR analysis of the estrogenic activity of terpenoids isolated from Ferula plants. *SAR QSAR Environ. Res.* **2007**, *18*, 663–673.
38. Ahmed, L.; Rasulev, B.; Turabekova, M.; Leszczynska, D.; Leszczynski, J. Receptor and ligand-based study of fullerene analogues: Comprehensive computational approach including quantum-chemical, QSAR and molecular docking simulations. *Org. Biomol. Chem.* **2013**, *11*, 5798–5808.
39. De Oliveira, D.B.; Gaudio, A.C. BuildQSAR: A new computer program for QSAR analysis. *Quant. Struct. Act. Relat.* **2003**, *19*, 599–604.
40. Hayriye, Y.; Yahya, G.; Zulbiye, O.; Gokce, A.; Safak Ozhan, K. 4D-QSAR study of $p56^{lck}$ protein tyrosine kinase inhibitory activity of flavonoid derivatives using MCET method. *Bull. Korean Chem. Soc.* **2012**, *32*, 4352–4360.
41. Yilmaz, H.; Güzel, Y.; Boz, M.; Türkmenoğlu, B. Pharmacophore and functional group identification of 4,4'-dihydroxydiphenylmethane as bisphenol-A (BSA) derivative. *Trop. J. Pharm. Res.* **2014**, *13*, 117–129.
42. Roy, K.; Mitra, I.; Kar, S.; Ojha, P.; Das, R.N.; Kabir, H. Comparative studies on some metrics for external validation of QSPR models. *J. Chem. Inf. Model.* **2012**, *52*, 396–408.
43. Ojha, P.K.; Mitra, I.; Das, R.N.; Roy, K. Further exploring r_m^2 metrics for validation of QSPR models dataset. *Chemom. Intell. Lab. Syst.* **2011**, *107*, 194–205.
44. Gutman, I.; Rücker, C.; Rücker, G. On walks in molecular graphs. *J. Chem. Inf. Comput. Sci.* **2001**, *41*, 739–745.
45. Devinyak, O.; Zimenkovsky, B.; Lesyk, R. Biologically active 4-thiazolidinones: A review of QSAR studies and QSAR modeling of antitumor activity. *Curr. Top. Med. Chem.* **2012**, *12*, 2763–2784.
46. Sharmaa, B.K.; Singha, P.; Sarbhaia, K.; Prabhakar, Y.S. A QSAR study on serotonin 5-HT6 receptor ligands: Indolyl and piperidinyl sulfonamides. *SAR QSAR Environ. Res.* **2010**, *21*, 369–388.

12

Phosphorothioate DNA Stabilized Fluorescent Gold and Silver Nanoclusters

Daniel S. Weadick and Juewen Liu *

Department of Chemistry, Waterloo Institute for Nanotechnology, Waterloo, ON N2L 3G1, Canada; E-Mail: danweadick@gmail.com

* Author to whom correspondence should be addressed; E-Mail: liujw@uwaterloo.ca

Academic Editor: Stephen Ralph

Abstract: Unmodified single-stranded DNA has recently gained popularity for the templated synthesis of fluorescent noble metal nanoclusters (NCs). Bright, stable, and biocompatible clusters have been developed primarily through optimization of DNA sequence. However, DNA backbone modifications have not yet been investigated. In this work, phosphorothioate (PS) DNAs are evaluated in the synthesis of Au and Ag nanoclusters, and are employed to successfully template a novel emitter using T_{15} DNA at neutral pH. Mechanistic studies indicate a distinct UV-dependent formation mechanism that does not occur through the previously reported thymine N3. The positions of PS substitution have been optimized. This is the first reported use of a T_{15} template at physiological pH for AgNCs.

Keywords: DNA; phosphorothioate; luminescence; nanoclusters

---

1. Introduction

The strong fluorescent intensity, relative stability, low toxicity and small size of noble metal nanoclusters make them an attractive replacement for organic fluorophores and semiconductor quantum dots in biological analysis, biological imaging and environmental monitoring [1–6]. The use of single-stranded DNA (ssDNA) as a synthetic template has gained popularity due to its commercial availability, biocompatibility, sequence dependent emission, and molecular recognition properties [2,4].

Despite the development of bright DNA-templated nanoclusters with emission wavelengths that range from the UV to NIR regions, our understanding of their structure, synthetic mechanism, and origin of fluorescence is currently limited. Past work has focused on the development and characterization of novel fluorescent clusters through the control of a variety of synthetic parameters, including nucleotide sequence, pH, redox conditions, and metal-DNA stoichiometry [3,4]. However, as of yet, DNA backbone modifications have not been examined.

Phosphorothioate DNA (PS DNA) is a DNA variant in which one of the non-bridging phosphate oxygen atoms is replaced by sulphur as shown in Figure 1. These modified oligonucleotides have become popular for designing antisense nucleic acids with high nuclease resistance and lipid bi-layer permeability, as well as for elucidating metal binding sites and ribozyme mechanisms [7–9]. More recently it has been employed in the assembly and positioning of metallic nanoparticles, and semi-conductor quantum dots, as well as in the development of chemical biology probes [10–14].

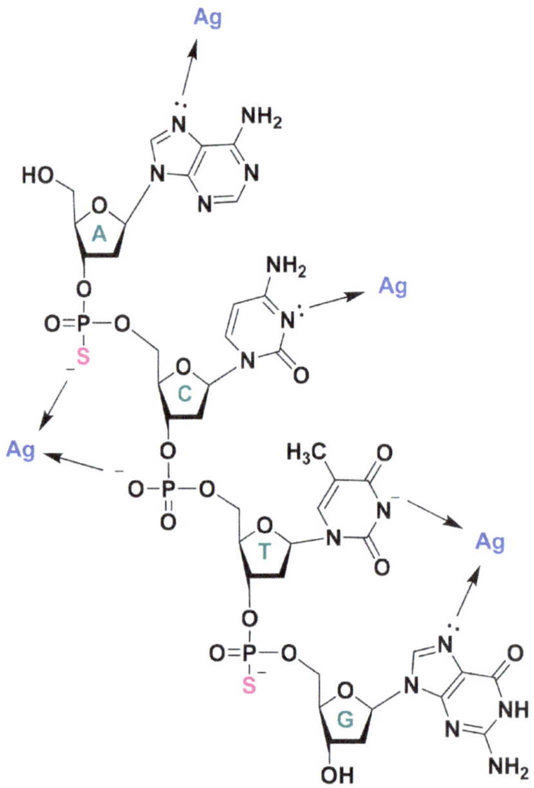

Figure 1. Schematic illustrating the potential sites of interaction between single stranded DNA and silver ions. Nucleobase interactions shown are based on previous literature reports. In this work, we explore the effect of backbone phosphorothioate modifications. Note that the thymine nitrogen needs to deprotonate for metal coordination.

Previously, small thiol compounds such as glutathione have been shown to be effective in stabilizing fluorescent gold and silver nanoclusters [6,15]. There are also reports that thiol can both increase and decrease the emission intensity of nanoclusters templated by DNA [6,16]. Compared to a normal thiol modification, PS is much more cost-effective and can be introduced at multiple locations [7]. This substitution gives PS nucleic acids increased hydrophobicity, membrane permeability, oxidative stability, nuclease resistance and affinity for soft metal ions [7,17,18]. It has also been shown to induce conformational changes, which may be explained by the increased atomic radius of sulfur, and redistribution of electron density in the phosphodiester bond [19,20]. Despite these changes, PS modification leaves DNA charge density and nucleobase pK_a unchanged [18,21]. It is expected that PS DNA may confer novel optical properties and provide insight into the structure and properties of DNA-nanoclusters. The aim of this work will be to synthesize PS DNA stabilized fluorescent metal NCs, and investigate the effect of phosphorothioate modification of the DNA backbone on optimal synthetic conditions, as well as the physical, optical and chemical properties of templated metal NCs.

2. Results and Discussion

2.1. DNA Sequence Screening for AgNCs

The structure of a single-stranded DNA is shown in Figure 1. A noble metal ion or cluster may interact with the phosphate backbone via electrostatic interaction. Gold and silver typically interact with phosphate weakly. However, with a PS modification, an enhanced backbone interaction is expected. At the same time, they may also coordinate with DNA via the endocyclic nitrogens and exocyclic carbonyl-groups of the nucleobases as well as through interactions with the π-system of the aromatic rings [22]. While no advantageous properties were detected using PS-DNA in AuNC synthesis (see Table S1 and Figure S1), some benefits were seen with silver.

In this synthesis, 15-m DNA was used for all AgNC synthesis. Base protonation has previously been shown to have a dramatic impact on template fluorescence [23,24]. Subsequently, C_{15}, A_{15} and T_{15} homopolymers (see Table 1) were evaluated at a solution pH at least one unit greater than and less than the nucleobase pK_a. The reduction was achieved with the standard $NaBH_4$ at a ratio of $NaBH_4{:}Ag^+ = 1{:}1$, and the samples were observed under UV excitation (Figure 2A). Emission spectra were also recorded for each (see Figure S2). The behavior of the unmodified phosphodiester (PO)-DNA is consistent with the previous literature reports. For example, unmodified T_{15} produced strong fluorescence only at high pH, while unmodified C_{15} stabilized AgNCs emitted strongly at neutral pH. In these cases, PS modification negatively impacted the fluorescent properties of the clusters produced. However, improved fluorescence was observed for PS templates under two conditions. A_{15}-PS was able to produced weak, but slightly improved visual emission at both low and neutral pH (Figure 2, Figure S2C,D) and more impressively, PS modification enabled the formation of a previously unreported T_{15}-templated violet emitter at neutral pH, resulting in increased overall visual fluorescence. PS-modification of C_{15} at pH 3 resulted in a slight shift in emission wavelength, but no enhancement in fluorescence (Figure S2A).

Table 1. DNA sequences used as templates in nanocluster synthesis (from 5'-terminus).

DNA Name	Sequence and Modifications
A_{15}-PO	AAAAAAAAAAAAAAA
A_{15}-PS	A*A*A*A*A*A*A*A*A*A*A*A*A*A*A
C_{15}-PO	CCCCCCCCCCCCCCC
C_{15}-PS	C*C*C*C*C*C*C*C*C*C*C*C*C*C*C
T_{15}-PO	TTTTTTTTTTTTTTT
T_{15}-PS	T*T*T*T*T*T*T*T*T*T*T*T*T*T*T
T_{15}-PS1	TTTTTTTT*T*T*T*T*T*T*T
T_{15}-PS2	T*TT*TT*TT*TT*TT*TT*TT

* = PS modification.

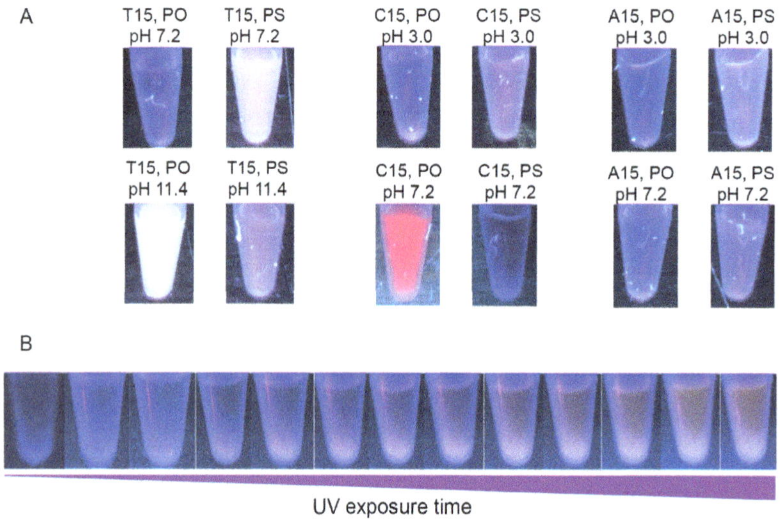

Figure 2. (**A**) Fluorescence photographs of phosphodiester (PO)- and phosphorothioate (PS)-modified 15-m DNA stabilized fluorescent AgNCs excited by a handheld UV lamp. Images were taken after exposure of samples to short wavelength UV light for approximately 5 min. Samples T_{15} pH 7.2 and 11.4, C_{15} pH 3, and A_{15} pH 3 and 7.2 were imaged under exposure to long wavelength UV light (365 nm). Samples of C_{15} pH 7.2 were imaged under short wavelength UV light (254 nm). Wavelength range of exposure did not impact the relative intensity of the two (PS and PO) samples; (**B**) Image sequence showing the increase in visible fluorescence of a T_{15}-PS AgNC sample prepared at pH 7.2 under continuous exposure to short wavelength UV light. Exposure time from left to right is as follows (mm:ss): 00:00, 00:20, 00:46, 01:20, 01:44, 01:58, 02:12, 02: 24, 02:54, 03:18, 03:56, 05:38, 06:08. Sample was prepared using 15 μM DNA, 120 μM $AgNO_3$, 120 μM $NaBH_4$, and 25 mM pH 7.2 phosphate buffer.

PS modification may have been detrimental to previously templated strong emitters for several reasons. PS modification may either stabilize the formation of distinct, non-fluorescent species, or it may simply be quenching the fluorescence of the clusters produced with unmodified templates. Thiol compounds have previously been found to quench the fluorescence of some DNA-AgNCs via electron donation from the sulfur lone-pair to a low lying d-orbital on silver [25]. In one case, using C_{12} DNA fluorescence was actually enhanced by biothiols, and changes circular dichromism spectra indicated a formation of a more compact DNA structure in addition to formation of S-Ag bonds [26]. However, the present data indicates that this effect does not occur with PS modification.

Since our aim is to find sequences where the PS DNA produced stronger fluorescence than the corresponding PO DNA. T_{15} DNA was chosen to be studied in more detail.

2.2. Different PS Compositions

For the above work, PS modification was introduced at each phosphate position. To further optimize DNA PS composition, two partially phosphorothioated T_{15} templates were evaluated to determine the effect of number and position of PS modifications on cluster fluorescence. One template (T_{15}-PS1, Table 1) possessed seven sequential PS linkages at the 3'-end of the template, while the other (T_{15}-PS2) had the same number of PS modifications but on every other linkage such that they were uniformly dispersed along the T_{15} strand. Samples templated with T_{15}-PO, T_{15}-PS1, T_{15}-PS2, and T_{15}-PS were prepared using 15 μM DNA, 120 μM $AgNO_3$ and 120 μM $NaBH_4$ in 25 mM phosphate buffer at both pH 7.2 and pH 11.4. The samples were incubated in dark for 2 h after adding $NaBH_4$ and then exposed to UV light for 10 min before measuring the spectra.

Across the four templates and two pH values, two major fluorescent bands were observed with varying intensity (Figure 3A,B). A violet emitter, λ_{EM} Max ~ 380 nm, λ_{EX} Max ~ 325 nm, was favored by neutral pH (7.2) and PS-modified templates, and a yellow-orange emitter λ_{EM} Max ~ 580 nm, λ_{EX} Max ~ 266 nm was favored by high pH (11.4) and unmodified PO-templates. Since the N3 position of thymine is deprotonated across this range (pK_a ~ 9.9), binding through this position may be involved in the formation of the yellow-orange emitter, but not the violet emitter. This may indicate that the violet emitter is produced through a pH insensitive interaction with the PS backbone, but is overcome by formation of the more favorable yellow-orange emitter when enabled via deprotonation at the N3 position.

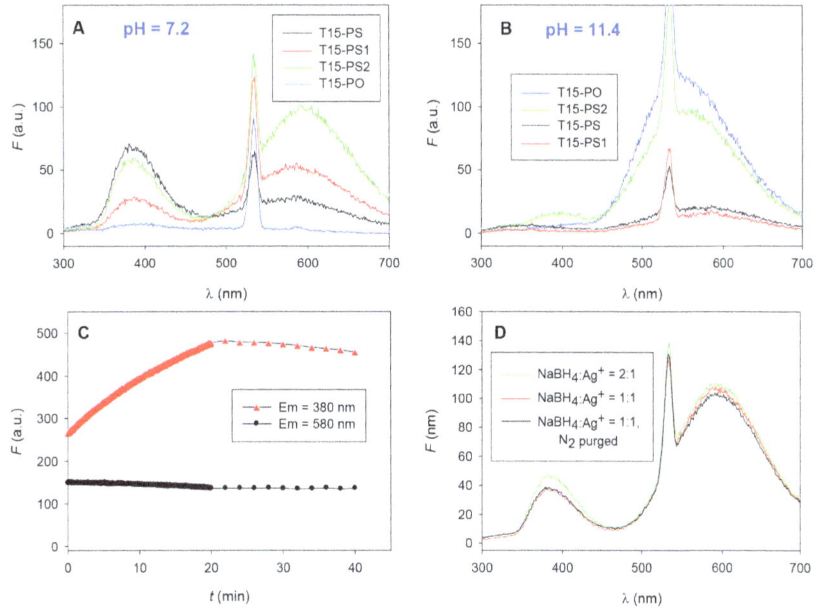

Figure 3. Emission spectra (λ_{EX} 266 nm) of fully-, partially-, and non-phosphorothioate modified T_{15} DNA template at (**A**) pH 7.2 and (**B**) pH 11.4; Samples were prepared using 15 μM DNA, 120 μM $AgNO_3$ and 120 μM $NaBH_4$ in 25 mM phosphate buffer at pH 7.2 or 11.4 and incubated in the dark for 2 hours. Spectra were measured following exposure under a handheld UV lamp at short wavelength for 10 min. Note that an excitation wavelength of 266 nm was chosen for its ability to produce emission from both bands, and favors emission from the yellow-orange band; (**C**) Emission intensities of T_{15}-PS-templated sample at 380 nm and 580 nm of samples under continuous excitation in fluorescence spectrophotometer at 266 nm using a 20 nm excitation slit and 1 s averaging time, followed by storage in the dark with a 1 s measurement taken every 2 min. UV exposure stopped after 20 min. The sample was prepared using 30 μM DNA, 240 μM $AgNO_3$ and 240 μM $NaBH_4$; (**D**) T_{15}-PS2, pH 7.2 AgNC samples prepared using different $NaBH_4$ concentrations and the effect of N_2 purging to remove oxygen. The sharp peak at 532 nm is due to the low efficiency of monochromators to block light at half frequency of the excitation wavelength.

One feature of note is that regardless of the degree or pattern of PS modification, only two distinct emitters are formed. This suggests that PS modification alters the favorability of forming a distinct species, rather than simply altering the template geometry or electronic environment surrounding the cluster which would be expected to cause a more gradual change in fluorescent properties. T_{15}-PS2 with alternating PS modifications was able to template both PO- and PS-favored species and produced the greatest overall fluorescence. As a result, it was used to investigate the conditions leading to the formation of the two fluorescent species.

2.3. UV-Light Activated Fluorescence

It needs to be noted that all the above T_{15}-PS DNA data shown above were collected after exposing the samples to UV light. When we observed under excitation by short wavelength handheld UV light, T_{15}-PS-DNA-templated samples, were initially non-fluorescent, but evolved increasing fluorescence upon continuous exposure over a period of approximately 5 min as seen in Figure 2B. To quantitatively evaluate the effect of light exposure, fluorescent intensity (λ_{EM} 380 nm and 580 nm) of each sample was measured under continuous excitation at 266 nm in the fluorescence spectrophotometer using a 20 nm slit width and 1 s averaging time (Figure 3C). The photo-induced emission was the most pronounced for the violet, 380 nm peak, templated by T_{15}-PS DNA at neutral pH. The photo induced emission effect was also evaluated for the T_{15}-PO template, but little increase was observed at high pH, and no fluorescence at all was observed at neutral pH (data not shown). This suggests that the effect is strongly favored by PS DNA, potentially due to the additional binding geometries or aggregation states enabled by this modification. After 20 min, the UV exposure was stopped and the peak also stopped increasing.

The observation that both emission bands either increase or remain constant under UV exposure suggests that the photo-induced emission does not result from a direct photo-oxidative conversion. This inter-conversion was further investigated later, by preparing samples using two times the concentration of $NaBH_4$ as well as using a nitrogen purged solution. Oxygen has previously been shown essential in the formation of certain partially oxidized species [24]. However, in this case neither condition has any significant effect on the relative distribution of the two species (Figure 3D), suggesting both species might be fully reduced.

2.4. Effect of Heat and Light on Kinetics of Cluster Formation

This is the first reported incidence of photo-induced fluorescence in a DNA-AgNC. Light has however, been found essential in the formation of luminescent gold particles as formed in the presence of adenine derivatives [27]. It was suggested in that study suggested that light may be necessary to induce a re-organization of gold atoms into a productive geometry, and it was found that heating could achieve the same goal. In order to investigate the mechanism of producing the two emission bands and to identify reproducible synthetic conditions, a few experiments were carried out. A T_{15}-PS2 template, possessing alternating PS linkages, and a pH of 7.2 was chosen for reaction conditions due to their ability to produce both emission bands with sufficient intensity. The evolution of fluorescence was measured as a function of incubation time under exposure to the following conditions: short wavelength UV light, ambient fluorescent light, dark at room temperature, and dark at 75 °C. Comparison of the conditions producing the two bands suggests unique formation mechanisms for each.

For the yellow-orange emitter (λ_{EM} 595 nm) UV light (black, Figure 4B) appeared to provide activation energy in the formation of the fluorescent species, rapidly increasing the reaction rate compared to that in the dark (red, Figure 4B), but having no effect on absolute intensity. This activation energy could also be supplied by heating at 75 °C (green, Figure 4B), but not by ambient fluorescent light (yellow, Figure 4B) which slightly reduced the emission intensity (21%) and had no appreciable effect on the reaction rate.

In the case of the violet emitter (λ_{EM} 385 nm), UV light was essential (black, Figure 4A), and essentially no fluorescence was observed under ambient light, or in the dark, with or without heat (yellow, green and red, Figure 4A). This suggests that UV irradiation is responsible for this short wavelength emission and the sulfur atom is critical for this activation. The exact mechanism of action remains to be explored. Under UV light, the violet band (black, Figure 4A) took longer (~40 min) to reach maximum intensity, compared to the yellow-orange band (black, Figure 4B) which reached an intensity maximum after only 5 min, but also exhibited greater photostability, maintaining 50% of its emission intensity approximately 230 min after reaching its peak compared to only approximately 30 min for the yellow-orange band. Although UV exposure was essential, it was not sufficient, and in the absence of $NaBH_4$, only very faint violet emission was observed, even after 5 h exposure to short wavelength short wavelength UV light (data not shown).

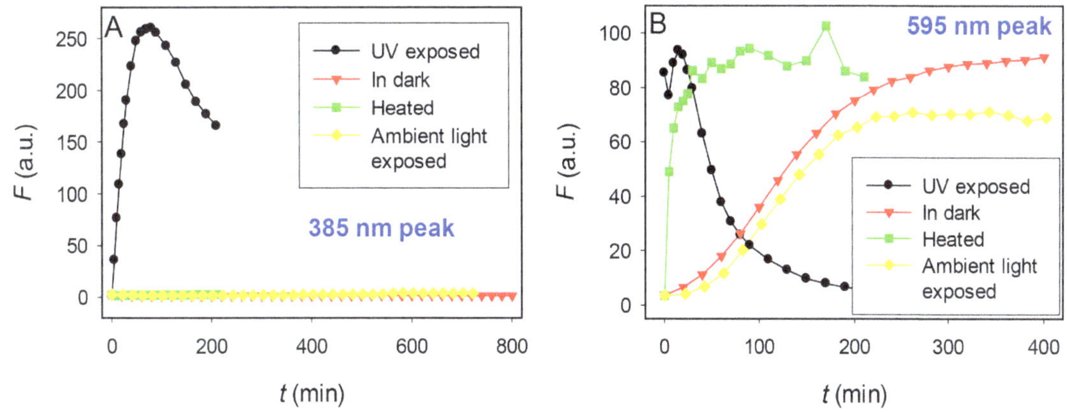

Figure 4. Evolution of T_{15}-PS2-AgNC fluorescence from the (**A**) violet band (λ_{EM} 385 nm) and (**B**) yellow-orange band (λ_{EM} 595 nm) as a function of incubation under UV exposure (**black**), in the dark at room temperature (**red**), while heated at 75 °C in the dark (**green**), and under exposure to ambient fluorescent light (**yellow**). Samples were prepared using 15 µM DNA, 120 µM $AgNO_3$ and 120 µM $NaBH_4$.

3. Experimental Section

3.1. Reagents

Oligonucleotides were purchased from Eurofins MWG Operon (Huntsville, AL, USA), or Integrated DNA Technologies (Coralville, IA, USA) and were purified via standard desalting by the manufacturer.

Citric acid monohydrate (Sigma Aldrich, St. Louis, MO, USA, 99.0%), citric acid trisodium dihydrate (Amresco, Solon, OH, USA, reagent grade), HEPES (Amresco, Solon, OH, USA, high purity grade), sodium phosphate monobasic dihydrate (Fisher, Waltham, MA, USA, 100.0%), sodium phosphate dibasic heptahydrate (Fisher, Waltham, MA, USA, 99.0%–102.0%), sodium hydroxide (Amresco, Solon, OH,

USA, ACS grade), nitric acid (Fisher, Waltham, MA, USA, ACS reagent grade), tetrachloroauric acid trihydrate (Fisher, Waltham, MA, USA, ACS Reagent Grade, ≥49% Au Basis, equivalent of 98% $HAuCl_4$), silver nitrate (Alfa Aesar, Ward Hill, MA, USA, 99.9%) and sodium borohydride (EMD, Billerica, MA, USA, 98.0%) were used as is. All solutions were prepared using ultrapure water (18.2 MΩ·cm at 25 °C, Synergy UV water purification system, Millipore Co., Billerica, MA, USA).

3.2. AgNC Synthesis

Buffer, DNA and $AgNO_3$ were dissolved in ultrapure water and vortexed. After approximately 5 min, freshly prepared $NaBH_4$ was added. The reaction mixture was then again vortexed, and incubated for approximately 2 h in the dark at room temperature. Typical reagent concentrations for AgNC synthesis were as follows: 25 mM buffer, 15 μM DNA, 120 μM $AgNO_3$, and 120 μM $NaBH_4$. Where indicated, samples were then exposed to short wavelength UV light prior to characterization.

3.3. Fluorescence Spectroscopy

Visible Fluorescence. Fluorescence was assessed visually under excitation under a hand held UV lamp at 254 nm or 365 nm in a dark room and recorded using a digital camera (Canon PowerShot SD1200 IS). Samples were also observed under ambient light. Full excitation and emission spectra were typically measured without dilution using a Varian Eclipse Fluorescence Spectrophotometer (Agilent Technologies, Santa Clara, CA, USA). Emission spectra were first measured using an excitation wavelength of 300 nm. Excitation spectra were then measured for emission maxima observed, and finally, emission spectra were again recorded for the measured excitation maxima.

4. Conclusions

In this work, we have investigated the effect of PS-modified ssDNA templates in the synthesis of fluorescent gold and silver nanoclusters. While PS modification did not bring stronger emission in the synthesis of AuNCs, PS-modified T_{15} template enabled the formation of stronger violet and yellow-orange emitters at neutral pH whereas unmodified T_{15} had previously only produced fluorescence at high pH. The insensitivity of this emitter to base protonation and the favorability of PS modifications suggests that DNA-Ag interaction may occur through the PS backbone. However, the synthesis is not completely independent of base sequence, as strong fluorescence could not be achieved using any other templates. The present work is the first time that a poly-T template has been used successfully to template fluorescent species at neutral pH, potentially extending the use of T-rich templates for physiological conditions.

Acknowledgments

Funding for this work is from NSERC of Canada.

Author Contributions

This study was conceived by Juewen Liu. Individual experiments were designed, executed, and interpreted by Daniel Weadick. The manuscript was jointly written by both authors. Both authors have read and approved the final manuscript.

Conflicts of Interest

The authors declare no conflict of interest.

References

1. Díez, I.; Ras, R.H.A. Fluorescent silver nanoclusters. *Nanoscale* **2011**, *3*, 1963–1970.
2. Liu, J. DNA-stabilized, fluorescent, metal nanoclusters for biosensor development. *Trends Anal. Chem.* **2014**, *58*, 99–111.
3. Shang, L.; Dong, S.; Nienhaus, G.U. Ultra-small fluorescent metal nanoclusters: Synthesis and biological applications. *Nano Today* **2011**, *6*, 401–418.
4. Zhang, L.; Wang, E. Metal nanoclusters: New fluorescent probes for sensors and bioimaging. *Nano Today* **2014**, *9*, 132–157.
5. Wei, H.; Wang, Z.; Yang, L.; Tian, S.; Hou, C.; Lu, Y. Lysozyme-stabilized gold fluorescent cluster: Synthesis and application as Hg^{2+} sensor. *Analyst* **2010**, *135*, 1406–1410.
6. Yuan, Z.; Peng, M.; Shi, L.; Du, Y.; Cai, N.; He, Y.; Chang, H.T.; Yeung, E.S. Disassembly mediated fluorescence recovery of gold nanodots for selective sulfide sensing. *Nanoscale* **2013**, *5*, 4683–4686.
7. Deleavey, G.F.; Damha, M.J. Designing chemically modified oligonucleotides for targeted gene silencing. *Chem. Biol.* **2012**, *19*, 937–954.
8. Sekhon, G.S.; Sen, D.A stereochemical glimpse of the active site of the 8–17 deoxyribozyme from iodine-mediated cross-links formed with the substrate's scissile site. *Biochemistry* **2010**, *49*, 9072–9077.
9. Wang, S.; Karbstein, K.; Peracchi, A.; Beigelman, L.; Herschlag, D. Identification of the hammerhead ribozyme metal ion binding site responsible for rescue of the deleterious effect of a cleavage site phosphorothioate. *Biochemistry* **1999**, *38*, 14363–14378.
10. Tan, L.H.; Xing, H.; Lu, Y. DNA as a powerful tool for morphology control, spatial positioning, and dynamic assembly of nanoparticles. *Acc. Chem. Res.* **2014**, *47*, 1881–1890.
11. Ma, N.; Sargent, E.H.; Kelley, S.O. One-step DNA-programmed growth of luminescent and biofunctionalized nanocrystals. *Nat. Nanotechnol.* **2009**, *4*, 121–125.
12. Huang, P.J.J.; Liu, J. Sensing parts-per-trillion Cd^{2+}, Hg^{2+}, and Pb^{2+} collectively and individually using phosphorothioate DNAzymes. *Anal. Chem.* **2014**, *86*, 5999–6005.
13. Zhang, D.; Deng, M.; Xu, L.; Zhou, Y.; Yuwen, J.; Zhou, X. The sensitive and selective optical detection of mercury(II) ions by using a phosphorothioate DNAzyme strategy. *Chem. Eur. J.* **2009**, *15*, 8117–8120.

14. Lee, J.H.; Wernette, D.P.; Yigit, M.V.; Liu, J.; Wang, Z.; Lu, Y. Site-specific control of distances between gold nanoparticles using phosphorothioate anchors on DNA and a short bifunctional molecular fastener. *Angew. Chem. Int. Ed. Engl.* **2007**, *46*, 9006–9010.
15. Link, S.; Beeby, A.; FitzGerald, S.; el-Sayed, M.A.; Schaaff, T.G.; Whetten, R.L. Visible to infrared luminescence from a 28-atom gold cluster. *J. Phys. Chem. B* **2002**, *106*, 3410–3415.
16. Han, B.; Wang, E. DNA-templated fluorescent silver nanoclusters. *Anal. Bioanal. Chem.* **2012**, *402*, 129–138.
17. Xie, X.; Liang, J.; Pu, T.; Xu, F.; Yao, F.; Yang, Y.; Zhao, Y.L.; You, D.; Zhou, X.; Deng, Z.; *et.al*. Phosphorothioate DNA as an antioxidant in bacteria. *Nucleic Acids Res.* **2012**, *40*, 9115–9124.
18. Guga, P.; Koziołkiewicz, M. Phosphorothioate nucleotides and oligonucleotides—Recent progress in synthesis and application. *Chem. Biodivers.* **2011**, *8*, 1642–1681.
19. Cruse, W.B.; Salisbury, S.A.; Brown, T.; Cosstick, R.; Eckstein, F.; Kennard, O. Chiral phosphorothioate analogues of B-DNA. The crystal structure of Rp-d[Gp(S)CpGp(S)CpGp(S)C]. *J. Mol. Biol.* **1986**, *192*, 891–905.
20. Smith, J.S.; Nikonowicz, E.P. Phosphorothioate substitution can substantially alter RNA conformation. *Biochemistry* **2000**, *39*, 5642–5652.
21. Moody, E.M.; Brown, T.S.; Bevilacqua, P.C. Simple method for determining nucleobase pK_a values by indirect labeling and demonstration of a pK_a of neutrality in dsDNA. *J. Am. Chem. Soc.* **2004**, *126*, 10200–10201.
22. Liu, J. Adsorption of DNA onto gold nanoparticles and graphene oxide: Surface science and applications. *Phys. Chem. Chem. Phys.* **2012**, *14*, 10485–10496.
23. Ritchie, C.M.; Johnsen, K.R.; Kiser, J.R.; Antoku, Y.; Dickson, R.M.; Petty, J.T. Ag nanocluster formation using a cytosine oligonucleotide template. *J. Phys. Chem. C* **2007**, *111*, 175–181.
24. Sengupta, B.; Ritchie, C.M.; Buckman, J.G.; Johnsen, K.R.; Goodwin, P.M.; Petty, J.T. Base-directed formation of fluorescent silver clusters. *J. Phys. Chem. C* **2008**, *112*, 18776–18782.
25. Han, B.; Wang, E. Oligonucleotide-stabilized fluorescent silver nanoclusters for sensitive detection of biothiols in biological fluids. *Biosens. Bioelectron.* **2011**, *26*, 2585–2589.
26. Huang, Z.; Pu, F.; Lin, Y.; Ren, J.; Qu, X. Modulating DNA-templated silver nanoclusters for fluorescence turn-on detection of thiol compounds. *Chem. Commun.* **2011**, *47*, 3487–3489.
27. Lopez, A.; Liu, J. Light-activated metal-coordinated supramolecular complexes with charge-directed self-assembly. *J. Phys. Chem. C* **2013**, *117*, 3653–3661.

13

Penetration and Toxicity of Nanomaterials in Higher Plants

Giuseppe Chichiriccò * and Anna Poma

Department of Life, Health and Environmental Sciences, University of L'Aquila, Via Vetoio, I-67010 Coppito, L'Aquila, Italy; E-Mail: annamariagiuseppina.poma@univaq.it

* Author to whom correspondence should be addressed; E-Mail: chichiri@univaq.it

Academic Editor: Robert Tanguay

Abstract: Nanomaterials (NMs) comprise either inorganic particles consisting of metals, oxides, and salts that exist in nature and may be also produced in the laboratory, or organic particles originating only from the laboratory, having at least one dimension between 1 and 100 nm in size. According to shape, size, surface area, and charge, NMs have different mechanical, chemical, electrical, and optical properties that make them suitable for technological and biomedical applications and thus they are being increasingly produced and modified. Despite their beneficial potential, their use may be hazardous to health owing to the capacity to enter the animal and plant body and interact with cells. Studies on NMs involve technologists, biologists, physicists, chemists, and ecologists, so there are numerous reports that are significantly raising the level of knowledge, especially in the field of nanotechnology; however, many aspects concerning nanobiology remain undiscovered, including the interactions with plant biomolecules. In this review we examine current knowledge on the ways in which NMs penetrate plant organs and interact with cells, with the aim of shedding light on the reactivity of NMs and toxicity to plants. These points are discussed critically to adjust the balance with regard to the risk to the health of the plants as well as providing some suggestions for new studies on this topic.

Keywords: carbon nanomaterials (CNMs); carbon nanotubes (CNTs); cytotoxicity; multi-walled carbon nanotubes (MWCNTs); nanomaterials (NMs); nanoparticles (NPs); single-walled carbon nanotubes (SWCNTs)

---

1. Introduction

Plants play a key role in the flow of energy from the sun to other organisms and are thus fundamental for any terrestrial ecosystem. Their cellular lines from gamy (zygote) and from meiosis (spores) develop into diploid sporophytic and aploid gametophytic generations, respectively. Sporophytes have perpetually young tissues at their ends (meristems) to grow organs indefinitely and to develop predominant surfaces to absorb water from the soil and sunlight, and carbon dioxide from the atmosphere. In anthropized environments these surfaces are exposed to pollutants on a daily basis, the aerial ones providing a huge landing field for airborne particles and the subterranean ones a kind of drain for pollutants absorbed by the soil. Among pollutant particles [1], there are nanomaterials (NMs) consisting of nanoparticles (NPs) and carbon-based nanomaterials (CNMs). NPs are particles 1–100 nm in size. CNMs can vary in length from tens of nanometers to a few centimeters and consist of: (1) single layered cylinders, named single-walled carbon nanotubes (SWCNTs), usually not exceeding 2 nm in diameter; (2) multiple concentric layers of cylinders not exceeding 80 nm in diameter and named multi-walled carbon nanotubes (MWCNTs); (3) fullerenes, which are hollow spheres less than 1 nm; and (4) graphene, which is a carbon allotrope made of a single layer of atoms arranged in repeating structures of hexagon shape as honeycomb. Functionalized NMs can be obtained by linking a number of chemical functional groups to their surface [2]. NPs may arise from natural (volcanic eruption), accidental (plants combustion), and planned (engine cars) activities, and both NPs and CNMs are produced in the laboratory by man for their potential use in the fields of agriculture, chemistry, biology, medicine, and energy [3,4]. Nanoparticles (nanogold positively charged) have even been used as useful probes to gain insight into the mechanisms of the plasma membrane internalization (endocytosis) in plant cells [5]. In addition, there is growing interest in the use of surface-functionalized NMs as absorbers of heavy metals from polluted environments [6]. A number of studies draw particular attention to the vulnerability of animals and plants to NMs [7–9], but the penetration and translocation in plants and the toxicity mechanisms are still poorly understood. Plant organs are typically protected by coatings and cell walls, which, notwithstanding their sealing capability, may be overcome by NMs. Once these penetrate into the plant, harmful effects may be produced, as well as effects advantageous for man. Here we give an updated survey on the cellular pathways for uptake and translocation of NMs and their effects in higher plants.

2. Vegetative System

The epidermis foliar (Figure 1) is made up of cells with the exposed walls waterproofed by hydroxylated fatty acids (cutin and waxes), which form a cuticle membrane. Epidermis is provided with stomata ranging from 100 to 1300 per mm^2, consisting of two guard cells, which, through expansion (turgor), form a pore between them ranging from 3 to 12 μm in width and 10–30 μm in length, for gas exchange. Eichert and Goldbach [10], according to the polar pore model [11], estimated 2–2.4 nm as the exclusion limit of the pore radius for polar and ionic solutes to penetrate the cuticle, while for diffusion via the stomatal surface the pore radius was quite variable and always exceeded 20 nm. The amplitude of the stomatal pores practically usable for the passage is not measurable; it is controlled by a number of factors such as the water layers and bacteria layers, among others.

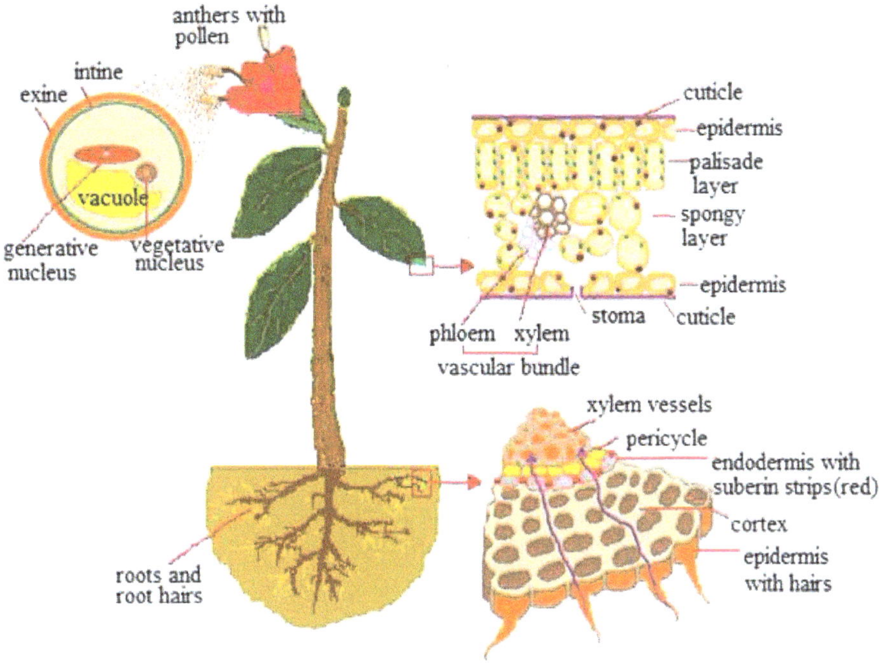

Figure 1. Pathways by which nanomaterials (NMs) are absorbed: exine apertures, cuticle, stomata, and root hairs.

The pathway for penetrating the cell is through the openings of the primary cell wall, consisting of a polysaccharidic-proteic structure—which, depending on the pectic component, is more or less porous [12]; the pore sizes range from 3.5 to 20 nm [13–15] and more often are around 5 nm. The progression to protoplasm (symplastic way) is made possible by cell membrane-embedded protein carriers and ionic channels, or by membrane invagination (endocytosis), which forms vesicles around the passenger. The transport cell to cell is made dynamic by cytoplasmic channels, the plasmodesmata, which are 20–50 nm in diameter at the midpoint and usually let inside small particles, around 3 nm [1], but the exclusion limits are subject to variations as endogenous proteins mediate crossing. Below the foliar epidermis, the photosynthetic palisade tissue (Figure 1) provides small intercellular spaces which, together with the cell walls, form the apoplastic pathway. Through the symplastic and apoplastic pathways, small particles can reach the photosynthate conducting system (phloem vessels), made up of living sieve-cells devoid of a nucleus and most organelles, while being connected to one another and to the surrounding tissues by wall sieve pores of 0.2–0.4 μm.

Underground, the roots are permeable only near the tips, where the epidermal cells extend out into hair roots (Figure 1); the remaining tracts (exodermis) are waterproofed by the suberin. Within the root the cortical apoplastic way is obstructed around the vascular system by a cell layer with radial walls impregnated by suberin (endodermis) (Figure 1), so the path to the vessels is through the tangential walls of the endodermis (symplastic way). However, the lateral roots enable an apoplastic path from the emergence zone up to the vessels [1]. Vessels consist of emptied cellular elements longitudinally connected to one another and to the surrounding cells with pits wider than 1 μm [16–18], so it is possible for small particles absorbed with the water to move passively within them.

The stem and its branches are wrapped by a tissue of suberized cells that replaces the epidermis; it is tiered by the radial growth of the stem. The annual and biennial plantlets do not grow radially, so their stems remain enveloped by epidermis provided with stomata.

3. Uptake and Toxicity of NMs

Many plants have been tested for their vulnerability to NMs, although only a few of them have been studied extensively with regard to their uptake and pathways for diffusion and internalization. For an exhaustive record, see the Tables in [8].

3.1. Leaf

When considering pollution, the first thing that comes to mind is the landing of particles on the leaves followed by penetration and/or obstruction of the stomatal pores, neglecting the potentiality of the particles to traverse the epidermal cuticle. To evaluate this, it must be kept in mind that in horizontally oriented leaves the stomata are usually in the epidermis of the dorsal surface, where airborne particles can scarcely be accumulated.

In *Vicia faba*, Eichert *et al.* [19] observed with confocal microscopy that polymeric NPs (suspensions of modified polystyrene particles) 43 nm in diameter penetrated the stomatal leaf pores, even if only sporadically and solely through a part of the total stomata, while particles of 1.1 μm did not ever penetrate. These observations, which are in agreement with previous studies cited above [10], made it possible to consider a size of 43 nm as the size exclusion limit predictable for the stomata penetration by nanoparticles. Birbaum *et al.* [20] used mass spectrometry and electronic and confocal microscopy and found that CeO_2–NPs with an average size of 37 nm when administered as spray (total exposure 0.4 g NPs) or in solution (10 ppm NPs) were retained by the leaves of *Zea mays* (50 μg of cerium per gram of leaf) with no sign of translocation to the stem, but the contribution of the stomata and the possible adsorption or incorporation of NPs was not ascertained. Kurepa *et al.* [21] treated seedlings of *Arabidopsis thaliana* grown on agar medium with TiO_2–nanoconjugates (2.8 ± 1.4 nm), and by using electron and X-ray fluorescence microscopy reported a penetrating ability of NPs into the epidermis and underlying palisade tissue, which suggested a contribution of the stomata and endocytotic vesicles in the absorption. Further studies by means of mass spectroscopy and electron microscopy analysis provided evidence of the foliar uptake following aerial treatments. Watermelon plants grown in pots and having large stomata and vessels were used by Wang *et al.* [22] for spraying with NPs (Fe_2O_3, TiO_2, MgO, ZnO); these initially were 27.3–46.7 nm in diameter and increased considerably in the suspension but significantly reduced during the spray treatment. Particles that did not exceed the diameter of 100 nm penetrated the foliar stomata and were translocated from leaves to stems and roots through the sieve elements. Taran *et al.* [23] used non-ionic colloidal solutions of NPs (Fe, Zn and Mn) in winter wheat to test their concentration in seedlings either arising from pretreated seeds, or sprayed with NPs after growth. They provided evidence for the absorption of Mn and Zn from the foliar epidermis and for translocation of NPs in seedlings that were pretreated. Cucumber plants hydroponically cultured were aerially treated by Hong *et al.* [24] with CeO_2–NPs of 8 nm ± 1nm (primary diameter) and of 231 nm ± 16 nm (hydrodynamic diameter) either in the form of powder or in suspension. Nanoparticles in both forms penetrated the foliar epidermis, but only the powder treatments succeeded in translocating to stems and roots. Larue *et al.* [25] sprayed the leaves of *Lactuga* with the salt $AgNO_3$, and with Ag–NPs, which were both round (38.6 nm in diameter) and non-round (38.2 nm × 57.8 nm) and had hydrodynamic diameters of 47.9 nm ± 29.2 nm. They provided detailed evidence for the cuticolar and stomatal uptake of NPs and translocation up into the vascular tissue through pathways, which seemed to be both apoplastic and symplastic. Moreover, they suggested transformation cycles within the plant involving the binding of Ag^+ ions to thiol groups and the conversion of Ag^+ ions in Ag–NPs, starting from dissolution of both the salt $AgNO_3$ and Ag–NPs. Larue *et al.* [26] had previously described leaf penetration by TiO_2–NPs in wheat and rapeseed.

Carbon-based nanomaterials (CNMs) have potential properties for penetration through leaves and translocation to the roots [27]; their foliar uptake is, however, not well documented. Studies on leaf cells are prevalently based on *in vitro* cultures of cells enzymatically treated for removing walls (protoplasts) before culturing, and describe membrane penetration via endocytosis, as reported by Shen *et al.* [28] in *Oryza* and *Arabidopsis* (see also in the following section). Such a penetration pathway was first reported by Liu *et al.* [29] in suspension cultures of intact tobacco cells line BY-2 (from seedling callus) treated with single-walled carbon nanotubes (SWCNTs, length <500 nm) bound to FITC dye for observations with confocal microscopy. Recently Gilardo *et al.* [30], by applying advanced methods to the leaves and to excised chloroplasts of spinach, proved the ability of SWCNTs to cross the stomata foliar as well as to penetrate chloroplasts and accumulate on thylakoids and stroma.

3.1.1. Toxicity of Metal- and Metal Oxide-NPs

The vulnerability of the foliar epidermis allowed for the estimation of the NPs' effects on the photosynthetic activity of chloroplasts. In *Spinacia oleracea*, Hong *et al.* [31] observed that treatment with TiO_2–NPs activated a photochemical reaction, and subsequent studies [8] suggested a mechanism based on the penetration of the TiO_2–NPs into the chloroplast, followed by binding to the photosystem II and activation of the basic reaction—that is, the charge separation. On the contrary, TiO_2–NPs (2.8 nm ± 1.4 nm) produced microtubule disorganization in epidermal and stomatal cells of wheat seedling leaves [32]. The influence of NPs in the magnetic form was studied by Rãcuciu and Creangã [33] by growing plantlets of *Zea mays* arising from seeds in a culture medium supplemented with magnetite (Fe_3O_4–NPs) particles of 8 nm in size, as a ferrofluid suspension. The plantlet growth and the levels of the chlorophylls a, chlorophylls b and carotenoids were stimulated by 10–50 μL/L of Fe_3O_4–NPs and inhibited by higher concentrations, while the chlorophylls a/b ratio and the photosynthesis process were decreased either by low or high concentrations of NPs. These results highlighted the potentiality of Fe_3O_4–NPs to affect the photosynthetic machinery, suggesting a role of the iron oxides (from magnetite) in the photosynthesis malfunction, and probable interference with the plasma membrane ion channels. Some years later [34] Wang *et al.* observed that Fe_3O_4–NPs induced oxidative stress in the shoots and roots of *Lolium perenne* and *Cucurbita mixta*, and to a greater extent than in treatment with Fe_3O_4 bulk particles, despite the fact that X-ray absorption spectroscopic analyses excluded internalization of the Fe_3O_4–NPs. Parsons *et al.* [35], in leaves of hydroponic seedlings of mesquite treated with 0.10 g $Ni(OH)_2$–NPs, found 400–803 mg/kg dry weight of Ni, with no effect on chlorophyll production.

3.1.2. Toxicity of CNMs

Carbon-based nanomaterials (CNMs), owing to their numerous potential applications, have been increasingly used for phytotoxic tests since their ability to enter the cell wall and plasma membrane was ascertained [29]. Studies on leaf cell cultures in *Arabidopsis* [36] and *Oryza sativa* [37] showed that treatments with multi-walled carbon nanotubes (MWCNTs) produced in the first plant a decrease of superoxidedismutase (SOD) activity associated with a decay of chlorophyll production, and in the second plant an increase of reactive oxygen species (ROS) and apoptotic processes. It is notable that symptoms of oxidative stress (ROS accumulation) and a dose-dependent programmed cell death were found in the above plant species by Shen *et al.* [28] after treating both protoplasts and integral leaves with SWNTs; these authors also provided some evidence in favor of the internalization of nanotubes through endocytosis-like processes. Recently, Giraldo *et al.* [30] found that SWCNTs, once having penetrated the membranes of spinach chloroplasts, increased the flow of electrons and photosynthetic activity, most likely by a stimulating action on the uptake of light with wavelengths of the near-infrared. Moreover, SWCNTs were shown to be sensitive to nitric oxides (NOx), suggesting that plants enclosing nanotubes could be used as detectors for NOx. Otherwise, Santos *et al.* [38] showed a toxic action of Fullerene C60 on the aquatic plant *Lemna gibba*, which manifested itself in a decrease of photosynthetic activity and plant growth.

The leaves were also shown to be sensitive to graphene oxides (GO). Cabbage, spinach, and tomato leaves decreased in size when the seedlings were treated with GO at 500–2000 mg/L, and the leaves of tomato and cabbage also decreased in number at concentrations of 1000–2000 mg/L [39].

3.2. Roots

Most data come from experiments on young plants developing from seeds, so the seed coat absorbability and the effects of NMs on germination were also proved. The studies on wheat, maize, spinach, zucchini, rapeseed, and some desert plants [23,40–45] showed the ability of metal-NPs to penetrate seeds without affecting germination, and some [23] reported the NPs' distribution in the corresponding seedlings. However, in *Arabidopsis thaliana* the TiO_2–nanoconjugates smaller than 5 nm remained stuck to the seed mucilage and failed to penetrate [21].

Many plant species are known to absorb NPs by the roots and translocate them in stems and leaves [46], depending on the physicochemical features of NPs, the type of plant, and the growth medium. Experiments in hydroponic maize cultures [15] pointed out that colloidal suspensions of TiO_2–NPs (1 g/L with sizes too large (30 nm) for the rhyzoderm cell wall pores (6.6 nm) prejudiced the hydraulic conductivity of the primary roots by a physical action on the wall pores that affected the transpiration and growth of the leaves. The same plants grown in pots and irrigated with TiO_2–NPs at 1 g/L concentration suffered much less from the treatment. Magnetite nanoparticles of 20 nm (Fe_3O_4–NPs) were able to penetrate roots and translocate to leaves of pumpkin plants grown in an aqueous medium [47]. In agar-medium cultures of *Arabidopsis thaliana* seedlings the TiO_2–nanoconjugates of 2.8 ± 1.4 nm in diameter succeeded in root cell penetration up to inside vacuoles and the nucleus [21]. Likewise, $Ni(OH)_2$–NPs, both uncoated (8.7 nm) and coated (2.5 nm before and 0.9 nm after synthesis), penetrated roots in hydroponic seedlings of mesquite and were transformed into an Ni(II)–organic acid complex, which moved to the leaves [35]. Experiments of Zhang *et al.* [48] in cucumber evidenced root absorption of CeO_2–NPs with hydrodynamic diameters of 40.2 nm ± 7.2 nm in deionized water, and 691.7 nm ± 26 nm in the nutrient solution.

Some data are available on treatments in soil. Du *et al.* [49] treated wheat plants with TiO_2–NPs and ZnO–NPs (20–100 nm in diameter) under field conditions using outdoor lysimeters. They observed with TEM that TiO_2–NPs either agglomerated and adhered to the cell walls of the root periderm, or particles of 50 ± 10 nm, probably derived from those of 20 ± 5, penetrated the primary epidermis roots and the cortex via apoplast up into cell vacuoles, whereas ZnO–NPs seemed to dissolve and then penetrate the cells in the form of Zn^{2+} ions. In *Zea mays* [20], no evidence was found for root absorption from soil irrigated with solutions of CeO_2–NPs having a 37 nm average size. Other experiments using ICP-OES Spectroscopy, X-ray fluorescence, and confocal microscopy reported that the CeO_2–NPs with primary diameter 8 nm ± 1 nm and hydrodynamic diameter 1373 nm ± 32 nm, when added to the soil where corn plants were growing, penetrated the roots and translocated in the shoots along a path that seemed to be apoplastic [50,51]. The studies also pointed out that organic substances in the soil influenced the mobility of NPs, and alginates favored nanoparticle accumulation in the roots and translocation to shoots.

For data on the root uptake and sizes of NPs, see the next section.

Carbon nanotubes (CNTs) may either accumulate on the ryzoderm surface or penetrate its cell walls and, if they are compatible with the size, they may be translocated in aerial plant tissues or get trapped during translocation. Accumulation of MWCNTs and SWCNTs was observed on the root surface of several crop plants including tomato, wheat, cabbage, lettuce, carrot, cucumber, onion, and rice [52–55]. Canas *et al.* [52] observed in several plants that CNTs did not penetrate the roots and produced effects on the root growth—adverse in tomato and advantageous in onion and cucumber—whose mechanisms of action were not fully clarified. At variance was the reported ability of CNTs to produce pores on the tomato seed coat that favored water uptake as well as seed germination and growth of seedlings [56]. Assuming CNTs are able to induce new water entry points on the seed coat, one would expect a similar capacity on the root rhyzoderm instead of a suppression of water uptake. This point, repeatedly debated by several authors, is still waiting to be cleared up.

Evidence for the absorption of carbon-based nanomaterials (CNMs) and translocation to aerial parts, including fruits, were provided by Smirnova *et al.* [57] in *Onobrychis arenaria* seedlings with electron microscopy (TEM), and by Khodakovskaya *et al.* [53] in tomato seedlings with TEM, Raman photothermal, and photoacoustic methods. The first authors used cylindrical MWCNTs 2 μm long with an external diameter of 20–70 nm. The second used: (1) MWCNTs with diameter 10–35 nm and length of 6 μm; (2) SWCNTs with diameter 0.86–2.22 nm and length of a few microns; and (3) graphene with a structure of 2–5 nm in thickness and a diameter of 100–120 nm.

Magnetic carbon-coated nanoparticles ranging from 5 to 50 nm with an average hydrodynamic diameter of 200 nm were used to treat plantlets of sunflower, tomato, pea, and wheat [58]. They were absorbed by the roots and then translocated to the xylem vessels up to the leaves (in wheat up to the trichomes), with a different efficiency depending on the species.

Lin *et al.* [55] after treating rice seeds and plants with CNMs suspended in natural organic matter, found a negligible uptake of MWCNTs, while fullerene C_{70} was found in seeds, seedlings, and the adult plants. Initially the fullerene was present more in the seeds and seedling roots than in the stem and leaves, while in adult plants it was only present neighboring the vascular system and leaves, suggesting uptake and substantial translocation within the plant. Recently, studies on bitter melon [59] substantiated, by Fourier Transform Infrared Spectroscopy (FTIR) analysis, the capacity of fullerol, a fullerene derivative, to penetrate roots and translocate to all parts of the plant including flowers and fruits. The smallest diameters of fullerol were 1.5–5.0 nm; they increased notably by increasing the concentration.

The mechanisms for CNMs' internalization into the plant are poorly understood, but some knowledge comes from the following studies. In *Catharanthus* plants, Liu *et al.* [29] found that SWCNTs labeled with fluorescein isothiocyanate were able to cross cell walls and the cell membrane through endocytosis, and then penetrate into the vacuole. For this ability, nanotubes have been considered as potential carriers of chemical substances through the cell and its substructures. Later, studies [28] in deprived wall cells (protoplasts) of rice and *Arabidopsis* described the formation of endocytosis-like structures after treatment with SWCNTs. More recent studies in *Nicotiana* and *Catharanthus* [60–63], in addition to confirming the endocytotic method of the nanotubes, reported their ability to penetrate the nucleus, plastids, and vacuoles, and to induce organelle recycling. In *Catharanthus* cell cultures treated with SWCNTs of 4.5 nm × 0.15–1.5 nm. Serag *et al.* [63] by ICP Scanning Raster Image Correlation Spectroscopy, found cytoplasmic material and SWNTs were both incorporated in the vacuole, which was interpreted as a mechanism of autophagy induced by damage to the cytoplasm.

Graphene oxides (GO) may be absorbed by roots. Studies on seeds and seedlings of *Arabidopsis thaliana* treated with GO of sizes 40–60 nm before and of 192 nm ± 24 nm after inclusion in the nutritive fluid, and a thickness of about 1.0 nm, were recently reported [64]. The authors confirmed the absorption of graphene by the roots and pointed out its accumulation in the air roots, rhyzoderma, and parechima root cells, with no evident signs of translocation except to the cotyledon cells.

3.2.1. Toxicity of Metal- and Metal Oxide-NPs

NPs can explicate a cytotoxic action on roots either after penetrating its cells and their organelles, or by inducing chemical changes in the surrounding medium or, more simply, by hindering the absorbing function of the roots.

Yang and Watts [65] treated plants of *Zea mays*, *Cucumis sativus*, *Brassica oleracea*, *Glycine max*, and *Daucus carota* with Al_2O_3–NPs and observed reduced root elongation due to toxic action related to the surface characteristics of the particles. Root degrowth had been previously reported in other plant species treated with low concentrations (2 mg/L) of Al_2O_3 in the form of phenanthrene-coated NPs [66]. Lin and Xing [67] treated hydroponic cultures of *Lolium perenne* with both Zn^{2+} ions and ZnO–NPs and found that ZnO–NPs were able to cross root cell walls, penetrate the cells, and reach vascular tissue via the endodermis, but the translocation to shoots was limited and much lower than that of Zn^{2+}. The uptake damaged the epidermal and cortical cells and decreased the plant biomass, with major effects caused by the Zn ions with respect to ZnO–NPs treatment. The same authors previously reported in other plant species a toxic effect of ZnO–NPs on root growth by using high concentrations (2000 mg/L) [68]. In hydroponic zucchini cultures, Stampoulis *et al.* [42], by using metal-oxide-NPs solutions and their corresponding bulk materials, found that Ag (size 100 nm) and Cu (size 50 nm) were toxic only in the form of nanoparticles, significantly reducing the biomass and plant transpiration as compared with their corresponding controls and bulk materials. The toxic effect as determined for Ag at concentrations of 100–1000 mg/L was dose dependent; it was in part related to the release of Ag ions from NPs and in part to a direct interaction of NPs with cells. The level of Ag within shoot cells, assayed with Inductively Coupled Plasma–Mass Spectrometry (ICP-MS), was much higher after Ag–NPs exposure than after bulk exposure.

Some experiments were carried out using soil or sand as medium. Du *et al.* [49] added TiO_2–NPs and ZnO–NPs (predominant sizes 20 nm ± 5 nm and 40 nm ± 10 nm respectively) into the soil where wheat plants grew, and observed with TEM that both NPs decreased the activity of proteases, catalase, and peroxidase enzymes in the soil, but only the small TiO_2–NPs (20 nm) penetrated the rhyzoderm of the primary roots and cortex cells up into vacuoles. The ZnO–NPs were not found in the roots, probably due to the dissolution and release of Zn^+ ions that had penetrated the root. All these observations were correlated with the loss of plant biomass. Dimkpa *et al.* [69] used ZnO–NPs and CuO–NPs in sand-grown wheat plants and observed accumulation in the shoots of CuO, Cu-sulfur complex, and Zn-phosphate, which induced oxidative stress (ROS increase) with inhibition of root growth. Transformations of ZnO–NPs had been previously reported in other species including some desert plants [43], where seed treatment with ZnO–NPs (8 nm) did not affect germination but the seedling roots showed a trend of reduced length and transformed the NPs into Zn-nitrate, Zn-phosphate, and Zn-citrate. Recently Kouhi *et al.* [45] evaluated in rapeseed seedlings the toxicity of: (1) ZnO–NPs <50 nm; (2) ZnO–microparticles (MPs); and (3) Zn^{2+}, and found that Zn^{2+} inhibited root elongation more than MPs and the MPs more than NPs. However, at high concentrations (250, 500 mg/L the three different treatments had the same toxic effects. The authors hypothesized that the dissolution of the different forms of Zn in contact with the roots played a key role in the toxicity. Further studies on hydroponic maize plants with ZnO–NPs and Zn^{2+} ions [70] suggested that both the Zn^{2+} added in the medium and the Zn^{2+} ions deriving from dissolution of NPs, as well as the NPs that escaped dissolution, entered the roots and were precipitated and entrapped in the form of phosphate, so only a few of the NPs reached the vascular tissue.

Well known for its trend in the ionic form to be biotransformed into nanoparticles is silver [25,71,72]; its biosynthesis finds application in various fields [73]. Treatment with Ag–NPs in the grass *Lolium multiflorum* [74] produced silver accumulation in the roots and in the shoots. This was higher than in treatment with the salt $AgNO_3$, which damaged the epidermal and cortical root cells and inhibited the seedlings' growth with a greater effect when using NPs of 6 nm as compared with those of higher sizes. The treatment with $AgNO_3$ or supernatants of the ultracentrifuged Ag–NPs solutions produced no damage. The inverse relationship between the toxicity of Ag–NPs and their size was subsequently observed in *Vicia faba* seedlings [75]. However, in *Zea mays* and *Brassica oleracea* var. *capitata* [44] the treatment with Ag–NPs and ZnO–NPs (hydrodynamic diameters 11 nm ± 0.7 nm) affected the seed germination, root meristem, and root development with toxic effects smaller than those produced by the corresponding ion salts. The above different responses suggested that: (1) Ag–NPs made Ag ions available, especially when in colloidal form; (2) Ag–NPs and Ag ions had a differential absorption by the cell; and (3) Ag–NPs and the corresponding ions operated with different mechanisms of action.

Biogenic formation of metal–NPs was also reported for gold in alfalfa [76] and poplar [77] plants, and for platinum in mustard [78]. In poplar the gold in the form of Au^{3+} ions was absorbed by the roots and transformed into Au–NPs, and in the form of Au–NPs was absorbed without dissolving in Au^{3+} ions. From the roots, Au–NPs moved inside the vascular system to leaves and in the cells of both organs they were in the plasmodesmata, in the cytoplasm, and inside organelles. As a result, the cells were damaged in contrast to what had previously been observed in *Brassica juncea* seedlings [79], where the uptake of Au–NPs following the treatment spray promoted growth and seed yield.

Some metal–NPs considered to be stable, such as CeO_2–NPs, make plants susceptible to their accumulation, with different effects. In tomatoes Wang *et al.* [80] reported that relatively low concentration (0.1–1 mg/L of CeO_2–NPs accumulated in the stem, leaves, and fruits without negative effects, and no toxic effects were observed in wheat and pumpkin by either Schwabe *et al.* [81], or Zhao *et al.* [50], despite the fact that nanoparticles were successfully absorbed by the roots. Conversely, Rico *et al.* [82] observed that rice grains from plants grown in soil treated with CeO_2–NPs reduced their nutritional value as regards minerals, fatty acids, proteins, starch, and antioxidants, some of which are a possible consequence of altered gene expression. Moreover, Zhang *et al.* [48], in hydroponic cultures of cucumber and soybeans treated with CeO_2–NPs, reported absorption and transformation of cerium. This partly accumulated as $CePO_4$ precipitates in the roots because of phosphate availability, and partly translocated in the stem and leaves where they became Ce-carboxylases. Subsequently, Zhang *et al.* [83] found that in *Lactuga* treatment with CeO_2–NPs, especially when using the smaller NPs (7 nm), increased the levels of ROS, the peroxidation products (MDA), and the antioxidants (SOD, POD), and these were associated with cell death and inhibition of root growth. The toxic effects were attributed to the transformation of Ce from 4+ to 3+ with Ce^{3+} ions release.

Other NPs were reported to change the formula of the core after their absorption, such as La_2O_3 to $LaPO_4$ in *Cucumis sativus*, $Ni(OH)_2$ to Ni^{2+} in *Prosopis* sp. and to Ni(II)-organic acid complex in mesquite, but without affecting the plants [8,35]. However, the dissolution of La_2O_3–NPs and Yb_2O_3–NPs by secretion of the plant roots made the treatments toxic [83,84].

With regard to the interactions of metal–NPs with root cell substructures, few reports are available. In *Arabidopsis* the accumulation of TiO_2–nanoconjugates into subcellular compartments did not cause evident damage, also suggesting their application as carriers of short oligonucleotides or peptides into the nucleus [21]. Conversely, in the same plant species treatment with TiO_2–NPs produced disorganization of the microtubules and isotropic growth of epidermal cell roots, which were ascribed to either the physic interaction of NPs with tubulin or the indirect effect through increasing ROS [32]. Oxidative stress induced by TiO_2–NPs had also been previously observed in the root of *Allium* [85].

Fluorescent NMs associated with quantum dots (QDs), largely used in animals, were applied in a number of plants [86], and more recently Navarro *et al.* [87] reported for CdSe/ZnS–QDs the induction of oxidative stress in *Arabidopsis* plants.

3.2.2. Genotoxicity of NPs

Tissues mitotically active (root meristem) of several plant species were analyzed for cytogenetic abnormalities induced by NPs. Roots of *Allium cepa*, *Glycine max*, and *Nicotiana tabacum* were exposed to: (1) Ag–NPs; (2) ZnO–NPs and CeO_2–NPs; and (3) TiO_2–NPs, for cytological analyses [88], assays with random amplified polymorphic DNA [89], and tests with comet and DNA laddering techniques [85], respectively. The root growth inhibition in the treated plants was associated with typical errors in cell division and chromosome behavior such as bridges, early chromosome separation, multiple breaks, and micronuclei release, as well as DNA damage. Similar mitotic aberrations and DNA alterations were also observed in *Zea mays* and *Vicia narbonensis* treated with TiO_2–NPs [90], as well as in *Vicia faba* seedlings treated with Ag–NPs [75]. The same cytogenetical changes were previously described by Rãcuciu and Creangã [41] in the root meristem of *Zea mays* treated with magnetic NPs. Oxidative damages to DNA were observed by Atha *et al.* [91] in seedlings of radish and ryegrass after treatment with CuO–NPs and the effects were evident also at low NPs suspension concentrations (10 mg/L).

3.2.3. Toxicity of CNMs

Studies reported advantageous, disadvantageous, or no evident effects of nanotubes on plants both at phenothypic and genotypic levels. A stimulative effect, prevalently of MWCNTs, on seedling root elongation and/or seed germination was observed in onion and cucumber [52], wheat [92], mustard [93], and tomatoes [53,56,94,95]. In tomatoes the MWCNTs penetrated the seed coat and seedling roots through new wall pores and enhanced the capacity of the plant to uptake water, with positive effect on germination and seedling growth. At concentrations of 50–200 μg/mL they improved the vegetative and the reproductive plant development up to double the rate of flowering and fruiting. The stimulation of root emergence and root growth was recently observed in *in vitro* cultures of *Rubus adenotrichos* treated with functionalized SWCNTs [96].

Concerning fullerene and its derivatives, Kole *et al.* [59] observed that in bitter melon the uptake of fullerol (water-soluble) by the roots and its translocation via the stem to leaves, flowers, and fruits, enhanced the plant biomass yield by 54%, its water content by 24%, and the fruit yield by 128%, and even increased the rate of the medical use of components (anticancer) in the fruit. This study also confirmed the transmission of CNMs to the next generation coming from the treated seeds; such a capacity had previously been reported by Lin *et al.* [55] in seeds and plants of rice in the first and second generations, and in these plants the concentration of fullerene (C_{70}) was higher in the aerial parts than in the roots, in agreement with its translocation.

No appreciable effect of MWCNTs on seed germination was observed in various crop plants (corn, cucumber, radish, rape, lettuce) [68].

As regards the inhibitory effect, Canas *et al.* [52] observed inhibition of the root growth in tomato and lettuce, although the same treatment on onion and cucumbers stimulated growth. These effects were due to interactions with the root surface as CNTs were not found within the root. As regards graphene, the absorption of GO (0.5–5.0 nm) from the roots of *Vicia faba* seedlings had both beneficial and toxic effects depending on the concentrations; toxic concentrations were of the order of 1600 mg/L and induced oxidative stress and an increase of electrolyte leakage to the detriment of the seedling growth [97]. In suspension cultures of T87 cell line obtained from seedlings of *Arabidopsis thaliana*, GO penetrated cells through endocytotic pathways and at low concentrations not exceeding 80 mg/L induced oxidative stress and damages to the nucleus and mitochondria, which produced cell death [98]. Similar adverse effects had been observed previously in the roots of cabbage and tomato seedlings by using GO at 500–2000 mg/L [39]. Begum and Fugetsu [99] treated spinach plants with MWCNTs and these were observed within the plant tissues to produce an increase of ROS and mechanisms of cell death. The toxicity was ascribed to oxidative stress induced by MWCNTs, as the toxic effects were reversed by adding antioxidants to the MWCNTs treatments.

Interesting results were recently reported by Hu *et al.* [100] in wheat plants after combined treatments of GO and Arsenic. GO and As administered alone were not toxic, but if they were given together, even at a concentration of 0.1 mg/L, they induced oxidative stress and alterations in cellular metabolisms (carbohydrates, amino acids, secondary metabolites) and cell structures, most likely because GO exalted the cell permeability to As, which could therefore damage the cells.

3.2.4. Genotoxicity of CNTs

The genotoxicity of CNTs, referred to as any interaction with the gene expression, has been studied in a few plants. In tomato seedlings Khodakovskaya *et al.* [53], by means of advanced physical methods (Raman, Phototherma, and Photoacoustic methods), described a nanotube-dependent activation, both in roots and leaves, of either stress-related genes, among which were those switched on by pathogens, or the gene controlling the water channels on the plasma membranes. Seeds and seedlings responded to the gene upregulation by increasing germination and growth. Growth induced by the upregulation of genes controlling the cell division, cell wall formation, and the water flow through the membrane was also observed in tobacco cells cultured with MWCNTs [101]. In roots of *Allium cepa*, Ghosh *et al.* [102] observed cytological errors and damages to DNA associated with an apoptosis process, as a probable consequence of internalization of the nanotubes. Most recently Yan *et al.* [103], by means of PCR analysis, immunostaining technique, and electron microscopy, provided evidence that in *Zea mays* the penetration and accumulation of SWCNTs inside roots may change the expression of genes controlling the seminal root and the hairs' root growth, of benefit to the first and disadvantage to the second.

4. Reproductive Systems

Both female and male reproductive generations (gametophytes) of higher plants are exposed with their coats and secretions to the reception and absorption of airborne particles. They are so small as to develop male (pollen) inside the pollen sacs, which open at maturity, and for the female (the primary endosperm or embryo sac) inside the egg, which opens apically with a canal. Angiosperm ovules are inside the pistil, which consists of the ovary and an elongated secretory trait, often channeled, which enlarges at the tip (the stigma) for receiving pollen. Gymnosperms lack pistils and have cones with naked ovules that, by means of a drop of secretion, capture pollen grains and lead them inside as along with any other airborne particle of a few microns in size. Pollen (Figure 1) encloses its cells with an inner wall and an outer wall that is reduced or absent in some dots (pores) or tracts (furrows) [104] to facilitate extrusion of the tube that transports the sperms to ovules. Lipids, proteins, and carbohydrates are spread on the pollen surface for interacting with secretions of the female structures.

5. Penetration and Toxicity of Metal- and Metal Oxide-NPs

The studies dealing with the vulnerability of the reproductive systems to NPs are limited to a few reports [105,106]; these, however, are rich in data. These authors studied the effects of Pd–NPs [105] and Ag–NPs [106] on *in vitro* cultured kiwi pollen by using electron microscopy, the SIMAA 6000 instrument for determining palladium, the Atomic Absorption spectrophotometer (AAanalyst 300) for silver, and X-ray resonance fluorescence analysis for the calcium ion Ca^{2+}. Both NPs were able to damage the plasma membrane and deplete endogenous calcium, with the result of decreasing the capacity of pollen for germination and elongation within the tube. Pd–NPs penetrated pollen faster and more deeply than $PdCl_2$. Ag–NPs increased the levels of ROS, differently to the silver free ion, which was more active than Ag–NPs in damaging the plasma membrane and preventing germination.

A recent study showed that tomato plants treated with relatively low concentrations of CeO_2–NPs produced, through sexual reproduction, poorly developing progeny with evident signs of oxidative stress and an enhanced capacity to accumulate Ceria [107].

These data are somewhat worrisome, substantiating the presumptive vulnerability of pollen based on discontinuous wall exine and intense metabolic activity concentrated in the vegetative cell producing pollen tube, and even showing transgenerational effects consequential to treatments with NMs.

6. Conclusions and Perspectives

The receptivity of plants to NMs goes beyond the predictable effects based on their structural traits, as well as the vessels and sieve elements for displacement of NMs upwards and downwards. NMs can overcome the barrier cuticle and then are able to penetrate the pores of the wall that are less selective than the cutin so as to allow the passage of NMs larger than the pores themselves [49,51,108]. Across the cuticle there are two alternative paths of diffusion, the lipophilic path and the polar path, made up of aqueous narrow pores (0.5–2 nm in diameter) whose length and sinuosity are unknown [10,109,110]. How the NMs pass through these paths is yet to be discovered. Across the cell membrane the NMs pass either by diffusion, via carriers, *de novo* formed membrane pores, or by energy-dependent pathways (endocytosis), as well as by linking to chemicals for cell internalization, so mimicking the behavior of biological compounds. It is more than likely that NMs can activate and dynamize processes by which cells become permeable; after all, it is well known that cell substructures may amplify their function when opportunely stimulated, *i.e.*, when the plasmodesmata are in contact with some viral particles [111]. From this perspective, the NMs might open the door to environmental toxicants by carrying them into cells. This hypothesis is substantiated by the increased uptake of pesticides observed in plants (zucchini, tomatoes, and soybeans) treated with C60 [112,113], as well as by recent studies [100] on combined treatments of graphene and arsenic, which suggest a role for graphene in the transport of As into cells.

Within cells the NMs may interact with the molecules structurally and non-structurally through mechanisms still poorly understood. As a result, the level of highly reactive molecular species of oxygen (ROS) is likely to increase, with high potential to damage any biomolecule and activate even programmed cell death mechanisms; of special note are the damages that NPs can inflict on DNA and chromosome behavior. The typical cellular response is an increase of antioxidant enzymes (SOD,

CAT, APX, *etc.*), so the changes in the cellular rate of ROS and anti-ROS and peroxidation products allow us to assess the degree of toxicity of NMs and the detoxifying ability of the plant cells. Oxidative stress induced by NMs can have long-term effects in some plant cells and in the case of upregulation of certain genes it can benefit plant growth and thus should be considered as potentially exploitable in agriculture [9]. Furthermore, it is known that the mitogen potentiality of ROS is able to induce DNA replication and cell proliferation; this could be a mechanism related to the increase in growth observed in some plants treated with NMs and therefore might be regarded as a benefit. Regardless, a number of favorable effects exploitable by man have been produced by NMs treatment [9] and perhaps major benefit for plants may come from NPs with their own antimicrobial and antifungal activity for use against pathogens [114,115].

When considering the reactive potential of NMs, several physicochemical traits should be emphasized without neglecting those acquired by residual impurities and exposure to air (airborne). Independently of the size, shape, surface structure, electronic structure, and charge, a basic characteristic is the nano-dimension, which allows a greater exposure of the atoms and a higher surface reactivity, albeit there is a wide range of reactivity and effects produced for the different types of nanoparticles. It can be expected that the smaller the surface area and the higher the zeta potential, the greater will be the reactive potential; there is no prediction about cell penetration and toxicity, especially as NPs are subject to various transformations and aggregation in the presence of natural organic matter [116]. Transformations include dissolution with release of metal ions, which are a factor of toxicity; chelation into complexes, which is within the detoxification mechanisms adopted by the plants; precipitation by chemicals arising from the growth environment or root secretion; and other reactions with numerous functional groups (hydroxyl, carboxyl, sulfhydryl, amino, *etc.*) that NPs encounter in the cellular pathways. In the case of CeO_2–NPs, Cerium transforms into Ce^{3+} within the plant and in this form is considered toxic for cells [83]. Keeping in mind that Ce^{3+} tends to bind oxygen to return to Ce^{4+}, we can hypothesize a switch of Cerium from 4+ to 3+ and the reverse in the cells, depending on the availability of oxygen; this could play a major role in oxidative stress. Conversely, in hydroponic cultures the ions of Ce^{3+} are precipitated in the roots by the phosphates that abound in the nutrient solution, and in the shoots are carboxylated by functional organic groups of the cells [48], producing unreactive forms. In addition to interactions of NPs with trophic and biological environments, we must not neglect those between different NPs hitherto known for Fe–NPs and Zn–NPs in the form of antagonism [23]. It is noteworthy that some metal ions within the plant are susceptible to transformation into NPs through redox reactions controlled by biomolecules, as recently reported by Kuppusamy *et al.* [117]. Hence, it is likely that within the plant a variable amount of metal ions resulting from NPs may again revert to NPs, in agreement with the previous observations of Larue *et al.* [25]. If the biosynthesis is to prevail to bio-dissolution, the NPs must concentrate and persist within the plant; their long-term effects are not known, while speculative applications regarding the biosynthesis of heavy and noble metal nanoparticles (gold, platinum, silver, zinc) are well known [117].

When comparing with animals, plant cells seem to be less damaged by NMs [7], and a role may be played by transmembrane transport via endocytosis, which is recurrent in animals—unlike in plant cells, where the cell wall and the pressure of turgor are an obstacle. Moreover, the large vacuolar system of plant cells may function as a detoxifying agent against NPs; the autophagy aforementioned [63] provides evidence. This activity, however, contrasts with the biosynthesis of NPs promoted by some vacuolar secondary metabolites [117]. It thus remains a matter of interest pending toxicological studies on differentiated cells (vacuolated) and undifferentiated cells (lacking vacuoles). An important point encompassing plant and animal cells is the potential of the plants to export NMs through their life cycle and the food chain, so, as already underlined by several authors [27,80], they could become carriers of chronic toxicity. Recent studies on watermelons and cucumbers [22,24] substantiate the ability of NMs to pass from the atmosphere into the leaf epidermis and its stomata, and then distribute inside the plant. Growing interest in the development of facilities for monitoring the impact of nano-materials on trophic and biological environments results from the self-sustaining mesocosm, which simulates natural ecosystems [118,119].

The emerging conclusions are that the studies on plants substantiate the reactive potential of nanomaterials; they are, however, based on different methods applied to different plant species, so the results lack homogeneity and are sometimes controversial and not ascribable to general models of interaction. A major limitation comes from the use of nanomaterials produced, characterized, and stored with different procedures that may confer different properties [120,121]; in particular, CNMs may retain active residuals of their precursors, such as Ni and Fe, and superficial pouches that expose the residuals. The methods of administration should ensure the control of the size and number concentration of NMs, as experienced by Wang *et al.* [22]. To shed light on the size limit for cell penetration, it is useful to know the sizes of NMs (hydrodynamic diameter) in the administration medium. The experiments with hydroponic plants and *in vitro* cell cultures, although they do not preserve the identity and integrity of the *in situ* plants, are useful and valuable for comparative purposes, not forgetting that adverse effects on plant cells may result from depletion of the micronutrients that interact with NMs. Furthermore, the techniques for detecting NMs within tissues and testing their toxicity to cells may produce artifacts, so special accuracy in preparing and manipulating NMs is needed in order to avoid misinterpretations; see [121].

The last point concerns the recurrent use of young plants, which limits the study to the vegetative phases of actively growing organs; it is hoped that studies will give more attention to life cycles and reproductive systems. The male and female generations producing gametes consist of a few cells with set haploid so they are easily vulnerable, especially during their pregamic interaction [122]. Closely related and to be evaluated thoroughly are the possible transgenerational effects of NMs.

While it remains important to study the basic mechanisms of the interaction of NMs with plants, our priority is to weigh up the potential benefits of NMs for plants and humans against the risk of exposing terrestrial ecosystems to particles that are small and potentially reactive.

Author Contributions

The authors both made substantial contributions to conception and design, acquisition of data, and analysis and interpretation.

Conflicts of Interest

The authors declare no conflict of interest.

References

1. Dietz, K.J.; Herth, S. Plant nanotoxicology. *Trends Plant Sci.* **2011**, *16*, 582–589.
2. Subbiah, R.; Veerapandian, M.; Yun, K.S. Nanoparticles: Functionalization and multifunctional applications in biomedical sciences. *Curr. Med. Chem.* **2010**, *17*, 4559–4577.
3. Treuel, L.; Eslahian, K.A.; Docter, D.; Lang, T.; Zellner, R.; Nienhaus, K.; Nienhaus, G.U.; Stauber, R.H.; Maskos, M. Physicochemical characterization of nanoparticles and their behavior in the biological environment. *Phys. Chem. Chem. Phys.* **2014**, *16*, 15053–15067.
4. Gao, Y.; Chen, K.; Ma, J.; Gao, F. Cerium oxide nanoparticles in cancer. *OncoTargets Ther.* **2014**, *7*, 835–840.
5. Onelli, E.; Prescianotto-Baschong, C.; Caccianiga, M.; Moscatelli, A. Clathrin-dependent and independent endocytic pathways in tobacco protoplasts revealed by labelling with charged nanogold. *J. Exp. Bot.* **2008**, *59*, 3051–3068.
6. Kong, X.; Yang, B.; Xiong, H.; Zhou, Y.; Xue, S.; Xu, B.; Wang, S. Selective removal of heavy metal ions from aqueous solutions with surface functionalized silica nanoparticles by different functional groups. *J. Cent. South Univ.* **2014**, *21*, 3575–3579.
7. Poma, A.; Colafarina, S.; Fontecchio, G.; Chichiriccò, G. Transgenerational Effects of NMs. In *Nanomaterials, Impacts on Cell Biology and Medicine*; Capco, D., Chen, Y., Eds.; Springer Science and Business Media: Dordrecht, Germany, 2014; Volume 811, pp. 235–254.
8. Miralles, P.; Church, T.L.; Harris, A.T. Toxicity, uptake, and translocation of engineered nanomaterials in vascular plants. *Environ. Sci. Technol.* **2012**, *46*, 9224–9239.
9. Husen, A.; Siddiqi, K.S. Phytosynthesis of nanoparticles: Concept, controversy and application. *Nanoscale Res. Lett.* **2014**, *9*, 229–252.
10. Eichert, T.; Goldbach, H.E. Equivalent pore radii of hydrophilic foliar uptake routes in stomatous and astomatous leaf surface—Further evidence for a stomatal pathway. *Physiol. Plant.* **2008**, *132*, 491–502.
11. Schreiber, L. Polar paths of diffusion across plant cuticles: New evidence for an old hypothesis. *Ann. Bot.* **2005**, *95*, 1069–1073.
12. Fleischer, A.; O'Neill, M.A.; Ehwald, R. The pore size of non-Graminaceous plant cell walls is rapidly decreased by borate ester cross-linking of the pectic polysaccharide rhamnogalacturonan II. *Plant Physiol.* **1999**, *121*, 829–838.
13. Carpita, N.; Sabularse, D.; Montezinos, D.; Delmer, D.P. Determination of the pore size of cell walls of living plants. *Science* **1979**, *205*, 1144–1147.
14. Tepfer, M.; Taylor, I.E.P. The permeability of plant cell walls as measured by gel filtration chromatography. *Science* **1981**, *213*, 761–763.
15. Asli, S.; Neumann, P.M. Colloidal suspensions of clay or titanium dioxide nanoparticles can inhibit leaf growth and transpiration via physical effects on root water transport. *Plant Cell Environ.* **2009**, *32*, 577–584.
16. Zwieniecki, M.A.; Holbrook, N.M. Bordered pit structure and vessel wall surface properties. Implications for embolism repair. *Plant Physiol.* **2000**, *123*, 1015–1020.
17. Sperry, J.S.; Hacke, U.G. Analysis of circular bordered pit function I. Angiosperm vessels with homogeneous pit membranes. *Am. J. Bot.* **2004**, *91*, 369–385.

18. Pittermann, J.; Choat, B.; Jansen, S.; Stuart, S.A.; Lynn, L.; Dawson, T.E. The Relationships between xylem safety and hydraulic efficiency in the Cupressaceae: The evolution of pit membraneorm and function. *Plant Physiol.* **2010**, *15*, 1919–1931.
19. Eichert, T.; Kurtz, A.; Steiner, U.; Goldbach, H.E. Size exclusion limits and lateral heterogeneity of the stomatal foliar uptake pathway for aqueous solutes and water-suspended nanoparticles. *Physiol. Plant.* **2008**, *134*, 151–160.
20. Birbaum, K.; Brogioli, R.; Schellenberg, M.; Martinoia, E.; Stark, W.J.; Günther, D.; Limbach, L.K. No evidence for cerium dioxide nanoparticle translocation in maize plants. *Environ. Sci. Technol.* **2010**, *44*, 8718–8723.
21. Kurepa, J.; Paunesku, T.; Vogt, S.; Arora, H.; Rabatic, B.M.; Lu, J.; Wanzer, M.B.; Woloschak, G.E.; Smalle, J.A. Uptake and distribution of ultra-small anatase TiO_2 alizarin red S nanoconjugates in *Arabidopsis thaliana*. *Nano Lett.* **2010**, *10*, 2296–2302.
22. Wang, W.-N.; Tarafdar, J.C.; Biswas, P. Nanoparticle synthesis and delivery by an aerosol route for watermelon plant foliar uptake. *J. Nanopart. Res.* **2013**, *15*, doi:10.1007/s11051-013-1417-8.
23. Taran, N.; Batsmanova, L.; Konotop, Y.; Okanenko, A. Redistribution of elements of metals in plant tissues under treatment by non-ionic colloidal solution of biogenic metal nanoparticles. *Nanoscale Res. Lett.* **2014**, *9*, 354–357.
24. Hong, J.; Peralta-Videa, J.R.; Rico, C.; Sahi, S.; Viveros, M.N.; Bartonio, J.; Zhao, L.; Gardea-Torresdey, J.L. Evidence of translocation and physiological impacts of foliar applied CeO_2 nanoparticles on Cucumber (*Cucumis sativus*) plants. *Environ. Sci. Technol.* **2014**, *48*, 4376–4385.
25. Larue, C.; Castillo-Michel, H.; Sobanska, S.; Bureau, S.; Barthès, V.; Ouerdane, L.; Carrière, M.; Sarret, G. Foliar exposure of the crop *Lactuca sativa* to silver nanoparticles: Evidence for internalization and changes in Ag speciation. *J. Hazard. Mater.* **2014**, *261*, 98–106.
26. Larue, C.; Veronesi, G.; Flank, A.M.; Surble, S.; Herlin-Boime, N.; Carriere, M. Comparative uptake and impact of TiO_2 nanoparticles in wheat and rapeseed. *J. Toxicol. Environ. Health A* **2012**, *75*, 722–734.
27. Deng, Y.; White, J.C.; Xing, B. Interaction between engineered nanomaterials and agricultural crops: Implications for food safety. *J. Zhejiang Univ. Sci. A* **2014**, *15*, 552–572.
28. Shen, C.-X.; Zhang, Q.F.; Li, J.; Bi, F.-C.; Yao, N. Induction of programmed cell death in *Arabidopsis* and rice by single wall carbon nanotubes. *Am. J. Bot.* **2010**, *97*, 1602–1609.
29. Liu, Q.; Chen, B.; Wang, Q.; Shi, X.; Xiao, Z.; Lin, J.; Fang, X. Carbon nanotubes as molecular transporters for walled plant cells. *Nano Lett.* **2009**, *9*, 1007–1010.
30. Giraldo, J.P.; Landry, M.P.; Faltermeier, S.M.; Mc Nicholas, T.P.; Iverson, N.M.; Boghossian, A.A.; Reuel, N.F.; Hilmer, A.J.; Sen, F.; Brew, J.A.; *et al*. Plant nanobionics approach to augment photosynthesis and biochemical sensing. *Nat. Mater.* **2014**, *13*, 400–408.
31. Hong, F.; Zhou, J.; Liu, C.; Yang, F.; Wu, C.; Zheng, L.; Yang, P. Effect of nano-TiO_2 on photochemical reaction of chloroplast of spinach. *Biol. Trace Elem. Res.* **2005**, *105*, 1–3.
32. Wang, S.; Kurepa, J.; Smalle, J.A. Ultra-small TiO_2 nanoparticles disrupt microtubular networks in *Arabidopsis thaliana*. *Plant Cell Environ.* **2011**, *34*, 811–820.
33. Rãcuciu, M.; Creangã, D.E. TMA-OH coated magnetic nanoparticles internalized in vegetal tissue. *Rom. J. Phys.* **2007**, *52*, 395–402.

34. Wang, H.; Kou, X.; Pei, Z.; Xiao, J.Q.; Shan, X.; Xing, B. Physiological effects of magnetite (Fe_3O_4) nanoparticles on perennial ryegrass (*Lolium perenne*) and pumpkin (*Cucurbita mixta*) plants. *Nanotoxicology* **2011**, *5*, 30–42.
35. Parsons, J.G.; Lòpez, M.L.; Gonzales, C.M.; Peralta-Videa, J.R.; Garrea-Torresdey, J.L. Toxicity and biotransformation of uncoated and coated nickel hydroxide nanoparticles on mesquite plants. *Environ. Toxicol. Chem.* **2010**, *29*, 1146–1154.
36. Lin, C.; Fugetsu, B.; Su, Y.; Watari, F. Studies on toxicity of multi-walled carbon nanotubes on *Arabidopsis* T87 suspension cells. *J. Hazard. Mater.* **2009**, *170*, 578–583.
37. Tan, X.M.; Lin, C.; Fugetsu, B. Studies on toxicity of multi-walled carbon nanotubes on suspension rice cells. *Carbon* **2009**, *47*, 3479–3487.
38. Santos, S.M.A.; Dinis, A.M.; Rodrigues, D.M.F.; Peixoto, F.; Videira, R.A.; Jurado, A.S. Studies on the toxicity of an aqueous suspension of C60 nanoparticles using a bacterium (gen. *Bacillus*) and an aquatic plant (*Lemna gibba*) as *in vitro* model systems. *Aquat. Toxicol.* **2013**, *142–143*, 347–354.
39. Fugetsu, B.; Begum, P. Graphene Phytotoxicity in the Seedling Stage of Cabbage, Tomato, Red Spinach, and Lettuce. In *Carbon Nanotubes—From Research to Applications*; ISBN: 978-953-307-500-6; Bianco, S., Ed.; InTech: Rijeka, Croatia. Available online: http://www.intechopen.com/books/carbon-nanotubes-fromresearch-to-applications/graphene-phytotoxicity-in-the-seedling-stage-of cabbage-tomato-red-spinach-and-lettuce (accessed on 23 May 2015).
40. Zheng, L.; Hong, F.; Lu, S.; Liu, C. Effect of nano-TiO_2 on strength of naturally aged seeds and growth of spinach. *Biol. Trace Elem. Res.* **2005**, *104*, 83–91.
41. Rãcuciu, M.; Creangã, D.E. Cytogenetical changes induced by β-cyclodextrin coated nanoparticles in plant seeds. *Rom. J. Phys.* **2009**, *54*, 125–131.
42. Stampoulis, D.; Sinha, S.K.; White, J.C. White assay-dependent phytotoxicity of nanoparticles to plants. *Environ. Sci. Technol.* **2009**, *43*, 9473–9479.
43. De la Rosa, G.; Lopez-Moreno, M.L.; Hernandez-Viescaz, J.; Montes, M.O.; Peralta-Videa, J.R.; Gardea-Torresdey, J.L. Toxicity and biotransformation of ZnO nanoparticles in the desert plants *Prosopis juliflora-velutina*, *Salsola tragus* and *Parkinsonia florida*. *Int. J. Nanotechnol.* **2011**, *8*, 492–506.
44. Pokhrel, L.R.; Dubey, B. Evaluation of developmental responses of two crop plants exposed to silver and zinc oxide nanoparticles. *Sci. Total Environ.* **2013**, *452–453*, 321–332.
45. Kouhi, S.M.; Lahouti, M.; Ganjeali, A.; Entezari, M.H. Comparative phytotoxicity of ZnO nanoparticles, ZnO microparticles, and Zn^{2+} on rapeseed (*Brassica napus* L.): Investigating a wide range of concentrations. *Toxicol. Environ. Chem.* **2014**, *96*, 861–868.
46. Rico, C.M.; Majumdar, S.; Duarte-Gardea, M.; Peralta-Videa, J.R.; Gardea-Torresdey, J.L. Interaction of nanoparticles with edible plants and their possible implications in the food chain. *J. Agric. Food Chem.* **2011**, *59*, 3485–3498.
47. Zhu, H.; Han, J.; Xiao, J.Q.; Jin, Y. Uptake, translocation, and accumulation of manufactured iron oxide by pumpkin plants. *J. Environ. Monit.* **2008**, *10*, 713–717.
48. Zhang, P.; Ma, Y.H.; Zhang, Z.Y.; He, X.; Zhang, J.; Guo, Z.; Tai, R.Z.; Zhao, Y.L.; Chai, Z.F. Biotransformation of ceria nanoparticles in cucumber plants. *ACS Nano* **2012**, *6*, 9943–9950.

49. Du, W.; Sun, Y.; Ji, R.; Zhu, J.; Wu, J.; Guo, H. TiO_2 and ZnO nanoparticles negatively affect wheat growth and soil enzyme activities in agricultural soil. *J. Environ. Monit.* **2011**, *13*, 822–828.
50. Zhao, L.; Peralta-Videa, J.R.; Varela-Ramirez, A.; Castillo-Michel, H.; Li, C.; Zhang, J.; Aguilera, R.J.; Keller, A.A.; Gardea-Torresdey, J.L. Effect of surface coating and organic matter on the uptake of CeO_2–NPs by corn plants grown in soil: Insight into the uptake mechanism. *J. Hazard. Mater.* **2012**, *225–226*, 131–138.
51. Zhao, L.; Peralta-Videa, J.R.; Peng, B.; Bandyopadhyay, S.; Corral-Diaz, B.; Osuna-Avila, P.; Montes, M.O.; Keller, A.A.; Gardea-Torresdey, J.L. Alginate modifies the physiological impact of CeO_2 nanoparticles in corn seedlings cultivated in soil. *J. Environ. Sci.* **2014**, *26*, 382–389.
52. Canas, J.E.; Long, M.; Nations, S.; Vadan, R.; Dai, L.; Luo, M.; Ambikapathi, R.; Lee, E.H.; Oslzyk, D. Effects of functionalized and non-functionalized single-walled carbon nanotubes on root elongation of select crop species. *Environ. Toxicol. Chem.* **2008**, *27*, 1922–1931.
53. Khodakovskaya, M.V.; de Silva, K.; Nedosekin, D.A.; Dervishi, E.; Biris, A.S.; Shashkov, E.V.; Galanzha, E.I.; Zharov, V.P. Complex genetic, photothermal, and photoacoustic analysis of nanoparticle-plant interactions. *Proc. Natl. Acad. Sci. USA* **2011**, *108*, 1028–1033.
54. Wild, E.; Jones, K.C. Novel method for the direct visualization of *in vivo* nanomaterials and chemical interactions in plants. *Environ. Sci. Technol.* **2009**, *43*, 290–294.
55. Lin, S.; Reppert, J.; Hu, Q.; Hudson, J.S.; Reid, M.L.; Ratnikova, T.A.; Rao, A.M.; Luo, H.; Ke, P.C. Uptake, translocation, and transmission of carbon nanomaterials in rice plants. *Small* **2009**, *5*, 1128–1132.
56. Khodakovskaya, M.V.; Dervishi, E.; Mahmood, M.; Xu, Y.; Li, Z.; Watanabe, F.; Biris, A.S. Carbon nanotubes are able to penetrate plant seed coat and dramatically affect seed germination and plant growth. *ACS Nano* **2009**, *3*, 3221–3227.
57. Smirnova, E.A.; Gusev, A.A.; Zaitseva, O.N.; Lazareva, E.M.; Onishchenko, G.E.; Kuznetsova, E.V.; Tkachev, A.G.; Feofanov, A.V.; Kirpichnikov, M.P. Multi-walled carbon nanotubes penetrate into plant cells and affect the growth of *Onobrychis arenaria* seedlings. *Acta Nat.* **2011**, *3*, 99–106.
58. Cifuentes, Z.; Custardoy, L.; de la Fuente, J.M.; Marquina, C.; Ibarra, M.R.; Rubiales, D.; Alejandro Pérez-de-Luque, A. Absorption and translocation to the aerial part of magnetic carbon-coated nanoparticles through the root of different crop plants. *J. Nanobiotechnol.* **2010**, *8*, doi:10.1186/1477-3155-8-26.
59. Kole, C.; Kole, P.; Randunu, K.M.; Choudhary, P.; Podila, R.; Ke, P.; Rao, A.M.; Marcus, R.K. Nanobiotechnology can boost crop production and quality: First evidence from increased plant biomass, fruit yield and phytomedicine content in bitter melon (*Momordica charantia*). *BMC Biotechnol.* **2013**, *13*, 37–46.
60. Serag, M.F.; Kaji, N.; Venturelli, E.; Okamoto, Y.; Terasaka, K.; Tokeshi, M.; Mizukami, H.; Ugent, K.B.; Bianco, A.; Baba, Y.; *et al.* Trafficking and subcellular localization of multiwalled carbon nanotubes in plant cells. *ACS Nano* **2011**, *5*, 493–499.
61. Serag, M.F.; Kaji, N.; Gaillard, C.; Okamoto, Y.; Terasaka, K.; Jabasini, M.; Tokeshi, M.; Misukami, H.; Bianco, A.; Baba, Y.; *et al.* A functional platform for controlled subcellular distribution of carbon nanotubes. *ACS Nano* **2011**, *5*, 9264–9270.

62. Serag, M.F.; Kaji, N.; Tokeshi, M.; Baba, Y. Introducing carbon nanotubes into living walled plant cells through cellulase-induced nanoholes. *RSC Adv.* **2012**, *2*, 398–400.
63. Serag, M.F.; Braeckmans, K.; Habuchi, S.; Kaji, N.; Bianco, A.; Baba, Y. Spatiotemporal visualization of subcellular dynamics of carbon nanotubes. *Nano Lett.* **2012**, *12*, 6145–6151.
64. Zhao, S.; Wang, Q.; Zhao, Y.; Rui, Q.; Wang, D. Toxicity and translocation of graphene oxide in *Arabidopsis thaliana*. *Environ. Toxicol. Pharmacol.* **2015**, *39*, 145–156.
65. Yang, L.; Watts, D.J. Particle surface characteristics may play an important role in phytotoxicity of alumina nanoparticles. *Toxicol. Lett.* **2005**, *158*, 122–132.
66. Aslani, F.; Bagheri, S.; Julkapli, N.M.; Juraimi, A.S.; Hashemi, F.S.G.; Baghdadi, A. Effects of engineered nanomaterials on plants growth: An overview. *Sci. World J.* **2014**, *2014*, doi:10.1155/2014/641759.
67. Lin, D.; Xing, B. Uptake and phytotoxicity of ZnO nanoparticles. *Environ. Sci. Technol.* **2008**, *42*, 5580–5585.
68. Lin, D.; Xing, B. Phytotoxicity of nanoparticles: Inhibition of seed germination and root growth. *Environ. Pollut.* **2007**, *150*, 243–250.
69. Dimkpa, C.O.; McLean, J.E.; Latta, D.E.; Manangon, E.; Britt, D.W.; Johnson, W.P.; Boyanov, M.I.; Anderson, A.J. CuO and ZnO nanoparticles: Phytotoxicity, metal speciation and induction of oxidative stress in sand-grown wheat. *J. Nanopart. Res.* **2012**, *14*, 1125–1140.
70. Lv, J.; Zhang, S.; Luo, L.; Zhang, J.; Yang, K.; Christie, P. Accumulation, speciation and uptake pathway of ZnO nanoparticles in maize. *Environ. Sci. Nano* **2015**, *2*, 68–77.
71. Gardea-Torresdey, J.L.; Gomez, E.; Peralta-Videa, J.; Parsons, J.G.; Troiani, H.E.; Yacaman, M.J. *Alfalfa sprouts*: A natural source for the synthesis of silver nanoparticles. *Langmuir* **2003**, *19*, 1357–1361.
72. Harris, A.T.; Bali, R. On the formation and extent of uptake of silver nanoparticles by live plants. *J. Nanopart. Res.* **2008**, *10*, 691–695.
73. Abdel-Aziz, M.S.; Shaheen, M.S.; El-Nekeety, A.A.; Abdel-Wahhab, M.A. Antioxidant and antibacterial activity of silver nanoparticles biosynthesized using *Chenopodium murale* leaf extract. *J. Saudi Chem. Soc.* **2014**, *18*, 356–363.
74. Yin, S.L.; Cheng, Y.; Espinasse, B.; Colman, B.P.; Auffan, M.; Wiesner, M.; Rose, J.; Liu, J.; Bernhardt, E.S. More than the ions: The effects of silver nanoparticles on *Lolium multiflorum*. *Environ. Sci. Technol.* **2011**, *45*, 2360–2367.
75. Abdel-Azeem, E.A.; Elsayed, B.A. Phytotoxicity of silver nanoparticles on *Vicia faba* seedlings. *N. Y. Sci. J.* **2013**, *6*, 148–156.
76. Gardea-Torresdey, J.L.; Parsons, J.G.; Gomez, E.; Peralta-Videa, J.; Troiani, H.E.; Santiago, P.; Yacaman, M.J. Formation and growth of Au nanoparticles inside live alfalfa plants. *Nano Lett.* **2002**, *2*, 397–401.
77. Zhai, G.; Walters, K.S.; Peate, D.W.; Alvarez, P.J.J.; Schnoor, J.L. Transport of gold nanoparticles through plasmodesmata and precipitation of gold ions in woody poplar. *Environ. Sci. Technol. Lett.* **2014**, *1*, 146–151.
78. Bali, R.; Siegele, R.; Harris, A.T. Biogenic Pt uptake and nanoparticle formation in *Medicago sativa* and *Brassica juncea*. *J. Nanopart. Res.* **2010**, *12*, 3087–3095.

79. Arora, S.; Sharma, P.; Kumar, S.; Nayan, R.; Khanna, P.K.; Zaidi, M.G.H. Gold-nanoparticle induced enhancement in growth and seed yield of *Brassica juncea*. *Plant Growth Regul.* **2012**, *66*, 303–310.
80. Wang, Q.; Ma, X.; Zhang, W.; Pei, H.; Chen, Y. The impact of cerium oxide nanoparticles on tomato (*Solanum lycopersicum* L.) and its implications for food safety. *Metallomics* **2012**, *4*, 1105–1112.
81. Schwabe, F.; Schulin, R.; Limbach, L.K.; Stark, W.; Bürge, D.; Nowack, B. Influence of two types of organic matter on interaction of CeO_2 nanoparticles with plants in hydroponic culture. *Chemosphere* **2013**, *91*, 512–520.
82. Rico, C.M.; Morales, M.I.; Barrios, A.C.; McCreary, R.; Hong, J.; Lee, W.-Y.; Nunez, J.; Peralta-Videa, J.R.; Gardea-Torresdey, J.L. Effect of cerium oxide nanoparticles on the quality of rice (*Oryza sativa* L.) grains. *J. Agric. Food Chem.* **2013**, *61*, 11278–11285.
83. Zhang, P.; Ma, Y.H.; Zhang, Z.Y.; He, X.; Li, Y.; Zhang, J.; Zheng, L.; Zhao, Y.L. Species-Specific toxicity of ceria nanoparticles to *Lactuga* plants. *Nanotoxicology* **2013**, *9*, 1–8.
84. Zhang, P.; Ma, Y.H.; Zhang, Z.Y.; He, X.; Guo, Z.; Tai, R.Z.; Ding, Y.Y.; Zhao, Y.L.; Chai, Z.F. Comparative toxicity of nanoparticulate/bulk Yb_2O_3 and $YbCl_3$ to Cucumber (*Cucumis sativus*). *Environ. Sci. Technol.* **2012**, *46*, 1834–1841.
85. Ghosh, M.; Bandyopadhyay, M.; Mukherjee, A. Genotoxicity of titanium dioxide (TiO_2) nanoparticles at two trophic levels: Plant and human lymphocytes. *Chemosphere* **2010**, *81*, 1253–1262.
86. Nair, R.; Poulose, A.C.; Nagaoka, Y.; Yoshida, Y.; Maekawa, T.; Kumar, D.S. Uptake of FITC labelled silica nanoparticls and quantum dots by rice seedlings: Effects on seed germination and their potential as biolabels for plants. *J. Fluoresc.* **2011**, *21*, 2057–2068.
87. Navarro, D.A.; Bisson, M.A.; Aga, D.S. Investigating uptake of water-dispersible CdSe/ZnS quantum dot nanoparticles by *Arabidopsis thaliana* plants. *J. Hazard. Mater.* **2012**, *211*, 427–435.
88. Kumari, M.; Mukherjee, A.; Chandrasekaran, N. Genotoxicity of silver nanoparticles in *Allium cepa*. *Sci. Total Environ.* **2009**, *407*, 5243–5246.
89. Lòpez-Moreno, M.L.; de la Rosa, G.; Hernàndez-Viezcas, J.A.; Castillo-Michel, H.; Botez, C.E.; Peralta-Videa, J.R.; Gardea-Torresdey, J.L. Evidence of the differential biotrasformation and genotoxicity of ZnO and CeO_2 nanoparticles on soybean (*Glycine max*) plants. *Environ. Sci. Technol.* **2010**, *44*, 7315–7320.
90. Ruffini, C.M.; Giorgetti, L.; Geri, C.; Cremonini, R. The effects of nano-TiO_2 on seed germination, development and mitosis of root tip cells of *Vicia narbodensis* L. and *Zea mays* L. *J. Nanopart. Res.* **2011**, *13*, 2443–2449.
91. Atha, D.H.; Wang, H.; Ptersen, E.J.; Cleveland, D.; Holbrook, D.R.; Jaruga, P.; Dizdaroglu, M.; Xing, B.; Nelson, C.B. Copper oxide nanoparticle mediated DNA damage in terrestrial plant models. *Environ. Sci. Technol.* **2012**, *46*, 1819–1827.
92. Wang, X.; Han, H.; Liu, X.; Gu, X.; Chen, K.; Lu, D. Multiwalled carbon nanotubes can enhance root elongation of wheat (*Triticum aestivum*) plants. *J. Nanopart. Res.* **2012**, *14*, 841–848.
93. Mondal, A.; Basu, R.; Das, S.; Nandy, P. Beneficial role of carbon nanotubes on mustard plant growth: An agricultural prospect. *J. Nanopart. Res.* **2011**, *13*, 4519–4528.

94. Villagarcia, H.; Dervishi, E.; de Silva, K.; Biris, A.S.; Khodakovskaya, M.V. Surface chemistry of carbon nanotubes impacts the growth and expression of water channel protein in tomato plants. *Small* **2012**, *8*, 2328–2334.
95. Khodakovskaya, M.V.; Kim, B.S.; Kim, J.N.; Alimohammadi, M.; Dervishi, E.; Mustafa, T.; Cernigla, C.E. Carbon nanotubes as plant growth regulators: Effects on tomato growth, reproductive system, and soil microbial community. *Small* **2013**, *12*, 115–123.
96. Flores, D.; Chacón, R.; Alvarado, L.; Schmidt, A.; Alvarado, C.; Chaves, J. Effect of using two different types of carbon nanotubes for blackberry (*Rubus adenotrichos*) *in vitro* plant rooting, growth and histology. *Am. J. Plant Sci.* **2014**, *5*, 3510–3518.
97. Anjum, N.A.; Singh, N.; Singh, M.K.; Sayeed, I.; Duarte, A.C.; Pereira, E.; Ahmad, I. Single-bilayer graphene oxide sheet impacts and underlying potential mechanism assessment in germinating faba bean (*Vicia faba* L.). *Sci. Total Environ.* **2014**, *472*, 834–841.
98. Begum, P.; Fugetsu, B. Induction of cell death by graphene in *Arabidopsis thaliana* (Columbia ecotype) T87 cell suspensions. *J. Hazard. Mater.* **2013**, *260*, 1032–1041.
99. Begum, P.; Fugetsu, B. Phytotoxicity of multi-walled carbon nanotubes on red spinach (*Amaranthus tricolor* L) and the role of ascorbic acid as an antioxidant. *J. Hazard. Mater.* **2012**, *243*, 212–222.
100. Hu, X.; Kang, J.; Lu, K.; Zhou, R.; Mu, L.; Zhou, Q. Graphene oxide amplifies the phytotoxicity of arsenic in wheat. *Sci. Rep.* **2014**, *4*, doi:10.1038/srep06122.
101. Khodakovskaya, M.V.; de Silva, K.; Dervishi, E.; Villagarcía, H. Carbon nanotubes induce growth enhancement of tobacco cell. *ACS Nano* **2012**, *6*, 2128–2135.
102. Ghosh, M.; Chakraborty, A.; Bandyopadhyay, M.; Mukherjee, A. Multi-walled carbon nanotubes (MWCNT): Induction of DNA damage in plant and mammalian cells. *J. Hazard. Mater.* **2011**, *197*, 327–336.
103. Yan, S.; Zhao, L.; Li, H.; Zhang, Q.; Tan, J.; Huang, M.; He, S.; Li, L. Single-walled carbon nanotubes selectively influence maize root tissue development accompanied by the change in the related gene expression. *J. Hazard. Mater.* **2013**, *246*, 110–118.
104. Punt, W.; Blackmore, S.; Nilsson, S.; Le Thomas, A. *Glossary of Pollen and Spore Terminology*; LPP Contributions Series No.1; LPP Foundation: Utrecht, The Netherlands, 1994.
105. Speranza, A.; Leopold, K.; Maier, M.; Taddei, A.R.; Scoccianti, V. Pd-Nanoparticles cause increased toxicity to kiwifruit pollen compared to soluble Pd (II). *Environ. Pollut.* **2010**, *158*, 873–882.
106. Speranza, A.; Crinelli, R.; Scoccianti, V.; Taddei, A.R.; Iacobucci, M.; Bhattacharya, P.; Chun Ke, P. *In vitro* toxicity of silver nanoparticles to kiwifruit pollen exhibits peculiar traits beyond the cause of silver ion release. *Environ. Pollut.* **2013**, *179*, 258–267.
107. Wang, Q.; Ebbs, S.; Chen, Y.; Ma, X. Trans-generational impact of cerium oxide nanoparticles on tomato plants. *Metallomics* **2013**, *5*, 753–759.
108. Chen, X.; Irani, N.G.; Friml, J. Clathrin-mediated endocytosis: The gateway into plant cells. *Curr. Opin. Plant Biol.* **2011**, *14*, 674–682.
109. Popp, C.; Burghardt, M.; Friedman, A.; Riederer, M. Characterization of hydrophilic and lipophilic pathways of *Hedera helix* L. cuticolar membrane: Permeation of water and uncharged organic compounds. *J. Exp. Bot.* **2005**, *56*, 2797–2806.

110. Kerstiens, G. Water transport in plant cuticles: An update. *J. Exp. Bot.* **2006**, *57*, 2493–2499.
111. Lucas, J.W. Plant viral movement proteins: Agents for cell-to-cell trafficking of viral genomes. *Virology* **2006**, *344*, 169–184.
112. De La Torre-Roche, R.; Hawthorne, J.; Deng, Y.; Xing, B.; Cai, W.; Newman, L.A.; Wang, Q.; Ma, X.; Hamdi, H.; White, J.C.; *et al.* Multiwalled carbon nanotubes and C_{60} fullerenes differentially impact the accumulation of weathered pesticides in four agricultural plants. *Environ. Sci. Technol.* **2013**, *47*, 12539–12547.
113. De la Torre-Roche, R.; Hawthorne, J.; Musante, C.; Xing, B.; Newman, L.A.; Ma, X.; White, J.C. Impact of Ag nanoparticle exposure on p,p′-DDE bioaccumulation by *Cucurbita pepo* (zucchini) and *Glycine max* (soybean). *Environ. Sci. Technol.* **2013**, *47*, 718–725.
114. Dimkpa, C.O.; McLean, J.E.; Britt, D.W.; Anderson, A.J. Antifungal activity of ZnO nanoparticles and their interactive effect with a biocontrol bacterium on growth antagonism of the plant pathogen *Fusarium graminearum. Biometals* **2013**, *26*, 913–924.
115. Husen, A.; Siddiqi, K.S. Carbon and fullerene nanomaterials in plant system. *J. Nanotechnol.* **2014**, *12*, 16–25.
116. Maurer-Jones, M.; Gunsolus, I.L.; Murphy, C.J.; Haynes, C.L. Toxicity of Engineered Nanoparticles in the Environment. *Anal. Chem.* **2013**, *85*, 3036–3049.
117. Kuppusamy, P.; Yusoff, M.M.; Govindan, N. Biosynthesis of metallic nanoparticles using plant derivatives and their new avenues in pharmacological applications—An updated report. *Saudi Pharm. J.* **2015**, doi:10.1016/j.jsps.2014.11.013.
118. Auffan, M.; Tella, M.; Santaella, C.; Brousset, L.; Paillès, C.; Barakat, M.; Espinasse, B.; Artells, E.; Issartel, J.; Masion, A.; *et al.* An adaptable mesocosm platform for performing integrated assessments of nanomaterial risk in complex environmental systems. *Sci. Rep.* **2014**, *4*, 5608–5615.
119. Cleveland, D.; Long, S.E.; Pennington, P.L.; Cooper, E.; Fulton, M.H.; Scott, G.I.; Brewer, T.; Davis, J.; Petersen, E.J.; Wood, L. Pilot estuarine mesocosm study on the environmental fate of silver nanomaterials leached from consumer products. *Sci. Total Environ.* **2012**, *421–422*, 267–272.
120. Herlekar, M.; Barve, S.; Kumar, R. Plant-mediated green synthesis of iron nanoparticles. *J. Nanopart.* **2014**, *2014*, doi:10.1155/2014/140614.
121. Petersen, E.J.; Henry, T.B.; Zhao, J.; MacCuspie, R.I.; Kirschling, T.L.; Dobrovolskaia, M.A.; Hackley, V.; Xing, B.; White, J.C. Identification and avoidance of potential artifacts and misinterpretations in nanomaterial ecotoxicity measurements. *Environ. Sci. Technol.* **2014**, *48*, 4226–4246.
122. Chichiriccò, G.; Picozzi, P. Reversible inhibition of the pollen germination and the stigma penetration in *Crocus vernus* ssp. *vernus* (Iridaceae) following fumigations with NO_2, CO_2, and O_3 gases. *Plant Biol.* **2007**, *9*, 730–735.

14

Thermal Emission Control via Bandgap Engineering in Aperiodically Designed Nanophotonic Devices

Enrique Maciá

Materials Physics Department, Faculty of Physics, University Complutense of Madrid, E-28040 Madrid, Spain; E-Mail: emaciaba@fis.ucm.es

Academic Editor: Lorenzo Rosa

Abstract: Aperiodic photonic crystals can open up novel routes for more efficient photon management due to increased degrees of freedom in their design along with the unique properties brought about by the long-range aperiodic order as compared to their periodic counterparts. In this work we first describe the fundamental notions underlying the idea of thermal emission/absorption control on the basis of the systematic use of aperiodic multilayer designs in photonic quasicrystals. Then, we illustrate the potential applications of this approach in order to enhance the performance of daytime radiative coolers and solar thermoelectric energy generators.

Keywords: thermal emission; aperiodic order; nanophotonics

---

1. Introduction

During the last few years, the search for materials exhibiting quite specific thermal emission properties has been spurred on by the necessity of attaining high-performance devices for light harvesting and thermal energy control. Thus, devices displaying either very low or very high emissivity values over relatively broad frequency windows are usually required. For instance, efficient daytime radiative coolers must radiate heat to outer space through a transparency window in the Earth's atmosphere ranging between 8 and 13 μm, but they should also reflect most incoming solar radiation within the UV and visible wavelengths [1]. Unfortunately, no known existing natural materials are able to perform this way, so one must synthesize them on purpose, and the original search turns into a materials engineering task.

A very convenient way to address this task consists in coating a suitable bulk material with a properly designed multilayered thin film, in order to exploit the beneficial optical properties contributed by each component in a synergetic way. Within this approach the coat can act either as a passive filter (when the refractive indices of the layers material are all real valued) or as an active emitter (when at least one of the layers material in the film has a complex valued refraction index). In both cases, the resulting thermal radiation spectrum of the composite structure can be substantially modified as compared to that corresponding to the original substrate alone, due to resonance effects within the multilayered coat [2].

These appealing results were originally obtained by considering a periodic sequence of layers in the thin film coat. Though the recourse to periodic order is quite convenient for the sake of simplicity, during the last decade it has been progressively realized that ordered structures can be suitably expanded to embrace not only periodic arrangements but aperiodic ones as well. In this way, the very notion of aperiodic order, that is, order without periodicity, has been explored to properly describe a growing number of physical systems, including metallic and dielectric multilayers [3–8]. Thus, well-known materials still remain in use, along with the previously gained technological expertise, as the attention is entirely focused on the (relatively low-cost) aspects related to device architecture optimization.

A typical example of an ordered aperiodic multilayer is provided by a structure consisting of a number of layers stacked according to a certain deterministic rule, for instance, one given by a substitution sequence. A simple example of such a nanostructured material is a two-component Fibonacci multilayer, where layers of two different materials (metallic, semiconductor, dielectric, ferroelectric) are arranged according to the celebrated Fibonacci substitution rule A→AB and B→A, whose successive application generates the sequence of layers A → AB → ABA → ABAAB → ABAABABA → ABAABABAABAAB → ... and so on. The number of layers in a sequence of order j is given by the Fibonacci number F_j, which is obtained from the recurrence relation $F_j = F_{j-1} + F_{j-2}$, $j \geq 2$, with $F_0 = F_1 \equiv 1$. Another example of aperiodic multilayer is provided by the substitution rule A → ABA and B → BBB, generating the fractal triadic Cantor sequence.

A key feature of these man-made materials is the presence of two kinds of order in the same sample at different length scales: at the atomic level we have the usual crystalline order determined by the periodic arrangement of atoms in each layer, whereas at longer scales we have the aperiodic order determined by the sequential deposition of the different layers. This aperiodic order is imposed during the growth process and can be precisely controlled. Since different physical phenomena have their own relevant physical scales, by properly matching the characteristic length scales we can efficiently exploit the aperiodic order we have introduced in the system, hence opening new avenues for technological innovation.

In particular, due to their characteristic highly fragmented frequency spectrum, aperiodic multilayers provide more full transmission peaks (alternatively, absorption bands) than periodic ones in a given frequency range for a given system length. This feature stems from the richer structural complexity of aperiodic sequences (related to the presence of quasiperiodic and/or self-similar order related fingerprints), naturally leading to the presence of more resonant frequencies stemming from multiple interference effects throughout the structure. The existence of more frequencies for operation becomes particularly useful when looking for selective enhancement (suppression) of thermal emission for some selected frequencies. Thus, by properly exploiting the *additional degrees of freedom* inherent to the aperiodic arrangements, one can design optical devices offering a broader and more flexible *design capability* than their periodic counterparts for certain specific applications [3–7].

2. Aperiodicity by Design

In the early years, most studied aperiodic multilayers were based on the systematic application of certain mathematically inspired substitution sequences in order to ascertain their fundamental properties. As the field of aperiodic order based photonics is coming to maturity a logical extension naturally appears: it consists in considering structures whose layers are intentionally arranged according to purposely defined aperiodic designs, able to tailor their physical properties to fit some specified requirement. In so doing, convenient aperiodic orderings are created by hand, generally following suitable optimization algorithms, thereby substantially enriching the variety of aperiodic orders of interest in both fundamental and applied science domains. We will refer to the systems obtained in this way as aperiodic systems by design [3,5].

This procedure is an example of what is generally known as a reverse engineering problem. In such cases, the ideal structures are found by numerically solving an optimization problem where a number of optimization parameters (e.g., the layer thicknesses or the incidence angle) are treated as independent variables under certain physical constraints (e.g., the requirement of a small enough total length value for the multilayer). Then, a merit function to be optimized is defined in terms of the physical magnitudes of interest in the considered device. For instance, if one is interested in optimizing the reflectance $R(\lambda)$ of a coating over a given spectral range one may use a function of the form:

$$f = \frac{1}{N}\sum_{\ell=1}^{N}\left[R_0 - R(\lambda_\ell)\right]^2 \tag{1}$$

where N is the total number of layers, R_0 is a reflectivity reference value, and λ_ℓ is the radiation wavelength propagating through the ℓ-th layer. Following this approach a number of Mo/Si aperiodic multilayers were designed and grown in order to attain a maximum reflectivity over a specified range of wavelengths or angles of incidence for their use as extreme-ultraviolet (XUV) radiation mirrors [9], or analyzers [10]. Similar approaches have been used to design efficient aperiodic W/Si multilayer mirrors for X-ray plasma diagnostic [11], or narrowband optical filters able to simultaneously tune multiple wavelengths [12]. More elaborated procedures, based on the application of the so-called genetic algorithm [13], have been successfully applied in order to design omnidirectional reflectors [14].

When trying to optimize some specific application by means of aperiodic order engineering a key question concerns the relationship between the structural order of the multilayer itself and its related physical properties. In multilayered optical devices the structural order naturally translates in terms of the refractive index spatial profile $n(z)$ function, where z is the stacking direction (Figure 1). Whereas aperiodic substitution sequences can only define relatively restricted types of aperiodic refractive index profiles, the recourse to analytical functions including suitable free parameters allows for a more general treatment, so that one can describe different kinds of refractive index distributions within a unified mathematical approach.

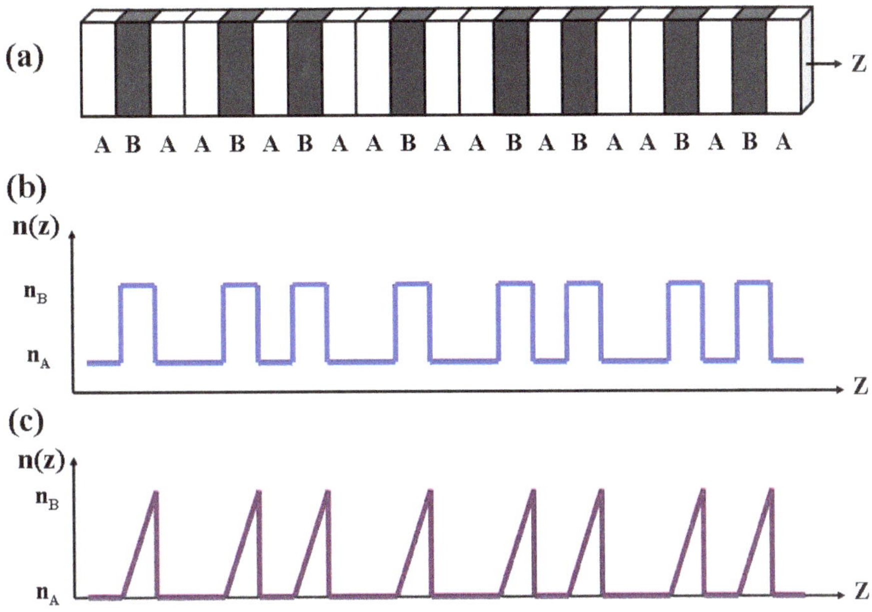

Figure 1. (**a**) Sketch of a Fibonacci dielectric multilayer grown along the *z* direction; (**b**) Refractive index profile *n*(*z*) for an arrangement of layers with constant refraction index; (**c**) Refractive index profile *n*(*z*) for an arrangement of layers with a linearly graded refraction index.

As an illustrative example one can consider a binary multilayer composed of layers whose refraction index values are assigned depending on the sign of the auxiliary function $V_\ell = \cos(\alpha\pi\ell^\nu)$ where ℓ labels the layer sequence, α and ν being real parameters, so that $n_\ell \to n_A$ if $V_\ell \leq 0$, and $n_\ell \to n_B$ otherwise [15]. For $\nu = 1$ and rational α one has a periodic refractive index profile, whereas for an irrational α value the multilayer sequence becomes quasiperiodic. In addition, for $\nu < 1$ (alternatively, $\nu > 1$) the refractive index profile is described by an aperiodic modulation whose wavelength progressively increases (decreases) as ℓ increases. Accordingly, by properly adjusting the values of the design parameters α and ν we can obtain a great diversity of aperiodic refractive index profiles. Thus, the parameter ν can be regarded as an exponent controlling the aperiodicity degree of the considered structure. Numerically obtained transmittance curves as a function of ν have been recently reported for multilayers based on this sequence in a systematic way (with $\alpha = \frac{\sqrt{5}-1}{2}$, the reciprocal of the golden mean). The obtained results show the existence of a characteristic ν value leading to averaged reflectivity values larger than those corresponding to both periodic and random structures of comparable size [16].

3. Thermal Emission Control

As it is well-known, even in the case of highly absorptive compounds, not all incident radiation is usually absorbed by a given material. This is because the material reflects some radiation at its surface. Therefore, most absorbing materials exhibit emissivity values lower than unity. An ingenious way to circumvent this shortcoming, which allows one to enhance the thermal emissivity of a given material at certain frequencies, was proposed on the basis of the structural design depicted in Figure 2. As it was previously discussed, the main idea is to coat the bulk substrate with a thin film made of alternating layers of different materials. In so doing, one aims to exploit quantum confinement effects in order to modify the emission processes through the multilayered structure, thereby properly engineering the thermal radiation characteristics of the overall system.

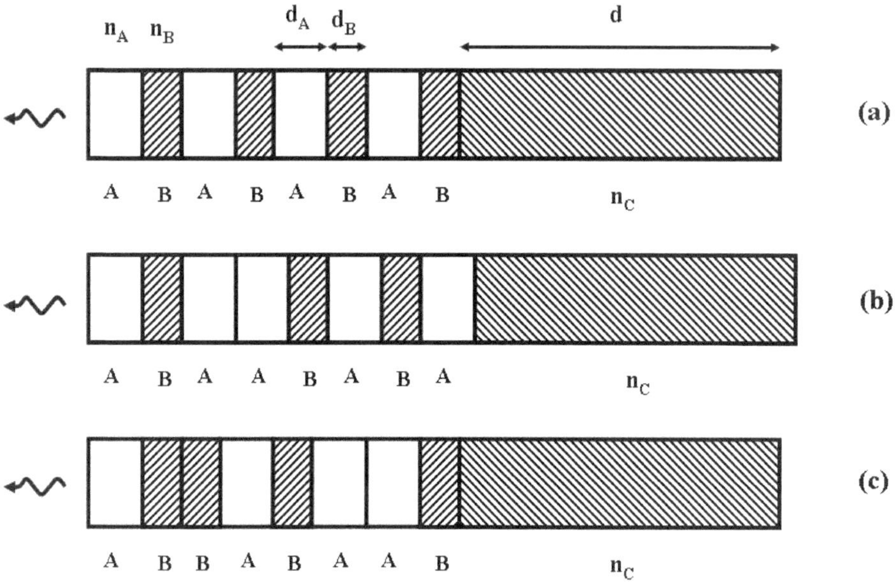

Figure 2. A multilayered stack (thin film) made of alternating layers of refractive indices n_A (white layers) and n_B (shadowed layers) coats a thick absorbing substrate of refraction index n_C (complex). The entire structure is embedded in a medium, taken to be air ($n = 1$). Different working regimes are obtained depending on the adopted layers distribution. In (**a**) the layers are periodically distributed, whereas in (**b**) and (**c**) they are respectively arranged according to the Fibonacci and Thue-Morse (A → AB, B → BA) sequences.

A convenient way to deal with this system within the framework of the transfer matrix technique relies on the evaluation of the so-called thermal optical power:

$$E(\omega) = 1 - R(\omega) - T(\omega) \tag{2}$$

where $R(\omega)$ and $T(\omega)$ are the reflectance and transmittance, respectively. $E(\omega)$ is physically interpreted as the ratio of the optical power emitted at frequency ω into a spherical angle element dΩ by a unit surface area of the thin film, to the power emitted by a blackbody with the same area at the same temperature. In this way, the power spectrum of the multilayered film located in front of an emitting hot surface (Figure 2) is given by the Equation (2):

$$P(\omega,\mathrm{T}) = E(\omega)\, P_B(\omega,\mathrm{T}) \tag{3}$$

where

$$P_B(\omega,T) = \frac{\hbar\omega^3}{2\pi^2c^2}\frac{1}{e^{\frac{\hbar\omega}{k_BT}}-1} \tag{4}$$

is the Planck's law [2]. Some illustrative results are shown in Figure 3.

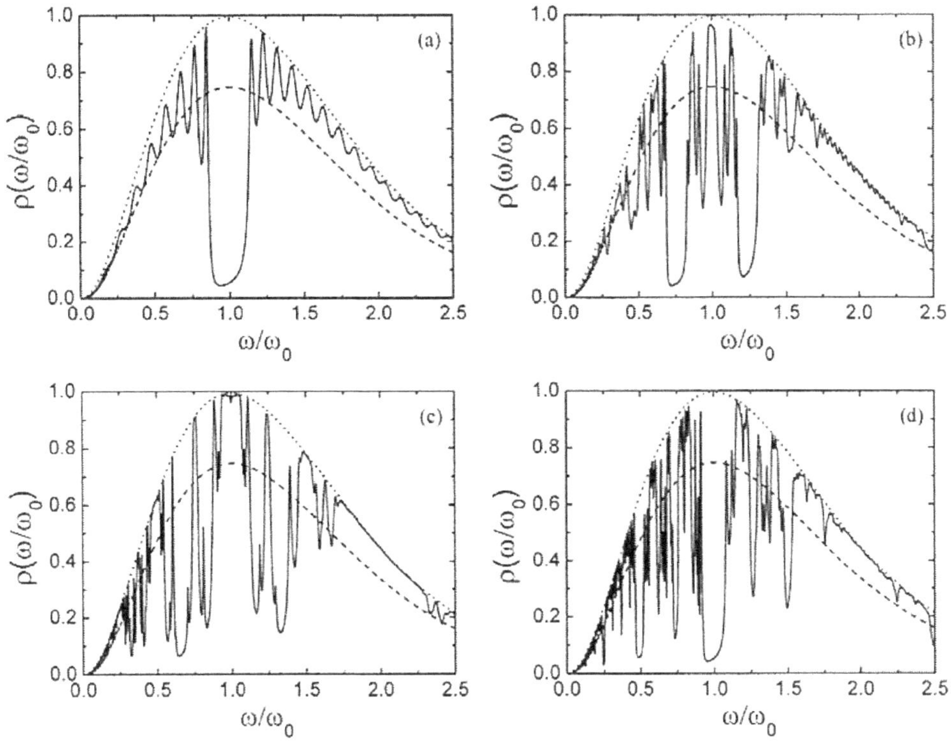

Figure 3. Thermal radiation spectra as a function of the reduced frequency under normal incidence conditions for the devices shown in Figure 2 with $n_A = 1.45$ (SiO_2) and $n_B = 1.0 + 0.01i$, where the thin film coat layers are respectively arranged according to the following sequences: (**a**) Periodic; (**b**) Fibonacci ($N = 377$). (**c**) Thue-Morse ($N = 512$); and (**d**) period-doubling ($N = 512$). The perfect blackbody thermal spectrum is given by the dotted curve, whereas the dashed curve gives the thermal spectrum of the substrate, with refractive index $n_C = 3 + 0.03i$. The temperature is chosen so that the blackbody (Wien) peak is aligned with the midgap frequency $\omega_0 = 2\pi c/\lambda_0$ ($\lambda_0 = 700$ nm). All the curves are properly normalized by this peak power. (Adapted from [17]. Copyright Courtesy of Eudenilson L. Albuquerque, 2007).

Let us first consider the case where a periodic film coating sits atop the heated substrate (Figure 3a). As we see the film significantly blocks heat radiation emitted by the substrate at the frequencies corresponding to the photonic crystal bandgap ($\omega/\omega_0 = 1$) as expected, but we also observe that the substrate's emission is enhanced from the gray-body level all the way up to the perfect blackbody rate at a number of frequencies corresponding to the pass-band transmission resonances of the multilayered film. This occurs because the thin film acts as an antireflective coating at these resonances. In this way, all the incident radiation tunnels through the multilayer structure into the substrate for these selected frequencies, so that the substrate effectively behaves as a perfectly absorbing blackbody in that case [2,17] A similar enhancement of the substrate's thermal emittance at certain resonance frequencies, accompanied by the corresponding inhibition at the stop-bands, is observed in the aperiodically arranged thin film coatings as well. In the case of the Fibonacci coating (Figure 3b) a characteristic trifurcation splitting can be clearly appreciated around a number of frequencies, and one finds a strong emittance within the spectral range corresponding to the midgap in the periodic case. In fact, the presence of allowed bands in certain forbidden regions of the periodic system is a characteristic feature of quasiperiodic systems [3–7].

An analogous pattern is observed in the thermal spectrum corresponding to the Thue-Morse thin film, whereas that corresponding to the period-doubling sequence is more similar to the periodic one around the $\omega/\omega_0 = 1$ spectral range. In all the aperiodic arrangements, however, one has a richer thermal emission spectrum, due to the highly fragmented nature of their transmission profiles. Quite interestingly, these spiky thermal emission spectra can be substantially smoothed (hence obtaining broader spectral ranges with enhanced emittance) by using metamaterials in the composition of aperiodic multilayer coats [17,18].

4. Possible Applications

4.1. Daytime Radiative Coolers

A passive cooler is a device which decreases its temperature below that of the ambient air without any electricity (or any other form of energy) input. Passive cooling was first demonstrated at night using a technique known as radiative cooling, in which a device exposed to the sky is used to radiate heat to outer space through a transparency window in the atmosphere ranging between 8 and 13 μm [19]. Daytime radiative cooling under direct sunlight was subsequently shown by properly extending the same fundamental technique using a photonic radiative cooler. To this end, a thin film multilayer composed of seven alternating layers of hafnium dioxide (high refraction index) and silicon dioxide (low refraction index) of varying thicknesses were deposited on top of a 200 nm silver substrate. The resulting coating was in turn deposited on top of a 750 μm thick, 200 mm diameter silicon wafer. The bottom four layers in the coating film have thicknesses within the range 10–70 nm, and assist in optimizing solar reflection in a manner akin to that used in periodic one-dimensional photonic crystals. The top three layers are one order of magnitude thicker (within the range 200–700 nm) and are primarily responsible for thermal radiation from the cooler device. Thus, it is the presence of two different kinds of layer sizes which ultimately accomplishes the two functional tasks required for the coating, namely, high solar reflectance (due to the shorter layers stack) and high infra-red emissivity within the 8–13 μm window (due to the longer layers stack). Both the total number and the proper thickness values of the alternating low/high refraction index layers was determined by using a numerical optimization method. Making use of this device a cooling of about 5 °C below ambient air temperature, amounting to 40 Wm^{-2} cooling power, has been reported under direct sunlight [1,20].

These promising figures were obtained by properly combining suitable material properties and interference effects in an integrated structure that collectively achieves high solar reflectance (97% at normal incidence) and strong thermal emission in the selected infra-red frequency window. Nevertheless, some room is still left for further improvement (say, full solar reflectance along with perfect transmission through the Earth's atmosphere infra-red windows), and one may wonder if these goals may be achieved by considering alternative layer sequence *orderings* for the different layers. Indeed, the possibility of attaining a broadband reflectance, covering a nearly full-visible spectrum window has been recently demonstrated in a TiO_2 nanotube-based photonic crystal whose lattice constant is *gradually* decreased according to an *arithmetic series* of the form $d_\ell = d_0 + (\ell - 1)d$, $\ell = 1, \ldots, N$, where d_0 is the thickness of the first layer and d is the common difference. In this way, a flattened reflectance peak reaching 100% intensity within the range 400–600 nm was numerically obtained by properly ensambling two different slabs containing $N = 8$ layers (with $d = 3.6$ nm, and $d_0 = 165.2$ nm or $d'_0 = 194$ nm, respectively) to form two $N = 16$ bilayers, which were in turn stacked to each other to get a $N = 32$ multilayer with a total length $L = 819.2$ nm [21]. Note that though the total length of this graded multilayer reflector amounts to about four times that corresponding to the solar reflector component used in the daytime cooler device in Reference [20] (*i.e.*, $L = 174$ nm), its characteristic *modular design* can readily explain its ability to ultimately outperform the latter one. In fact, this property can be readily understood by regarding the overall structure as a modular stack of shorter photonic crystals with different lattice parameters, each one reflecting light with a peak located at its respective Bragg resonance condition. Thus, by properly selecting the location and width of each module main reflectance peaks, one may design a much broader, almost flat, reflectance spectrum profile, naturally stemming from the different modules spectra superimposition.

Further elaborating on this physical picture it was numerically shown that an even wider full reflectance band (ranging from 380 to 780 nm) could be obtained by considering hierarchical photonic crystals consisting in geometrically distributed layers, where the aperiodic order of the $n(z)$ profile is given in terms of a *geometric series* rather than an arithmetic one [22]. In this study the influence of the angle of incidence and the light polarization was carefully analyzed, and the refraction index dispersion effects in the layers was explicitly taken into account, hence providing a quite realistic description. Geometric series naturally introduce a scale factor in the resulting refraction index profile, a feature they share with fractal structures characterized by self-similar symmetry. Accordingly, it is reasonable to expect that additional improvement in both the solar reflectance and the infra-red thermal emission properties could be attained in radiative cooling devices able to exploit the additional design degree of freedom provided by the scalability property characteristic of fractal multilayers geometries [23,24]. Thus, it has been shown that, for every type of fractal structure, the transmission spectrum of any given generation $k > 2$ contains embedded transmission spectra of all preceding generations $j = 1, 2, \ldots, k - 1$, the spectrum of every preceding generation being squeezed along the frequency axis by a scale factor s^{k-j} [25–28]. The progressive fragmentation of the frequency spectrum as k increases gives rise to narrower transmission bands, which in turn result from preferential localization of light-wave energy in certain resonant cavities present in the structure. In fact, one easily realizes that the triadic Cantor second generation structure can be regarded as a BBB cavity embedded by a pair of ABA mirrors. For higher generations the number of existing cavities grows by a factor of three and the related transmission peaks split accordingly.

Finally, an additional step in this route towards higher order designs, one may consider quasiperiodic structures, which combine the self-similar features typical of fractal geometries with the very possibility of including graded refraction index profiles. A suitable example is illustrated in Figure 1c and it consists in two kinds of layers: one has a constant refraction index (n_A) and the other has a linearly varying refraction index given by the expression,

$$n_B = n_A + \frac{n_B - n_A}{d_B} z \tag{5}$$

where d_B is the B layer thickness [29]. Of course, instead of simply linear one may consider an exponential or any other refraction index gradation laws as well [30]. In turn, layers A and B are arranged according to a sequence prescribed by a given substitution rule (*i.e.*, Fibonacci, Thue-Morse,

Period-doubling). In accordance with the obtained numerical results significantly broadband full reflectance profiles, covering a given spectral range, could be obtained by properly combining these aperiodic photonic crystals in modular designs.

4.2. Solar Cells and Solar Thermoelectric Generators

Thin film solar cells with thicknesses ranging from hundreds of nm to a few μm are attractive candidates for low-cost replacement of significantly thicker wafer-based devices, due to reduced costs of raw materials and processing. Light trapping photon management strategies rendering this cell optically thick enough despite its limited physical thickness is then crucial to attain high efficiency solar cells. To this end, the design of back-reflectors acting as diffraction gratings able to turn the incident light by an angle of 90° at specific frequencies (resonant photonic modes) that can be guided and thus absorbed along the plane of the cell have proved very useful devices. Though earlier designs were based in periodic layer sequences, more recent works have analyzed in detail the richness of quasiperiodic geometries, demonstrating that a nano-grating grown according to the Fibonacci sequence used as back-reflector in a thin film solar cell guarantees higher absorption enhancement with respect to the periodic counterpart over the whole incidence angle range, providing a slightly better performance of about 5% for angles less than 20° and for angles greater than 45° [31]. This improvement stems from the denser Bragg spectrum of Fibonacci lattices which allows to reach the phase match condition over more wavelengths, hence leading to a higher multiple resonance pattern for a given spectral range. This property also renders the quasiperiodic back-reflector more robust with respect to the incoming light incidence angle variations.

The aperiodic order based design strategy, used for metallic back reflectors to improve light absorption in thin film solar cells, can be readily extended to solar thermoelectric generators as well. In this case, the key element is a thin film heat absorber which collects the solar radiation concentrated through a suitable lens. In order to optimize the device efficiency one requires a wavelength-selective surface absorptance for this film to precisely match the solar spectrum distribution in the near UV, visible and near IR domains [32]. Whereas most experimental substrates considered up to now rely on specific materials, rather than engineered photonic crystals [33], the results reviewed in the previous sections clearly suggest that one may expect significant improvement in the spectral response of absorbing films based on aperiodic designs.

5. Conclusions

In this work we have reviewed the convenience of exploiting different aperiodic thin film designs for their use as coating materials in photonic devices requiring a high thermal emission spectrum combined with full reflectance properties at certain relatively broadband spectral windows. These aperiodic designs can be classified according to their increasing order of structural complexity. Thus, one can consider graded structures whose layer widths are determined by: (1) arithmetic or geometric series; (2) fractal self-similar patterns; (3) quasiperiodic sequences given by substitution rules; and (4) analytical functions able to encompass some of the previous refraction index profiles in a unified way. An important aspect supporting the use of these aperiodic order based designs concerns the purported robustness of their frequency spectra upon the unavoidable layer width fluctuations appearing during the growth of actual multilayers [7,28]. Accordingly, a deeper study of some of the proposed aperiodic designs may well deserve a closer experimental scrutiny. In doing so, it would also be interesting to consider coatings based on organic multilayers, as a suitable alternative to the inorganic materials (silicon, oxides, fluorides) mostly considered up to now [34–36].

Acknowledgments

The author warmly thanks Luca Dal Negro, Sergey V. Gaponenko, Juan Antonio Monsoriu, and Rue Wen Peng for many useful comments and for sharing different materials of common interest during the last decade. The author thanks Eudenilson L. Albuquerque for kindly providing one of the pictures which illustrate the text. The author is grateful to Victoria Hernández for a critical reading of the manuscript.

Conflicts of Interest

The author declares no conflict of interest.

References

1. Raman, A.P.; Anoma, M.A.; Zhu, L.; Rephaeli, E.; Fan, S. Passive radiative cooling below ambient air temperature under direct sunlight. *Nature* **2014**, *415*, 540–545.
2. Cornelius, C.M.; Dowling, J.P. Modification of Planck blackbody radiation by photonic band-gap structures. *Phys. Rev. A* **1999**, *59*, 4736–4746.
3. Dal Negro, L.; Lawrence, J.; Trevine, J.; Walsh, G. Aperiodic Order for Nanophotonics. In *Optics of Aperiodic Structures: Fundamentals and Device Applications*; Dal Negro, L., Ed.; Pan Stanford Publishing: Boca Raton, FL, USA, 2014; pp. 1–55.
4. Maciá, E. The Importance of being Aperiodic: Optical Devices. In *Optics of Aperiodic Structures: Fundamentals and Device Applications*; Dal Negro, L., Ed.; Pan Stanford Publishing: Boca Raton, FL, USA, 2014; pp. 57–90.
5. Maciá, E. Exploiting aperiodic designs in nanophotonic devives. *Rep. Prog. Phys.* **2012**, *75*, 036502.
6. Dal Negro, L.; Boriskina, S.V. Deterministic aperiodic nanostructures for photonics and plasmonics applications. *Laser Photonics Rev.* **2012**, *6*, 178–218.
7. Maciá Barber, E. *Aperiodic Structures in Condensed Matter: Fundamentals and Applications*; Taylor & Francis CRC: Boca Raton, FL, USA, 2009.
8. Maciá, E. The role of aperiodic order in science and technology. *Rep. Prog. Phys.* **2006**, *69*, 397–441.
9. Vishnyakov, E.A.; Kamenets, F.F.; Kondratenko, V.V.; Luginin, M.S.; Panchenko, A.V.; Pershin, Y.P.; Pirozhkov, A.S.; Ragozin, E.N. Aperiodic multilayer structures in soft X-ray radiation optics. *Quantum Electron.* **2012**, *42*, 143–152.
10. Wang, H.; Zhu, J.; Wang, Z.; Zhang, Z.; Zhang, S.; Wu, W.; Chen, L.; Michette, A.G.; Powell, A.K.; Pfauntsch, S.J.; *et al.* Broadband Mo/Si multilayer analyzers for the 15–17 nm wavelength range. *Thin Solid Films* **2006**, *515*, 2523–2526.

11. Champeaux, J.P.; Troussel, P.; Villier, B.; Vidal, V.; Khachroum, T.; Vidal, B.; Krumrey, M. Development and realization of non-periodic W/Si multilayer mirrors for 5–14 keV X-ray plasma diagnostic. *Nuclear Instrum. Methods Phys. Res. A* **2007**, *581*, 687–694.
12. Gu, X.; Chen, X.; Chen, Y.; Zeng, X.; Xia, Y.; Chan, Y. Narrowband multiple wavelengths filter in aperiodic optical superlattice. *Opt. Commun.* **2004**, *237*, 53–58.
13. Holland, J. *Adaptation in Natural and Artificial Systems*, University Michigan Press: Ann Arbor, MI, USA, 1975.
14. Jiang, L.; Zheng, G.; Shi, L.; Yuan, J.; Li, X. Broad omnidirectional reflectors design using genetic algorithm. *Opt. Commun.* **2008**, *281*, 4882–4888.
15. Griniasty, M.; Fishman, S. Localization by pseudorandom potentials in one dimension. *Phys. Rev. Lett*. **1988**, *60*, 1334–1338.
16. Nascimento, E.M.; de Moura, F.A.B.F.; Lyra, M.L. Suppressed transmission in aperiodically modulated multilayered dielectric structures. *Photonic. Nanostruct.* **2009**, *7*, 101–107.
17. De Medeiros, F.F.; Albuquerque, E.L.; Vasconcelos, M.S.; Maritz, P.W. Thermal radiation in quasiperiodic photonic crystals with negative refractive index. *J. Phys.* **2007**, *19*, 496212.
18. Maksimovic, M.; Jaksic, Z. Emittance and absortance tailoring by negative refractive index metamaterial-based Cantor multilayers. *J. Opt. A* **2006**, *8*, 355–362.
19. Gentle, A.R.; Smith, G.B. Radiative heat pumping from the earth using surface phonon resonant nanoparticles. *Nano Lett.* **2010**, *10*, 373–379.
20. Rephaeli, E.; Raman, A.; Fan, S. Ultrabroadband photonic structures to achive high-performance daytime radiative cooling. *Nano Lett.* **2013**, *13*, 1457–1461.
21. Guo, M.; Xie, K.; Wang, Y.; Zhou, L.; Huang, H. Aperiodic TiO_2 nanotube photonic crystal: Full-visible-spectrum solar light harvesting on photonic devices. *Sci. Rep.* **2014**, *4*, 6442.
22. Alagappan, G.; Wu, P. Geometrically distributed one-dimensional photonic crystals for light-reflection in all angles. *Opt. Express* **2009**, *17*, 11550–11557.
23. Gaponenko, S.V. *Introduction to Nanophotonics*; Cambridge University Press: Cambridge, UK, 2010; pp. 258–270.
24. Zhukovsky, S.V.; Lavrinenko, A.V.; Gaponenko, S.V. Optical Filters Based in Fractal and Aperiodic Multilayers. In *Optics of Aperiodic Structures: Fundamentals and Device Applications*; Dal Negro, L., Ed.; Pan Stanford Publishing: Boca Raton, FL, USA, 2014; pp. 91–142.
25. Zhukovsky, S.V.; Lavrinenko, A.V.; Gaponenko, S.V. Spectral scalability as a result of geometrical self-similarity in fractal multilayers. *Europhys. Lett.* **2004**, *66*, 455.
26. Gaponenko, S.V.; Zhukovsky, S.V.; Lavrinenko, A.V.; Sandomirskii, K.S. Propagation of waves in layered structures viewed as number recognition. *Opt. Commun.* **2002**, *205*, 49–57.
27. Chigrin, D.N.; Lavrinenko, A.V.; Yarotsky, D.A.; Gaponenko, S.V. Observation of total omnidirectional reflection from a one-dimensional dielectric lattice. *Appl. Phys. A* **1999**, *68*, 25–28.
28. Lavrinenko, A.V.; Zhukovsky, S.V.; Sandomirski, K.S.; Gaponenko, S.V. Propagation of classical waves in nonperiodic media: Scaling properties of an optical Cantor filter. *Phys. Rev. E* **2002**, *65*, 036621.
29. Singh, B.K.; Kumar, P.; Pandey, P.C. Tunable photonic band-gaps in one-dimensional photonic crystals containing linear graded index material. *Appl. Phys. B* **2014**, *117*, 947–956.

30. Singh, B.K.; Pandey, P.C. Influence of graded index materials on the photonic localization in one-dimensional quasiperiodic (Thue-Morse and Double-Periodic) photonic crystals. *Opt. Commun.* **2014**, *333*, 84–91.
31. Micco, A.; Ricciardi, A.; Pisco, M.; La Ferrara, V.; Mercaldo, L.V.; Delli Veneri, P.; Cutolo, A.; Cusano, A. Light trapping efficiency of periodic and quasiperiodic back-reflectors for thin film solar cells: A comparative study. *J. Appl. Phys.* **2013**, *114*, 063103.
32. Chen, G. Theoretical efficiency of solar thermoelectric energy generators. *J. Appl. Phys.* **2011**, *109*, 104908.
33. De Leob, M.T.; Chong, H.; Kraft, M. Solar thermoelectric generators fabricated on a silicon-on-insulator substrate. *J. Micromech. Microeng.* **2014**, *24*, 085011.
34. Ariga, K.; Yamauchi, Y.; Rydzek, G.; Ji, Q.; Yonamine, Y.; Wu, K.C.-W.; Hill, J.P. Layer-by-layer nanoarchitectonics: Invention, innovation, and evolution. *Chem. Lett.* **2014**, *43*, 36–68.
35. Costa, R.R.; Mano, J.F. Polyelectrolyte multilayered assemblies in biomedical technologies. *Chem. Soc. Rev.* **2014**, *43*, 3453–3479.
36. Joseph, N.; Ahmadiannamini, P.; Hoogenboom, R.; Vankelecom, F.J. Layer-by-layer preparation of polyelectrolyte multilayer membranes for separation. *Polym. Chem.* **2014**, *5*, 1817–1831.

15

Electronic Structure and Magnetism of Mn-Doped ZnO Nanowires

Fuchun Zhang [1], Dandan Chao [1], Hongwei Cui [1], Weihu Zhang [1] and Weibin Zhang [2,*]

[1] College of Physics and Electronic Information, Yanan University, Yanan 716000, China; E-Mails: yadxzfc@yau.edu.cn (F.Z.); yadxddc@163.com (D.C.); hweicui001@163.com (H.C.); ydzwh@163.com (W.Z.); wbzhang@dongguk.edu (W.Z.)

[2] Department of Physics, Dongguk University, Seoul 100-715, Korea

* Author to whom correspondence should be addressed; E-Mail: zmrright@163.com

Academic Editor: Lorenzo Rosa

Abstract: The geometric structures, electronic and magnetic properties of Mn-doped ZnO nanowires were investigated using density functional theory. The results indicated that all the calculated energy differences were negative, and the energy of the ground state was 0.229 eV lower than ferromagnetic coupling, which show higher stability in antiferromagnetic coupling. The calculated results indicated that obvious spin splitting phenomenon occurred near the Femi level. The Zn atoms on the inner layer of ZnO nanowires are easily substituted by Mn atoms along the [0001] direction. It was also shown that the Mn^{2+}-O^{2-}-Mn^{2+} magnetic coupling formed by intermediate O atom was proved to be caused by orbital hybridization between Mn *3d* and O *2p* states. The magnetic moments were mainly attributed to the unpaired Mn *3d* orbitals, but not relevant with doping position of Mn atoms. Moreover, the optical properties of Mn-doped ZnO nanowires exhibited a novel blue-shifted optical absorption and enhanced ultraviolet-light emission. The above results show that the Mn-doped ZnO nanowires are a new type of magneto-optical materials with great promise.

Keywords: ZnO; dilute magnetic semiconductors; nanowire; density functional theory; magnetic property

---

1. Introduction

Dilute magnetic semiconductors (DMSs) have attracted much attention due to their unique potential usage of both charge and spin of freedom of carriers in magneto-optical, magneto-electrical, and magneto-transport devices [1–4], Especially, oxide-diluted magnetic semiconductors (such as ZnO) have shown excellent piezoelectric and photoelectric properties, and great potential applications in spintronic devices [5–7]. Moreover, theoretical and experimental research has predicted Curie temperatures (T_C) above room temperature, high solubility of magnetic ions, and transparency for visible light in ZnO-based DMSs [8,9]. In recent years, great progress has been achieved in one-dimensional (1D) ZnO-based DMSs. Chang *et al.* [10] prepared $Zn_{1-x}Mn_xO$ nanowires (NWs) at 500 °C with a high doped content of 13%. Philipose *et al.* [11] also successfully prepared $Zn_{1-x}Mn_xO$ NWs, and the samples showed stable ferromagnetic (FM) and ultraviolet emission properties. Wang *et al.* [12] prepared $Zn_{1-x}Mn_xO$ NWs and no second phase was observed in the Mn-doped ZnO NWs. Theoreticians have found ZnO to be an excellent candidate host semiconductor for high-T_C ferromagnetism. Dietl *et al.* [2] made theoretical predictions about the higher Curie temperature of Mn-doped ZnO, Sato *et al.* [13] revealed the magnetic properties of *3d* TMs-doped ZnO by using the local density approximation (LDA). Marcel *et al.* [14] investigated the ferromagnetic (FM) and antiferromagnetic (AFM) properties of *3d* TMs-doped ZnO. Our previous studies have shown that potential FM ground states are more stable in V-doped and Fe-doped ZnO NWs [15,16], and have obvious half-metallic properties in Fe-doped ZnO NWs [16]. In addition, He *et al.* [17] have reported on the electronic and magnetic properties of Mn-doped ZnO nanotubes using density functional theory (DFT) with the generalized gradient approximation (GGA). However, many problems in the research of ZnO-based DMSs still remain both scarce and controversial regarding the influence of impurities on the electronic, optical, and magnetic properties. The magnetic coupling mechanism and origin of the ferromagnetism is still not clear, and experimental observations demonstrate controversial results about optical phenomenon in the optical absorption with regards to Mn-doped ZnO NWs. In order to clarify the effect of Mn doping on the electronic, optical, and magnetic properties of ZnO NW, it is necessary to perform first-principles methods based on DFT.

In this paper, the geometric structures, and electronic and magnetic properties of Mn-doped ZnO NWs were systematically investigated by first-principles methods, based on DFT. The applicable methods to adjust and control oxide DMSs were obtained by analysis of the magnetic coupling mechanism of Mn-doped ZnO NWs. We systematically study mechanisms of FM and AFM coupling on the electronic, optical, and magnetic properties. The main results provide theoretical guidance for preparing ZnO-based DMS materials of high quality and high-T_C.

2. Theoretical Models and Calculated Methods

The models for Mn-doped ZnO NWs are generated from a wurtzite ZnO (7 × 7 × 2) supercell structure along the [0001] direction (see Figure 1a). All atoms within the dashed line remained to build up the ZnO NW structure in Figure 1b. Our NW model contains 96 atoms ($Zn_{46}Mn_2O_{48}$). The vacuum region of ZnO NWs along [$01\bar{1}0$] and [$10\bar{1}0$] is 15 Å to avoid the effects of the interaction between Mn-doped ZnO NWs on the results, and we select the periodical structure along [0001] direction, as shown in Figure 1b. To investigate the magnetic coupling mechanism between the two Mn atoms, we replace two of the Zn atoms with Mn in ZnO NWs. The six possible magnetic coupling models are as shown in Figure 2, and Mn doping concentration is at 4.2%.

The calculations were carried out by using the first principles methods based on DFT, additionally, the Vienna *Ab initio* Simulation Package (VASP) was applied in our work [18]. The detailed parameters were set as follows, the exchange-correlation function was calculated based on GGA in the form of the Perdew-Burke-Ernzher (PBE) [19], the valence electronic configuration of O, Zn and Mn atoms were $2s^22p^4$, $4s^23d^{10}$ and $4s^23d^5$, respectively, the cut-off energy of plane wave was set at 420 eV, and the convergence in energy and force was less than 1×10^{-5} eV and 10^{-3} eV/Å, respectively, and the stress on the cell was less than 0.01 GPa. The Brillouin zone was sampled with a mesh of 1 × 1 × 16 k points, generated by the Monkhorst-Pack scheme.

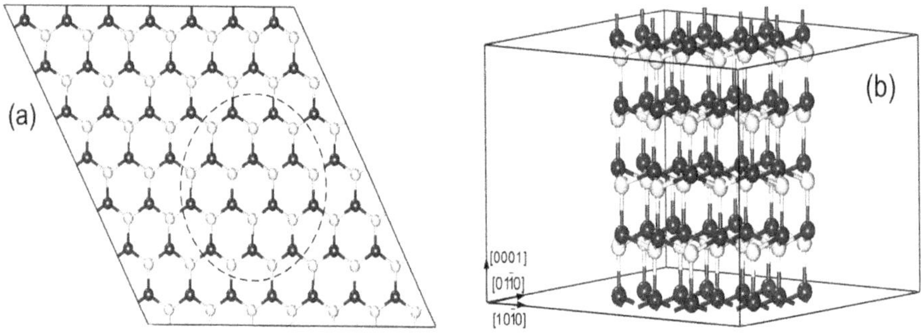

Figure 1. (**a**) The top view of the 7 × 7 × 2 ZnO supercell structure; (**b**) $Zn_{48}O_{48}$ nanowires (NWs) supercell along the [0001] direction (The white and black spheres represent O and Zn atoms).

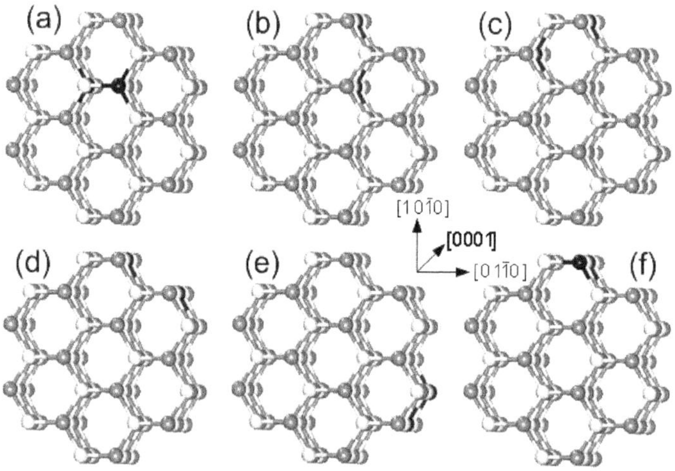

Figure 2. The structures of Mn-doped ZnO NWs. (**a**) Configuration I; (**b**) Configuration II; (**c**) Configuration III; (**d**) Configuration IV; (**e**) Configuration V; (**f**) Configuration VI. (The white, gray, and black spheres represent O, Zn, and Mn atoms, respectively).

3. Results and Discussion

To verify the geometric structure, and electronic and magnetic properties of the Mn-doped ZnO NWs, the geometric structures and electronic properties of pure $Zn_{48}O_{48}$ NW were calculated under full geometry relaxation. The total energy of the optimized ZnO NW was found to be 6.371 eV lower than that of the unoptimized one. The optimized Zn–O bond length along the [0001] direction was 1.899 Å on the outermost surface layer, which was 4.67% shorter than that of bulk ZnO. The optimized Zn–O bond lengths were 1.969 Å and 1.978 Å on the inner-layer NWs, and 1.15% and 0.7% shorter than that of bulk ZnO, respectively. The optimized Zn–O bond lengths along the $[01\bar{1}0]$ direction were between 1.955 Å and 1.962 Å on the outermost surface layer, their corresponding bond angles of ∠O–Zn–O (108.04°) and ∠Zn–O–Zn (108.04°) changed into ∠O–Zn–O (113.76°) and ∠Zn–O–Zn (105.81°). In addition, it was observed from Figure 3a that the total density of states (DOS) for the spin-up and spin-down were identical, namely, pure ZnO NW is a nonmagnetic material. The calculated band gap was 1.86 eV, which was significantly larger than that of our calculated bulk ZnO (0.97 eV).

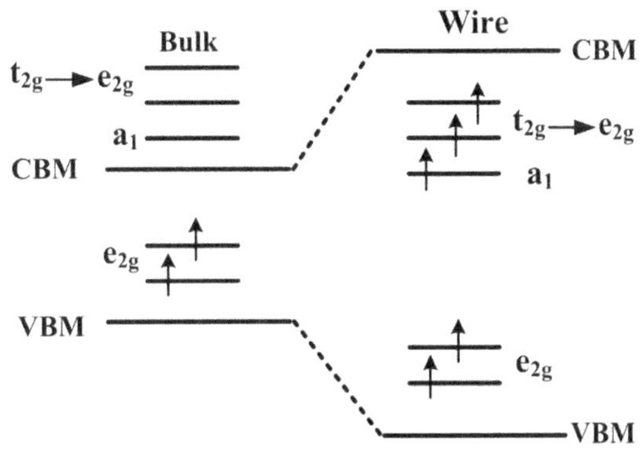

Figure 3. The schematic plot of *3d* level splitting in the ZnO bulk and NW.

Table 1 illustrates the magnetic coupling modes, geometric parameters, energy, bond lengths, and magnetic moments of Mn-doped ZnO NW, in which ΔE ($\Delta E = E_{AFM} - E_{FM}$) denotes energy difference after optimized treatments, ΔE_0 ($\Delta E = E_{AFM} - E_{FM}$) denotes the unoptimized energy difference, and the relative energy $\Delta\varepsilon$ denotes the energy difference corresponding to the ground state configuration. As shown in Table 1, the ΔE of Mn-doped ZnO NWs are all negative, which indicates that the AFM states are more stable than FM states. The configuration V corresponds to the ground state of Mn-doped ZnO nanowires, and its energy is lower than FM state energy by 0.229 eV. In this configuration, the two Mn atoms reside in the inner layer sites along the [0001] direction. As for configuration VI, the energy difference ΔE is only −0.007 eV, which indicates that AFM coupling is slightly more stable than FM coupling. In order to further investigate AFM interactions, we note that the corresponding energies of FM and AFM states are almost same as shown in Table 1 (ΔE=-0.007 eV), and the AFM coupling is weak with the increase of Mn–Mn bond length for configuration VI (d_{Mn-Mn} = 5.205 Å). Hence, it is prone to forming a paramagnetic state and spin-glass state. It is indicated that the Mn–O bond length may play a more important role than the Mn–Mn bond length in predicting the magnetic coupling, which suggests that the range of magnetic interactions between the two Mn atoms is rather short and independent of the Mn–Mn distance. While configurations II, III, and IV are in an AFM state for Mn-doped ZnO nanowires along other directions. Our calculated results are consistent with previously reported results [11,20–22]. Moreover, all magnetic moments of Mn atoms are 3.947–4.193 μ_B and mainly originate from *3d* electrons of Mn atoms. Oxygen atoms also show AFM spin polarization in nearest-neighbor Mn atoms, which produce a weak magnetic moments of −0.061–−0.089 μ_B. As for the relaxed geometric structure, the calculated results show that the Mn–O and Mn–Mn bond lengths are 1%–3% shorter than that of an unrelaxed one. The outer layer of ZnO NWs shows more intense relaxation than the inner one, therefore presenting high surface effects and quantum size effects.

Table 1. Magnetic coupling modes, energy, bond lengths, and magnetic moments.

Modes	ΔE (eV)	ΔE_0 (eV)	$\Delta\varepsilon$ (eV)	Coupling	d_{Mn-O} (Å)	d_{Mn-Mn} (Å)	Mn_1 (μ_B)	Mn_2 (μ_B)	O (μ_B)
I	−0.302	−0.052	0.721	AFM	1.914	3.134	4.171	−4.176	−0.061
II	−0.219	−0.053	0.208	AFM	1.819	3.111	4.134	−4.070	−0.069
I	−0.144	−0.047	0.405	AFM	1.916	3.060	3.947	−4.113	−0.074
IV	−0.139	−0.095	0.561	AFM	1.816	3.264	4.021	−4.104	−0.071
V	−0.229	−0.061	0.000	AFM	1.839	2.694	4.078	−4.067	−0.070
VI	−0.007	−0.003	0.117	AFM	1.823	5.205	4.193	−4.021	−0.089

To further study the origin of magnetic coupling between two Mn atoms, Figure 3 displays the schematic of splitting 3*d* level occupation in bulk ZnO and NW. For Mn-doped ZnO NW, the positions of Zn^{2+} ions are replaced by Mn^{2+} ions, and an Mn^{2+} (d^5 configuration) ion has an tetrahedral structure in a crystalline field. The 3*d* states of an Mn atom split into one upper, triply degenerate t_{2g} state (d_{xy}, d_{yz}, d_{xz}) and one lower, doubly degenerate e_g state ($d_{z^2-r^2}$, $d_{x^2-y^2}$). However, the space group of ZnO with a wurtzite structure degrades from a high symmetric T_d to a C_{3v} group in the NW, the t_{2g} states are no longer the original triply degenerate, but further divide into one doubly degenerate e_{2g} state and one singly a_1 state. In particular, O 2*p* states also splits into an e_g state and an a_1 state, due to the high symmetry of T_2 in O 2*p* orbital. Magnetic coupling between O 2*p* and Mn 3*d* states cause localization characteristics of the e_g states owing to the same symmetry. Figure 4 shows the FM and AFM mechanisms of the Mn-doped ZnO NW. According to the Hund rules and the Pauli exclusion principle, the two lower e_g levels located in the gap are fully occupied in the upper-spin bands, and pushed upward and downward by the same Δd to form an FM coupling when there is no energy input, as shown in Figures 3 and 4a. For the AFM coupling in Figure 4a, the upper-spin e_g level of Mn ion couples with that of another Mn ion with the same level, the occupied e_g levels are pushed up by Δd energy, and the unoccupied ones are pushed down by Δd energy due to the half-occupation of the e_g levels, as shown in Figure 4b, thus, the super-exchange interaction is smaller than the double exchange interaction. By increasing distance between the two Mn^{2+} ions, the AFM coupling of d-d state becomes weaker, and the doped system can form paramagnetic or spin-glass states, which is consistent with our analysis about geometric structure and stability in previous section. The Mn-doped ZnO nanowires show an AFM behavior, and are prone to form a zigzag chain of Mn–O–Mn atoms along the [0001] direction, which indicate that the doped systems have a good effect on the [0001] preferred orientation to prepare ZnO-based dilute magnetic semiconductors.

In order to further explain the magnetic coupling of Mn-doped ZnO NW, Figure 5 shows the total density of states (TDOS) and partial density of states (PDOS) of FM and AFM states. It can be seen that the TDOS of Mn-doped ZnO NW moves towards low energy (0.95 eV), and one can see that the spin-up Mn *3d* states pass through the Femi level for FM states, which confirms that the doped system is a half-metallic ferromagnet and exhibits 100% spin polarization. The strong FM coupling between the Mn *3d* and O *2p* states occurs at the top valence band (−1.0–0 eV) and the bottom valence bands (−4.0–−7.0 eV), the dispersive bonding t_{2g}^{b} state and the local nonbonding e_{g}^{e} state is formed at the valence band (−2–−6 eV) and near the Femi level, respectively. The antibonding t_{2g}^{a} state is formed at the conduction band (1.6–2.1 eV). In particular, one can see from Figure 5c that the role of the O atom mediates the magnetic coupling between Mn atoms. The magnetic coupling chains of Mn^{2+}–O^{2-}–Mn^{2+} would reveal a double exchange mechanism owing to impurity states in the gap region. On the other hand, the magnetic moments of Mn atoms are 4.012–4.193 μB and mainly come from the Mn *3d* orbitals, which are less than the theoretical value of 5 μB. However, small contributions to the magnetic moments come from O *2p* orbitals. As for AFM state, we find that the spin-up and spin-down DOS are asymmetrical, the spin-up electrons are more than the spin-down electrons. As the DOS of spin-up and spin-down states all pass through the Femi level, the AFM state is a possible metal ferromagnet. Spin-exchange splitting is also observed between Mn 3*d* and O 2*p* states, as shown in Figure 5e,f. Thus, the local bonding t_{2g}^{b} states are fully occupied between −1.5 eV and −1.0 eV. Whereas the nonbonding t_{2g}^{a} states are 4/5 occupied between −1.0 eV and the Femi level, the anti-bonding e_{g}^{e} states are full empty between 0.6 eV and 2.5 eV, which correspond to the Mn *3d* states. We note that the bandwidth of the bonding and antibonding band is greatly narrowed for AFM states, and the spin-orbital coupling and hybridization effects are significantly higher than FM states. Therefore, stable AFM states can be easily formed in Mn-doped ZnO NWs. This is mainly because the Mn^{2+} ($3d^5$) ion shows a half-full *3d* electronic configuration, the spin-up *3d* orbitals of Mn^{2+} atom are all occupied, and there are no excessive unoccupied orbitals for hybridization coupling with neighboring Mn^{2+} ions. As a result, *3d* orbital electrons of the neighboring Mn^{2+} are prone to filling the levels along the reverse parallel direction, which is in accordance with the mechanisms shown in Figures 3 and 4. In addition, the up-spin or down-spin orbitals are fully empty for the higher e_{2g} state and AFM coupling is favored. Namely, the AFM coupling can be formed more easily in Mn-doped ZnO NWs according to theoretical analysis, which is similar to the reported results [23,24].

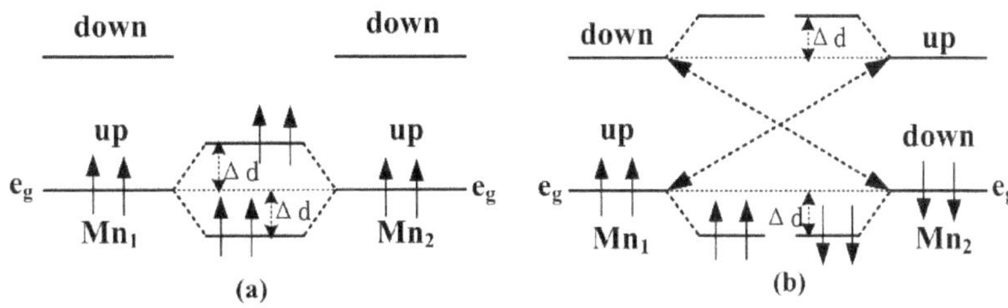

Figure 4. The ferromagnetic (FM) and antiferromagnetic (AFM) mechanisms of the Mn-doped ZnO NW. (**a**) FM coupling; (**b**) AFM coupling.

The optical absorption spectrum of Mn-doped ZnO NWs is illustrated in Figure 6. We applied the scissors operator ($E_{scissor}$ = 1.51 eV) to adjust the experimental and calculated band gap based on the experimental result (3.37 eV). To study the effect of Mn-doped on the optical properties, the absorption spectrum of pure ZnO NWs is also calculated. Some previous experimental studies [25–27] found that pure ZnO NW shows a strong absorption peak in the ultraviolet region as a result of near-band edge absorption. A wide green-yellow luminous bands form within the visible region of 550–600 nm. As shown in Figure 6, the calculated optical properties show that the absorption spectrum of the Mn-doped ZnO NWs are blue-shifted in the ultraviolet region, and the broad ultraviolet emission from 320 nm to 340 nm is observed in a doped system, which is related to electron transitions from the top valence band to the bottom conduction band. The calculated result is similar to the reported experimental optical absorption spectra of Mn-doped ZnO nanoparticles [28,29] and Mn-doped ZnO nanowires [30,31]. Additionally, one "sharp" absorption peak appears in the far-ultraviolet region near 90 nm, which are transitions from the *s–d* and *p–d* coupling of localized Mn 3*d* orbitals. The increase in the intensity of the UV peak is considered to be due to the participation of the 3*d* electrons of Mn [10,29,30], and is very similar to Co-doped ZnO nanowires [32]. The above-calculated results indicate a strong photo-response for ultraviolet-light luminescence. Namely, the Mn-doped ZnO NWs may be a new type of magneto-optical material with great promise.

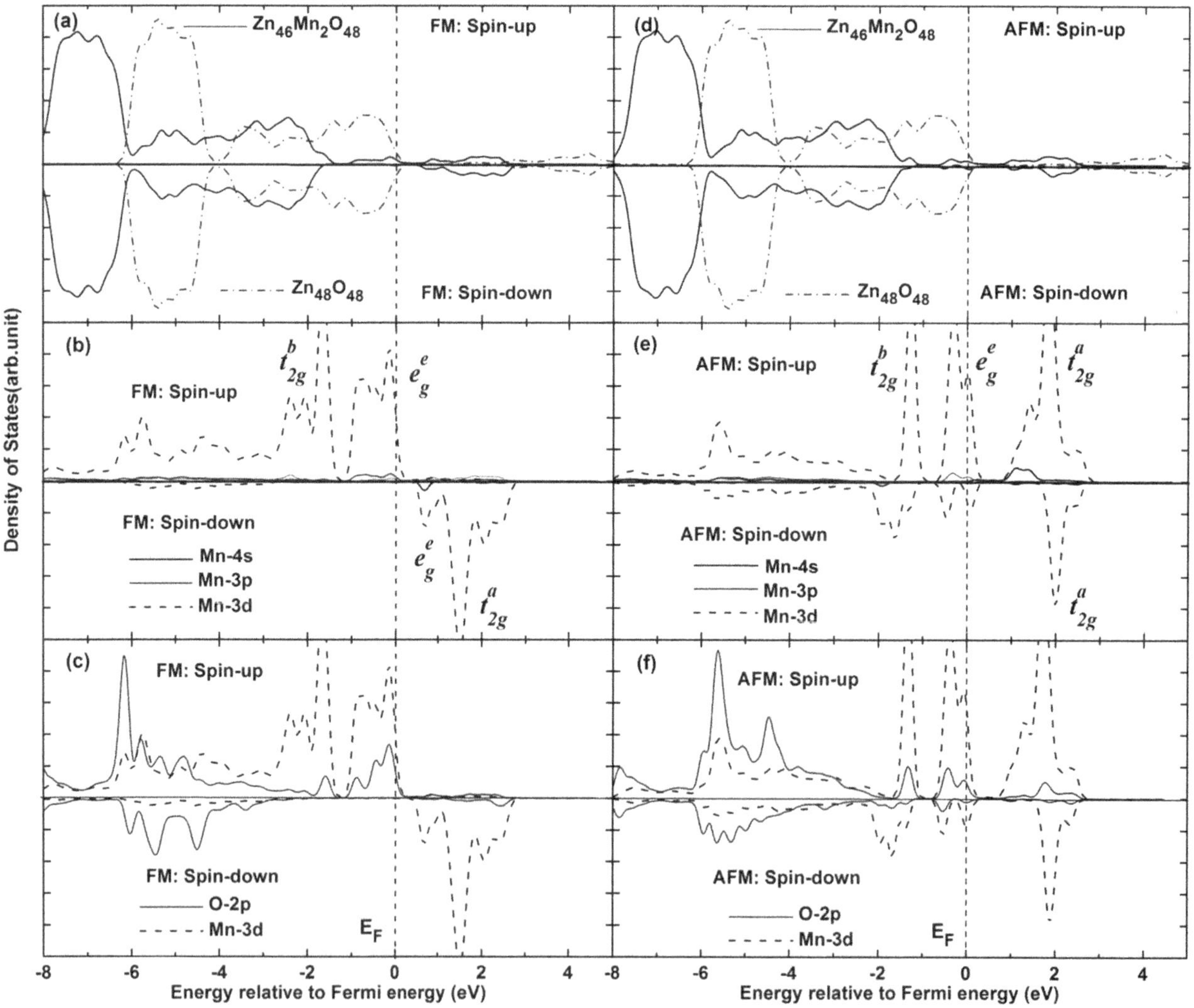

Figure 5. Total density of states (TDOS) and partial density of states (PDOS) of Mn-doped ZnO NW. (**a**) TDOS of pure $Zn_{46}O_{48}$ and Mn-doped ZnO nanowire for FM states; (**b**) PDOS of Mn atom for FM states; (**c**) PDOS of Mn 3*d* and O 2*p* for FM states; (**d**) TDOS of pure $Zn_{46}O_{48}$ and Mn-doped ZnO nanowire for FM states; (**e**) PDOS of Mn atom for AFM states; (**f**) PDOS of Mn 3*d* and O 2*p* for AFM states.

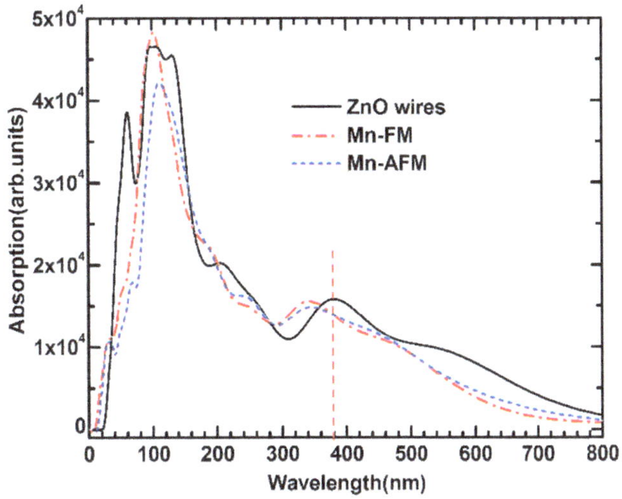

Figure 6. The optical absorption curves of FM and AFM for Mn-doped ZnO NW.

4. Conclusions

In summary, geometric structures, and electronic and magnetic properties of Mn-doped ZnO NWs were investigated by using first-principles methods based on DFT. The magnetic origination and magnetic coupling mechanism were analyzed in detail. The results indicate that AFM coupling is the most stable for six doping configurations, and *p–d* hybridization coupling occurred at the top valence bands between Mn 3*d* and O 2*p* states. The dispersive bonding state and local nonbonding state are formed in valence bands, and a local antibonding state is formed in conduction bands. Magnetic moments are originated mainly from Mn 3*d* orbitals. The theoretical results further reveal the half-metallic magnetic properties of FM coupling and metallic magnetic properties of AFM coupling. Our research will provide theoretical guidance for preparing ZnO-based DMSs with a high Curie temperature.

Acknowledgments

The authors thank all colleagues who contributed to the results described in this article and whose names appear in joint publications. This research was supported by the National Natural Science Foundation of Shaanxi Province, China (2014JM2-5058), the Scientific Research Program of Yan'an city, China (2013-KG03), the Science Innovation Training Program (2014107191081) and the Science Foundation of Yan'an University (YD2014-02).

Author Contributions

The FZ and WZ conceived and designed the study. The DC and HC performed the theoretical calculation and edited the manuscript. The FZ wrote the manuscript.

Conflicts of Interest

The authors declare no conflict of interest.

References

1. Ohno, H. Making nonmagnetic semiconductors ferromagnetic. *Science* **1998**, *281*, 951–954.
2. Dietl, T.; Ohno, H.; Matsukura, F.; Cibert, J.; Ferrand, D. Zener model description of ferromagnetism in zinc-blende magnetic semiconductors. *Science* **2000**, *287*, 1019–1022.
3. Pan, Z.W.; Dai, Z.R.; Wang, Z.L. Nanobelts of semiconducting oxides. *Science* **2001**, *281*, 1947–1949.
4. Huang, M.H.; Mao, S.; Feick, H.; Yan, H.Q.; Wu, Y.Y.; Kind, H.; Weber, E.; Russo, R.; Yang, P.D. Room-temperature ultraviolet nanowire nanolasers. *Science* **2001**, *292*, 1897–1899.
5. Ye, C.Q.; Li, M.Z.; Luo, J.; Chen, L.F.; Tang, Z.M.; Pei, J.; Jiang, L.; Song, Y.L.; Zhu, D.B. Photo-induced amplification of readout contrast in nanoscale data storage. *J. Mater. Chem.* **2012**, *22*, 4299–4305.
6. Chau, R.; Doyle, B.; Datta, S.; Kavalieros, J.; Zhang, K. Integrated nanoelectronics for the future. *Nat. Mater.* **2007**, *6*, 810–812.
7. Ma, X.; Tian, H. Bright functional rotaxanes. *Chem. Soc. Rev.* **2010**, *39*, 70–80.
8. Jin, Z.W.; Fukumura, T.; Kawasaki, M.; Ando, K.; Saito, H.; Sekiguchi, T.; Yoo, Y.Z.; Murakami, M.; Matsumoto, Y.; Hasegawa, T.; *et al.* High throughput fabrication of transition-metal-doped epitaxial ZnO thin films: A series of oxide-diluted magnetic semiconductors and their properties. *Appl. Phys. Lett.* **2001**, *78*, 3824.
9. Norberg, N.S.; Kittilstved, K.R.; Amonette, J.E. Synthesis of colloidal Mn^{2+}:ZnO quantum dots and high-TC FM nanocrystalline thin films. *J. Am. Chem. Soc.* **2004**, *126*, 9387–9398.
10. Chang, Y.Q.; Wang, D.B.; Luo, X.H.; Xu, X.Y.; Chen, X.H.; Li, L.; Chen, C.P.; Wang, R.M.; Xu, J.; Yu, D.P. Synthesis, optical, and magnetic properties of diluted magnetic semiconductor $Zn_{1-x}Mn_xO$ nanowires via vapor phase growth. *Appl. Phys. Lett.* **2003**, *83*, 4020–4022.
11. Philipose, U.; Naira, S.V. High-temperature ferromagnetism in Mn-doped ZnO nanowires. *Appl. Phys. Lett.* **2006**, *88*, 263101.
12. Zhang, X.M.; Zhang, Y.; Wang, Z.L. Synthesis and characterization of $Zn_{1-x}Mn_xO$ nanowires. *Appl. Phys. Lett.* **2008**, *92*, 162102.
13. Sato, K.; Katayams, H. Material design for transparent ferromagnets with ZnO-Based Magnetic Semiconductors. *Jpn. J. Appl. Phys.* **2000**, *39*, L555–L558.
14. Sluiter, M.H.F.; Kawazoe, Y.; Sharma, P.; Inoue, A.; Raju, A.R.; Rout, C.; Waghmare, U.V. First principles based design and experimental evidence for a ZnO-based ferromagnet at room temperature. *Appl. Phys. Lett.* **2005**, *94*, 187204.
15. Zhang, F.C.; Dong, J.T.; Zhang, W.H.; Zhang, Z.Y. Ferromagnetism of V-doped ZnO nanowires. *Chin. Phys. B* **2013**, *22*, 027503.
16. Zhang, F.C.; Zhang, W.H.; Dong, J.T.; Zhang, Z.Y. First-principles study of Fe-doped ZnO nanowires. *Chin. Phys. Lett.* **2011**, *28*, 126102.
17. He, A.L.; Wang, X.Q.; Fan, Y.Q.; Feng, Y.P. Electronic structure and magnetic properties of Mn-doped ZnO nanotubes: An *ab initio* study. *J. Appl. Phys.* **2010**, *108*, 084308.
18. Kresse, G.; Furthmüller, J. Efficient iterative schemes for *ab initio* total-energy calculations using a plane-wave basis set. *Phys. Rev. B* **1996**, *54*, 011169.

19. Kresse, G.; Joubert, D. From ultrasoft pseudopotentials to the projector augmented-wave method. *Phys. Rev. B* **1999**, *59*, 1758–1775.
20. Baik, J.M.; Lee, J.L. Fabrication of vertically well-aligned (Zn, Mn)O nanorods with room temperature ferromagnetism. *Adv. Mater.* **2005**, *22*, 2745–2748.
21. Zhu, Y.H.; Xia, J.B. Electronic structure of Mn-doped ZnO quantum wires: A mean-field theory study. *Phys. Rev. B* **2007**, *75*, 205113.
22. Pradhan, A.K.; Zhang, K.; Mohanty, S.; Dadson, J.B.; Hunter, D. High-temperature ferromagnetism in pulsed-laser deposited epitaxial (Zn, Mn)O thin films: Effects of substrate temperature. *Appl. Phys. Lett.* **2005**, *86*, 152511.
23. Lawes, G.; Risbud, A.S.; Ramirez, A.P.; Seshadri, R. Absence of ferromagnetism in Co and Mn substituted polycrystalline ZnO. *Phys. Rev. B* **2005**, *71*, 045201.
24. Prashant, K.S.; Ranu, K.D.; Choudhary, R.J.; Pandeya, A.C. Doping, strain, defects and magneto-optical properties of $Zn_{1-x}Mn_xO$ nanocrystals. *CrystEngComm* **2013**, *15*, 4438–4447.
25. Zhang, Z.H.; Qi, X.Y.; Jian, J.K.; Duan, X.F. Investigation on optical properties of ZnO nanowires by electron energy-loss spectroscopy. *Micron* **2006**, *37*, 229–233.
26. Kong, Y.C.; Yu, D.P.; Zhang, B.; Fang, W.; Feng, S.Q. Ultraviolet-emitting ZnO nanowires synthesized by a physical vapor deposition approach. *Appl. Phys. Lett.* **2001**, *78*, 407–410.
27. Chen, T.; Xing, G.Z.; Zhang, Z.; Chen, H.Y.; Wu, T. Tailoring the photoluminescence of ZnO nanowires using Au nanoparticles. *Nanotechnology* **2008**, *19*, 435711.
28. Hao, Y.M.; Lou, S.Y.; Zhou, S.M.; Yuan, R.J.; Zhu, G.Y.; Li, N. Structural, optical, and magnetic studies of manganese-doped zinc oxide hierarchical microspheres by self-assembly of nanoparticles. *Nano. Rese. Lett.* **2012**, *7*, doi: 10.1186/1556-276X-7-100.
29. Yang, L.W.; Wu, X.L.; Huang, G.S.; Qiu, T.; Yang, Y.M. *In situ* synthesis of Mn-doped ZnO multileg nanostructures and Mn-related Raman vibration. *J. Appl. Phys.* **2005**, *97*, 014308.
30. Phan, T.; Sun, Y.; Vincent, R.; Cherns, D.; Nghia, N.; Yu, S.-C. Optical properties of Mn-doped ZnO nanowires. *J. Korean Phys.* **2008**, *52*, 1633–1636.
31. Deka, S.; Joy, P.A. Synthesis and magnetic properties of Mn doped ZnO nanowires. *Solid State Commun.* **2007**, *142*, 190–194.
32. Chang, Y.Q.; Wang, P.W.; Ni, S.L.; Long, Y.; Li, X.D. Influence of Co content on raman and photoluminescence spectra of Co doped ZnO nanowires. *J. Mater. Sci. Technol.* **2012**, *28*, 313–316.

16

Targeting of Apoptotic Cells Using Functionalized Fe_2O_3 Nanoparticles

Moataz Mekawy [1,2,*], Atsushi Saito [3], Hiroaki Shimizu [4] and Teiji Tominaga [1]

[1] Department of Neurosurgery, Graduate School of Medicine, Tohoku University, 2-1 Seiryo-machi, Aoba-ku, Sendai, Miyagi 980-8575, Japan; E-Mail: tomi@nsg.med.tohoku.ac.jp

[2] Center of Materials Science, Zewail City of Science and Technology, Sheikh Zayed District, 12588, 6th of October City, Giza, Egypt

[3] Department of Neurosurgery, Aomori Prefecture Central Hospital, 2 1 1 Higashitsukurimichi, Aomori 0308553, Japan; E-Mail: satsushi2002@yahoo.co.jp

[4] Department of Neurosurgery, Graduate School of Medicine, Akita University, 1-1-1 Hondo, Akita 010-8543, Japan; E-Mail: hshim@nsg.med.akita-u.ac.jp

* Author to whom correspondence should be addressed; E-Mail: mmekawy@zewailcity.edu.eg

Academic Editor: Pablo del Pino

Abstract: Fe_2O_3 nanoparticles (NPs) have been synthesized and functionalized with SiO_2 and $-NH_2$ group, respectively. Conjugation to fluorescently-labeled poly-caspase inhibitor (SR-FLIVO) has been carried out for better cellular uptake studies of apoptosis arising from brain focal cerebral ischemia. Highest conjugation affinity to SR-FLIVO was found to be *ca.* 80% for Fe_2O_3-SiO-NH_2 functionalized nanoparticles (FNPs). Tracking of SR-FLIVO conjugated functionalized nanoparticles (SR-FLIVO-FNPs) *in vivo* and *in vitro* has been carried out and detected using microscopic techniques after histochemical staining methods. Experimental results revealed that SR-FLIVO-FNPs probe could passively cross the blood brain barrier (BBB) and accumulated within the apoptotic cell. Optimization of SR-FLIVO-FNPs probe can effectively promise to open a new era for intracellular drug delivery and brain diagnosis.

Keywords: functionalized nanoparticles (FNPs); fluorescently-labeled poly-caspase inhibitor (SR-FLIVO); apoptosis; brain diagnosis

---

1. Introduction

Apoptosis detection (denoted to a form of programmed cell death) has attracted much research in the past decade. It can be distinguished according to morphological, molecular, and biochemical changes occurred in the cell which is going to die. If apoptosis is diagnosed early and properly, there will be a high possibility for the prevention of its common contributing diseases. However, inappropriate regulation of apoptosis may play an important role for many pathological conditions in ischemia and stroke [1,2].

Previously, Annexin V has been addressed for apoptosis detection [3]; however, due to its very short life and the difficulty of reliably detecting it, the time point of assessment can be critical. In addition, Annexin V does not bind to all apoptotic tumor cells [4] and it also binds positively to normal and healthy bone marrow derived cells [5]. Furthermore, the inversion of phosphatidyl-serine may not be exclusively related to apoptosis and this adds to the background issues [6].

Distinguished signs for apoptosis are activation of caspases, DNA fragmentation, and membrane swelling [7]; therefore, caspase inhibitors play an important role in the early detection of apoptosis because the apoptotic cells have more active caspases than control cells.

Since the early reports of polycaspase inhibitor benzoxycarbonyl-Val-Ala-Asp-fluoromethyl ketone (Z-VAD-FMK) and its derivatives [8], they have attracted much attention. They are widely used because they can be bound specifically to the activated caspases, hence they can reflect detection of apoptosis. On the other hand, and as a progressive step, caspase inhibitors could be linked to organic dyes to facilitate the *in vivo* detection of apoptosis using enhanced fluorescence techniques [3].

Recently, nanotechnology based materials have been able to play an essential role not only in diagnosis but also in being an important platform for medical therapies. Several nanomaterials could be applicable for such purposes such as carbon nanotube [9], quantum dots [10], metallic [11], polymeric [12] and magnetic NPs [13] which have received considerable attention in the past decade to improve MRI diagnostics [14] taking into account their compatibility [15]. Thus, recent progress could lead to several models of magnetic NPs targeting for tumor imaging and therapy [16,17]. However, to date, the development of specific apoptotic-targeted NPs remains extremely challenging.

Several types of iron oxide based nanoparticles (NPs) have been used in the field of magnetic applicability. Only Fe_3O_4 and Fe_2O_3 had been approved by FDA to be used in clinical studies. However, Fe_2O_3 has better stability since Fe_3O_4 showed high tendency towards surface oxidation in the presence of oxygen and it is easily transformed to Fe_2O_3 even at room temperature [18–21].

The key factor that addresses the potential applicability of NPs in theranostics is their surface chemistry nature. Thus, surface functionalized nanoparticles are considered to be of high importance due to their small sizes with different surface natures and controlled penetration abilities which reflects their potential use in many biomedical applications such as *in vitro* cell separation, *in vivo* drug delivery, MRI contrasting agents and many others [22]. However, before discussing their applicability, the characteristics of nanoparticles should be examined carefully. Among those important characteristics, the ability to conjugate to different biomolecules, crossing the blood brain barrier (BBB), and accumulation inside the apoptotic cells for better modifications of the diagnosis techniques. In other words, it is important to examine the *in vivo* and *in vitro* cellular uptake of these nanoparticles to be applicable in the future as specific contrasting agents and/or drug delivery carriers.

In this study, we aim to synthesize a new probe for apoptotic cells detection based on functionalized Fe_2O_3 functionalized nanoparticles (FNPs). We have synthesized Fe_2O_3 core NPs shielded with silica-shell that is grafted with NH_2 group for facile conjugation to the fluorescently-labeled poly-caspase inhibitor, sulforhodamine (SR-FLIVO) and for better molecular imaging studies of apoptosis arising from brain focal cerebral ischemia.

SR-FLIVO [3,23] is a red fluorescent probe that has the ability to diffuse across the cell through the BBB forming covalent bonds with activated caspases, and thus detects *in vivo* apoptosis. However, any un-reacted SR-FLIVO is leached outside the brain via the normal blood circulation. By considering the advantages of this molecular design which can allow the fluorescence detections based on hestochemical and TUNEL staining for the accumulation of SR-FLIVO-FNPs probe, one can predict that this probe can work effectively for *in vivo* brain imaging studies and open the era for inter and intracellular drug delivery. To the best of our knowledge, this is the first study to address such objectives using SR-FLIVO-FNPs probe.

2. Results and Discussion

2.1. Surface Morphology

Transmission electron microscopy (TEM) experimental results (Figure 1) revealed that well resolved, mono-dispersed, and crystalline NPs and FNPs have been synthesized with average sizes of 6.3 ± 0.3, 10.8 ± 0.5, and 11.6 ± 0.6 nm for Fe_2O_3, Fe_2O_3-SiO_2, and Fe_2O_3-SiO-NH_2 NPs, respectively.

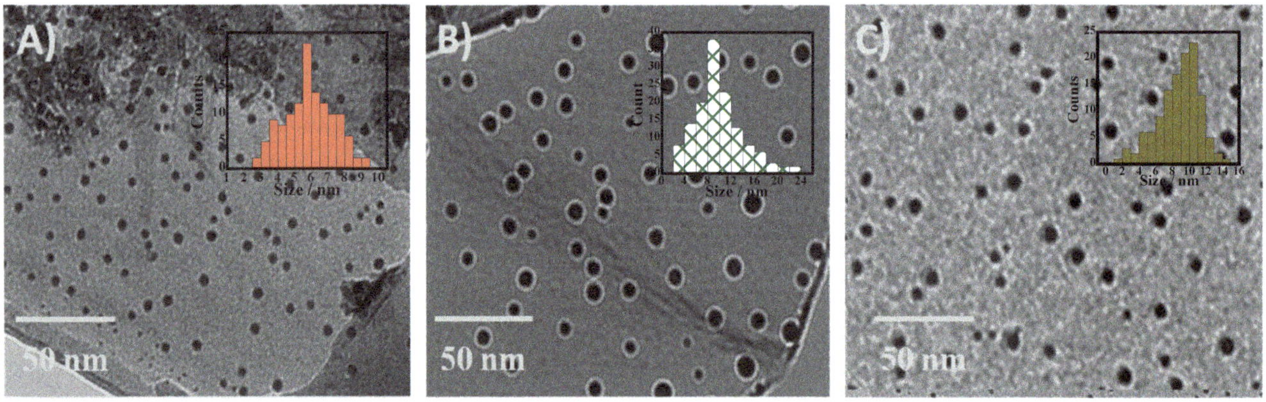

Figure 1. Transmission electron microscopy (TEM) and size distribution histograms for (**A**) Fe_2O_3; (**B**) Fe_2O_3-SiO_2; and (**C**) Fe_2O_3-SiO-NH_2 nanoparticles (NPs), respectively.

2.2. Conjugation to SR-FLIVO

The conjugation affinity which reflects the highest degree of FNPs surface coverage with SR-FLIVO was examined. Results shown in Figure 2 revealed that the conjugation affinity reached 80% for Fe_2O_3-SiO-NH_2 FNPs. Time course fluorescence emission spectra was recorded and a remarkable quenching in the fluorescence intensity was noticed until reaching equilibrium. This could be ascribed to the formation of covalent bonding between the surface NH_2 group of FNPs with the S=O group of SR-FLIVO [24–26]. This was confirmed by Fourier transform infrared spectroscopy (FTIR) spectra shown in Figure 3 which reveal the existence of a strong band at 1341 cm^{-1} which is ascribed to sulfonamide formation in SR-FLIVO-FNPs. Thus, it is considered among the candidate probes for apoptotic cells detection.

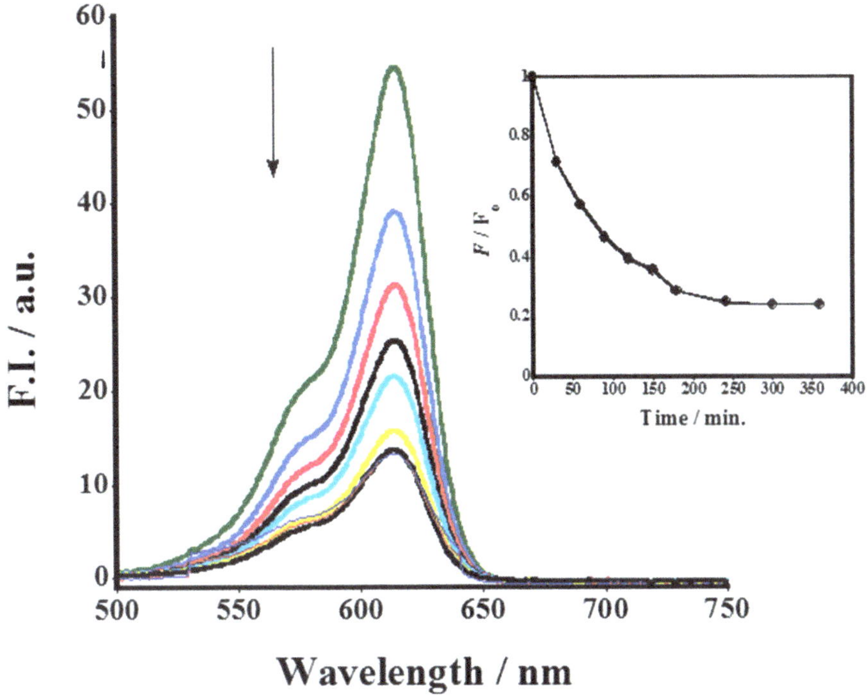

Figure 2. Time course fluorescence spectra measurement revealing the conjugation of Fe_2O_3-SiO-NH_2 functionalized nanoparticles (FNPs) to fluorescently-labeled poly-caspase inhibitor (SR-FLIVO). The inset shows that the conjugation efficiency reaches *ca.* 80%.

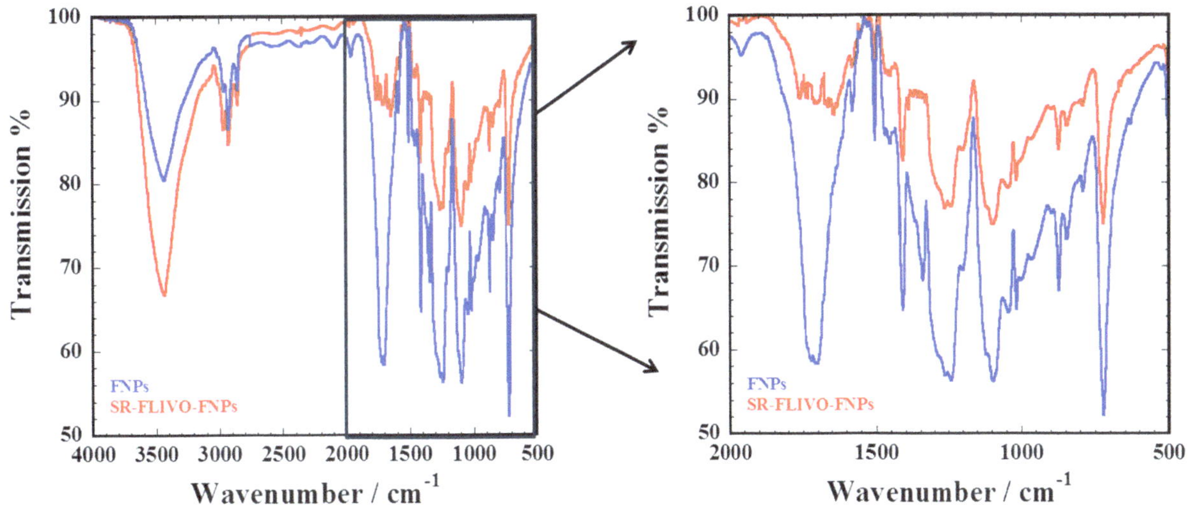

Figure 3. Fourier transform infrared spectroscopy (FTIR) spectra of FNPs (red) and SR-FLIVO-FNPs (blue).

FTIR spectral results show several bands that could be assigned as: bands at 633 and 791 cm^{-1} which are attributed to the C–F stretching vibration in FLIVO, a strong band at 725 cm^{-1} which is attributed to the symmetric stretching vibration of Si–O–Si, a band at 845 cm^{-1} which is attributed to the wag N–H, a band at 875 cm^{-1} which is attributed to the stretching vibration of Si–O–H, a band at 1016 cm^{-1} which is attributed to the weak antisymmetric vibration of Si–O, a band at 1045 cm^{-1} which is attributed to the stretching vibration of Si–O–C arising from APTMS layer [27], and bands at 1095, 1237 and 1262 cm^{-1} which are attributed to the C–N stretching vibration of aliphatic amines. A strong band at 1341 cm^{-1} is attributed to sulfonamide stretching vibration that indicates the contribution of S=O group of SR-FLIVO in conjugation with free NH_2 from FNPs. Bands at 1416 and 1458 cm^{-1} are attributed to the C–H bending vibration. A band at 1508 cm^{-1} is attributed to alkane H–C–H bending vibration. A band at 1580 cm^{-1} is attributed to the aliphatic N–H bending vibration. Weak shoulders appear at 1641, 1704, 1766 cm^{-1} that are believed to correspond to symmetric NH_3^+ deformation mode partly superimposed by CH_2 bending and formation of labile H–bonding at FNPs [28]. A strong band with weak deformation appears at 1704 cm^{-1} and is attributed to the C=O stretching vibration at SR-FLIVO. A weak band appears at 1962 cm^{-1} and is attributed to the overtone bending vibration of aromatic C–H at SR-FLIVO. Bands at 2850, 2930 cm^{-1} are attributed to the symmetric and asymmetric stretching of CH_3, respectively. A band at 3436 cm^{-1} is attributed to the stretching vibration of NH group.

2.3. TUNEL Staining

Figure 4 shows the enhancement of the fluorescence intensity of activated caspases at ischemic lesion, which is reflected with SR-FLIVO-NPs and terminal deoxynucleotidyl transferase dUTP nick end labeling (TUNEL) staining. In contrast, no enhancement was detected for the fluorescence intensity in the control (non-ischemic) side due to absence of activated caspases.

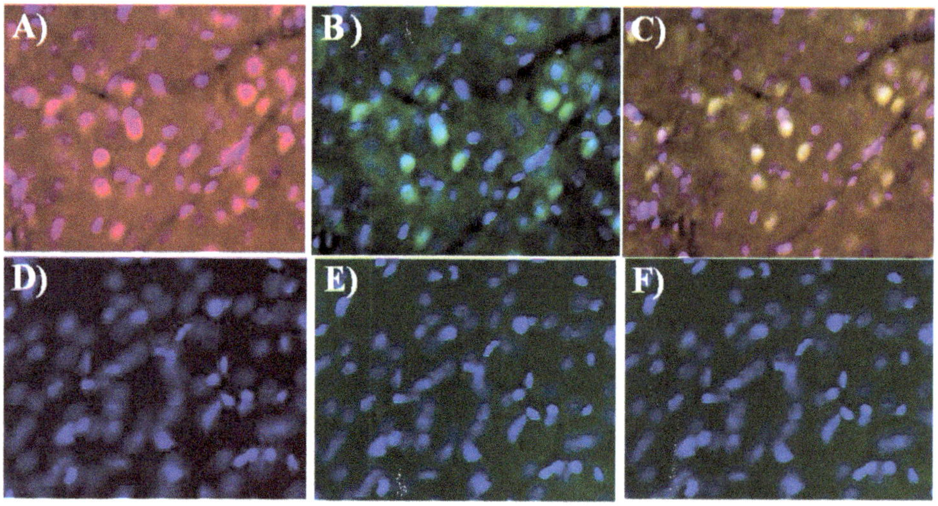

Figure 4. Fluorescence microscopy images after histochemical staining for ischemic lesion (upper raw) and non-ischemic lesion (lower raw) of SR-FLIVO-FNPs (**A**,**D**), terminal deoxynucleotidyl transferase dUTP nick end labeling (TUNEL) (**B**,**E**), and merged SR-FLIVO-FNPs/TUNEL staining (**C**,**F**) which show specific detection of apoptotic cells at the ischemic lesion in the rat brain after focal cerebral ischemia.

2.4. Intracellular Tracking

Cellular tissues were recorded using TEM after brain reperfusion of normal rats (n = 4), ischemic rats with and without injection of SR-FLIVO-FNPs (n = 4 for each group). Figure 5A shows the apoptotic cell before injection of SR-FLIVO-FNPs revealing surface blebbing which is considered a specific pattern of apoptosis. This is due to a deep cytoskeleton rearrangement, causing progressive changes in cell shape such as swelling and organelle distribution [29,30]. TEM micrographs showed that the SR-FLIVO-FNPs could passively crossing the BBB and deposited inside the apoptotic cells as shown in Figure 5B. This finding could also be confirmed using Prussian blue staining as shown in Figure 5D–F. These results reflect the *in vivo* tracking of SR-FLIVO-FNPs probe and the successful crossing of the weak compromised BBB due to vasogenic edema and the small size of the FNPs based probe used which can give a specific detection of apoptotic cells in the ischemic lesions.

For *in vivo* intracellular tracking, Fluorescence microscope images shown in Figure 6 revealed that positive expression has been recorded using SR-FLIVO-FNPs probe at the ischemic core. This could be explained due to the covalent binding with caspase as previously reported [3,8].

Moreover, Figure 7 shows the tracking and apoptotic cellular uptake of SR-FLIVO-FNPs probe after Prussian blue staining which confirms the intracellular probe uptake by apoptotic cells.

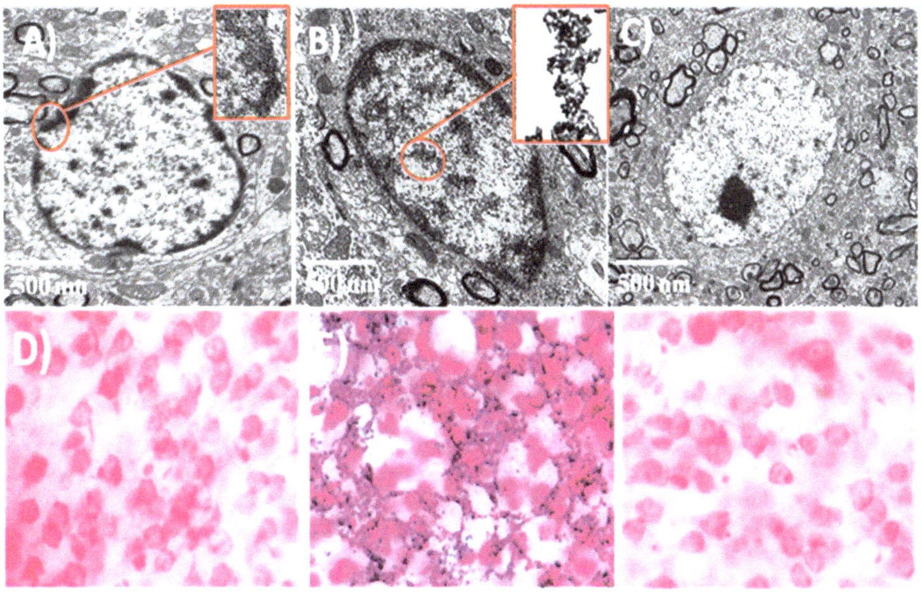

Figure 5. TEM micrographs for the apoptotic cells before (**A**) and after (**B**) injection of SR-FLIVO-FNPs. (**C**) control cell in non-ischemic lesion. Light microscope images after Prussian blue staining for (**D**) control rat brain after injection of SR-FLIVO-FNPs, (**E**) ischemic lesion of rat brain with localized SR-FLIVO-FNPs, and (**F**) non-ischemic lesion of rat brain after injection of SR-FLIVO-FNPs.

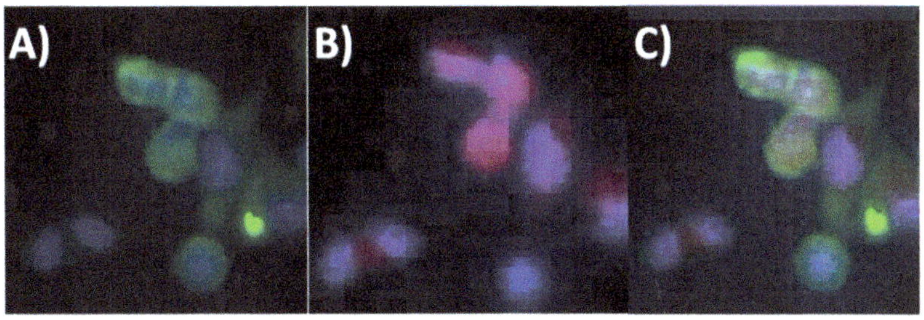

Figure 6. Fluorescence microscope images for apoptotic cells (**A**) TUNEL staining, (**B**) SR-FLIVO-FNPs staining and (**C**) merged TUNEL/SR-FLIVO-FNPs.

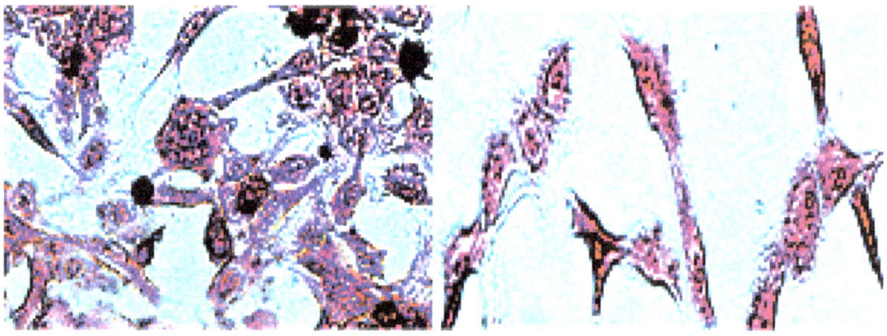

(A) (B)

Figure 7. Light microscope images after Prussian blue staining in ischemic lesion (**A**) which shows the accumulation of SR-FLIVO-FNPs, and non-ischemic lesion (**B**) in which no remarkable accumulation of NPs was observed, respectively.

3. Experimental Section

3.1. Synthesis and Surface Modification of NPs

Fe_2O_3 nanoparticles have been synthesized using microemulsion procedure as previously described with some modifications [31]. In a typical synthesis method, 5.0 g of 0.15 mM $FeCl_2{\cdot}4H_2O$ was mixed with 0.5 mL of 1.0 M HCl at room temperature for 5 min (Solution A). 1.2 g CTAB was mixed with 10 mL *n*-Octane till a clear solution appeared followed by addition of 10 mL 1-butanol and stirring at 40 °C for 20 min (Solution B). Finally, solutions A and B were mixed together under vigorous stirring for 30 min at 40 °C followed by drop wisely addition of 6.0 mL of 0.25 M NaOH and kept under stirring for another 20 min to produce microemulsion of Fe_2O_3 nanoparticles. Decantation and then centrifugation carried out and finally the NP supernatant was washed thoroughly with water and acetone followed by drying.

For better stability, facile conjugation to other molecules, and inhibition of NPs aggregation, formation of silica shell surrounded the core of Fe_2O_3 NPs carried out via co-condensation synthesis method using a starting composition of equi-molar $Na_2SiO_3{\cdot}9H_2O{:}FeCl_2{\cdot}4H_2O$ as silica and iron sources, respectively. Removal of the hexadecyltrimethylammonium bromide (CTAB) surfactant carried out using hot ethanol solvent extraction method in which NPs (Fe_2O_3 and Fe_2O_3-SiO_2) were immersed in hot ethanol at 50 °C for 20 h. Finally the surface grafting with NH_2 group for Fe_2O_3-SiO_2 NPs carried out using 3-Aminopropyltriethoxysilane (APTES) by immersing 1.0 g of the NPs in 20% APTES/Toluene mixture and allowed to interact for 12 h at 50 °C followed by washing thoroughly with ethanol and hexane.

Finally, the Fe_2O_3, Fe_2O_3-SiO_2 and Fe_2O_3-SiO-NH_2 (FNPs) were characterized using JEOL TEM (JEM 1400 Plus). For size measurements, a region of interest was chosen and sizes of 100 NPs or FNPs were statistically analyzed to calculate the final average size.

3.2. Conjugation of FNPs with SR-FLIVO

Conjugation to SR-FLIVO has been carried out where 1.0 mg of FNPs was interacted with 1.0 mL of SR-FLIVO and the conjugation affinity has been recorded against time using HITACHI

Fluorescence spectrometer (F-2500). Scheme 1 illustrates the synthesis and conjugation pathways of FNPs and SR-FLIVO.

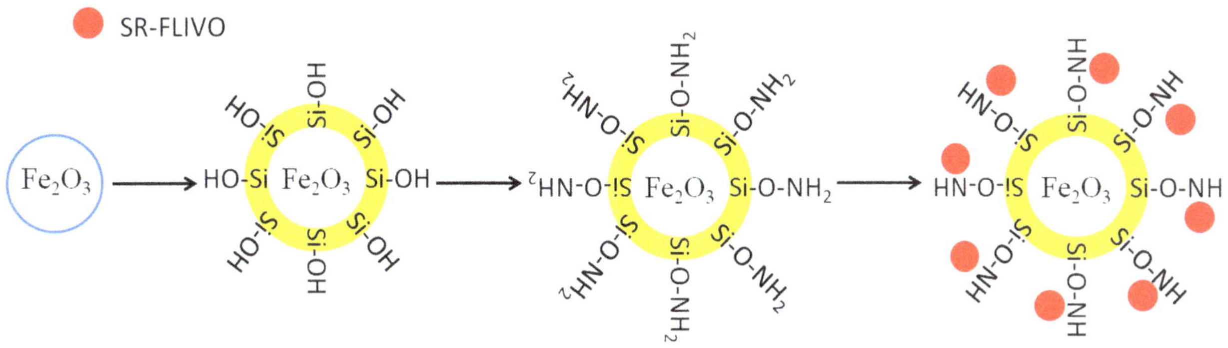

Scheme 1. Schematic diagram for the synthesis and conjugation pathways of FNPs and SR-FLIVO.

3.3. *In Vivo Study*

A rat model of temporary focal cerebral ischemia has been used [32]. Spraque-Dawley rats ($n = 4$ for each group under study) underwent temporary middle cerebral occlusion for 2 h by using an intraluminal suture. Since the caspase activities and apoptotic cells were detected between 12 and 24 h after the reperfusion in pilot studies, SR-FLIVO-FNPs were injected 18 h after the reperfusion. The rats were allowed to survive for the next 60 min and then were sacrificed. The brains were fixed with formalin and sectioned into 50 μm thick serial slices. Histochemical TUNEL staining was performed and investigated with Olympus Fluorescence microscope (BX51).

In addition, TEM was recorded for the cellular tissue to ensure if; the SR-FLIVO-FNPs can passively cross the blood brain barrier (BBB) for apoptotic cells targeting. Moreover, Cellular Prussian blue staining was carried out to confirm the existence of SR-FLIVO-FNPs.

3.4. *In Vitro Tracking*

For *in vitro* tracking of SR-FLIVO-FNPs, the 9L gliosarcoma cell line has been used for cell culture study. For each well, the number of cells was adjusted to be 1×10^6. Thus, 0.3 mM H_2O_2 has been used for *in vitro* apoptotic model preparation [33]. Therefore, cell morphology has been checked with time (Figure S1). An apoptotic model has been adjusted for 18 h then a solution of SR-FLIVO-FNPs was introduced to the apoptotic culture cells, and kept for additional 1 h, then washed thoroughly using D-PBS and finally TUNEL staining was performed followed by cellular recording using fluorescence microscope.

4. Conclusions

In summary, we have successfully developed a cell signaling probe that based on functionalized Fe_2O_3 NPs and fluorescent caspase inhibitor SR-FLIVO. This could be considered as a promising platform probe if it is well optimized to be used clinically as a contrasting agent in the future to specifically diagnose the apoptotic lesion and as a drug delivery vehicle; these potential uses are now under our consideration.

Author Contributions

First and second authors: designed and conducted the experiments and wrote the manuscript; third and fourth authors: contributed reagents/materials/analysis tools and gave advices to perform the needed experiments.

Conflicts of Interest

The authors declare no conflict of interest.

References

1. David, L.V.; Stanley, J.K. Cell death in development. *Cell* **1999**, *96*, 245–254.
2. Kim, K.; Lee, M.; Park, H.; Kim, J.H.; Kim, S.; Chung, H.; Choi, K.; Kim, I.S.; Seong, B.L.; Kwon, I.C. Cell-permeable and biocompatible polymeric nanoparticles for apoptosis imaging. *J. Am. Chem. Soc.* **2006**, *128*, 3490–3491.
3. Griffin, R.J.; Williams, B.W.; John, B.A.; Bischof, C.; Olin, M.; Gary, L.; Johnson, B.S.; Lee, B.W. Use of a fluorescently labeled poly-caspase inhibitor for *in vivo* detection of apoptosis related to vascular-targeting agent arsenic trioxide for cancer therapy. *Technol. Cancer Res. Treat.* **2007**, *6*, 651–654.
4. Dicker, D.T.; Kim, S.H.; Jin, Z.; El-Deiry, W.S. Heterogeneity in non-invasive detection of apoptosis among human tumor cell lines using annexin-V tagged with EGFP or Qdot-705. *Cancer Biol. Ther.* **2005**, *4*, 1014–1017.
5. Dillon, S.R.; Constantinescu, A.; Schlissel, M.S. Annexin V binds to positively selected B cells. *J. Immunol.* **2001**, *166*, 58–71.
6. Tait, J.F.; Smith, C.; Wood, B.L. Measurement of phosphatidylserine exposure in leukocytes and platelets by whole-blood flow cytometry with annexin V. *Blood Cells Mol. Dis.* **1999**, *25*, 271–278.
7. Clarke, P.G. Apoptosis: From morphological types of cell death to interacting pathways. *Trends Pharmacol. Sci.* **2002**, *23*, 308–309.
8. Berger, A.B.; Sexton, K.B.; Bogyo, M. Commonly used caspase inhibitors designed based on substrate specificity profiles lack selectivity. *Cell Res.* **2006**, *16*, 961–963.
9. Taghdisi, S.M.; Lavaee, P.; Ramezani, M.; Abnous, K. Reversible targeting and controlled release delivery of daunorubicin to cancer cells by aptamer-wrapped carbon nanotubes. *Eur. J. Pharm. Biopharm.* **2011**, *77*, 200–206.
10. Suriamoorthy, P.; Zhang, X.; Hao, G.; Joly, A.G.; Singh, S.; Hossu, M.; Sun, X.; Chen, W. Folic acid-CdTe quantum dot conjugates and their applications for cancer cell targeting. *Cancer Nanotechnol.* **2010**, *1*, 19–28.

11. Huang, X.; Jain, P.K.; El-Sayed, I.H.; El-Sayed, M.A. Gold nanoparticles: Interesting optical properties and recent applications in cancer diagnostics and therapy. *Nanomedicine* **2007**, *2*, 681–693.
12. Wong, C.; Styliano, P.T.; Cui, J.; Martin, J.; Chauhan, V.P. Multistage nanoparticle delivery system for deep penetration into tumor tissue. *Proc. Natl. Acad. Sci. USA* **2011**, *108*, 2426–2431.
13. Schleich, N.; Po, C.; Jacobs, D.; Ucakar, B.; Gallez, B. Comparison of active, passive and magnetic targeting to tumors of multifunctional paclitaxel/SPIO-loaded nanoparticles for tumor imaging and therapy. *J. Control. Release* **2014**, *194*, 82–91.
14. Iv, M.; Telischak, N.; Feng, D.; Holdsworth, S.J.; Yeom, K.W.; Daldrup-Link, H.E. Clinical applications of iron oxide nanoparticles for magnetic resonance imaging of brain tumors. *Nanomedicine* **2015**, *10*, 993–1018.
15. Hanini, A.; Schmitt, A.; Kacem, K.; Chau, F.; Ammar, S.; Gavard, J. Evaluation of iron oxide nanoparticle compatibility. *Int. J. Nanomed.* **2011**, *6*, 787.
16. Peng, X.H.; Qian, X.; Wang, A.Y.; Chen, Z.; Nie, S.; Shin, D.M. Targeted magnetic iron oxide nanoparticles for tumor imaging and therapy. *Int. J. Nanomed.* **2008**, *3*, 311–321.
17. Schleich, N.; Danhier, F.; Préat, V. Iron oxide-loaded nanotheranostics: Major obstacles to *in vivo* studies and clinical translation. *J. Control. Release* **2015**, *198*, 35–54.
18. Khandhar, A.P.; Ferguson, R.M.; Krishnan, K.M. Monodispersed magnetite nanoparticles optimized for magnetic fluid hyperthermia: Implications in biological systems. *J. App. Phys.* **2011**, *109*, 07B310.
19. Tartaj, P. *Encyclopedia of Nanoscience and Nanotechnology*; American Scientific Publishers: Stevenson Ranch, CA, USA, 2003; Volume 6, p. 823.
20. Atta, M.A.; Al-Lohedan, H.A.; Al-Hussain, S.A. Synthesis of stabilized myrrh-capped hydrocolloidal magnetite nanoparticles. *Molecules* **2014**, *19*, 11263–11278.
21. Cornel, R.M.; Schwertmann, U. *The Iron Oxides, Structure, Properties, Reactions and Uses*; VCH: Weinheim, Germany, 1996.
22. Lockman, P.R.; Koziara, J.M.; Mumper, R.J.; Allen, D.D. Nanoparticle surface charges alter blood-brain barrier integrity and permeability. *J. Drug Target.* **2004**, *12*, 635–641.
23. ImmunoChemistry Technologies LLC. Available online: http://www.immunochemistry.com/FLIVO.htm (accessed on 2 April 2014).
24. Smith, S.N.; Steer, R.P. The photophysics of Lissamine rhodamine-B sulfonyl chloride in aqueous solution: Implications for fluorescent protein—Dye conjugates. *J. Photochem. Photobiol. A* **2001**, *139*, 151–156.
25. Neves, C.; Byk, G.; Escriou, V.; Bussone, F.; Scherman, D.; Wils, P. Novel method for covalent fluorescent labeling of plasmid DNA that maintains structural integrity of the plasmid. *Bioconjug. Chem.* **2000**, *11*, 51–55.
26. Yamaguchi, A.; Watanabe, J.; Mekawy, M.M.; Fujiwara, R.; Morita, K.; Yamashita, T.; Amino, Y.; Chen, Y.; Logudurai, R.; Teramae, N. Extraction mechanisms of charged organic dye molecules into silica-surfactant nanochannels in a porous alumina membrane. *Anal. Chim. Acta* **2006**, *556*, 157–163.
27. El-Gamel, N.E.; Wortmann, L.; Arroub, K.; Mathur, S. Surface immobilization and release of sparfloxacin drug from SiO_2@Fe_2O_3 core-shell nanoparticles. *Chem. Commun.* **2011**, *47*, 10076.

28. Shimizu, I.; Okabayashi, H.; Taga, K.; Nishio, E.; O'Connor, C.J. Diffuse reflectance infrared Fourier transform spectral study of the thermal and adsorbed-water effects of a 3-aminopropyltriethoxysilane layer modified onto the surface of silica gel. *Vib. Spectosc.* **1997**, *14*, 113–123.
29. Zhoua, J.; Leuschnerb, C.; Kumarc, C.; Hormesc, J.F.; Soboyejo, W.O. Sub-cellular accumulation of magnetic nanoparticles in breast tumors and metastases. *Biomaterials* **2006**, *27*, 2001–2008.
30. Jendelová, P.; Herynek, V.; Urdzíková, L.; Glogarová, K.; Kroupová, J.; Andersson, B.; Bryja, V.; Burian, M.; Hájek, M.; Syková, E. Magnetic resonance tracking of transplanted bone marrow and embryonic stem cells labeled by iron oxide nanoparticles in rat brain and spinal cord. *J. Neurosci. Res.* **2004**, *76*, 232–243.
31. Chin, A.B.; Yaacob, I.I. Synthesis and characterization of magnetic iron oxide nanoparticles via w/o microemulsion and Massart's procedure. *J. Mater. Process. Technol.* **2007**, *191*, 235–237.
32. Longa, E.Z.; Weinstein, P.R.; Carlson, S.; Cummins, R. Reversible middle cerebral artery occlusion without craniectomy in rats. *Stroke* **1989**, *20*, 84–91.
33. Kitamura, Y.; Ota, T.; Matsuoka, Y.; Tooyama, I.; Kimura, H.; Shimohama, S.; Nomura, Y.; Gebicke-Haerter, P.J.; Taniguchi, T. Hydrogen peroxide-induced apoptosis mediated by p53 protein in glial cells. *GLIA* **1999**, *25*, 154–164.

Permissions

All chapters in this book were first published in Nanomaterials, by MDPI; hereby published with permission under the Creative Commons Attribution License or equivalent. Every chapter published in this book has been scrutinized by our experts. Their significance has been extensively debated. The topics covered herein carry significant findings which will fuel the growth of the discipline. They may even be implemented as practical applications or may be referred to as a beginning point for another development.

The contributors of this book come from diverse backgrounds, making this book a truly international effort. This book will bring forth new frontiers with its revolutionizing research information and detailed analysis of the nascent developments around the world.

We would like to thank all the contributing authors for lending their expertise to make the book truly unique. They have played a crucial role in the development of this book. Without their invaluable contributions this book wouldn't have been possible. They have made vital efforts to compile up to date information on the varied aspects of this subject to make this book a valuable addition to the collection of many professionals and students.

This book was conceptualized with the vision of imparting up-to-date information and advanced data in this field. To ensure the same, a matchless editorial board was set up. Every individual on the board went through rigorous rounds of assessment to prove their worth. After which they invested a large part of their time researching and compiling the most relevant data for our readers.

The editorial board has been involved in producing this book since its inception. They have spent rigorous hours researching and exploring the diverse topics which have resulted in the successful publishing of this book. They have passed on their knowledge of decades through this book. To expedite this challenging task, the publisher supported the team at every step. A small team of assistant editors was also appointed to further simplify the editing procedure and attain best results for the readers.

Apart from the editorial board, the designing team has also invested a significant amount of their time in understanding the subject and creating the most relevant covers. They scrutinized every image to scout for the most suitable representation of the subject and create an appropriate cover for the book.

The publishing team has been an ardent support to the editorial, designing and production team. Their endless efforts to recruit the best for this project, has resulted in the accomplishment of this book. They are a veteran in the field of academics and their pool of knowledge is as vast as their experience in printing. Their expertise and guidance has proved useful at every step. Their uncompromising quality standards have made this book an exceptional effort. Their encouragement from time to time has been an inspiration for everyone.

The publisher and the editorial board hope that this book will prove to be a valuable piece of knowledge for researchers, students, practitioners and scholars across the globe.

List of Contributors

Chenyang Xue
Key Laboratory of Instrumentation Science and Dynamic Measurement of Ministry of Education, North University of China, Taiyuan 030051, China
Science and Technology on Electronic Test and Measurement Laboratory, North University of China, Taiyuan 030051, China

Junyang Li
Key Laboratory of Instrumentation Science and Dynamic Measurement of Ministry of Education, North University of China, Taiyuan 030051, China
Science and Technology on Electronic Test and Measurement Laboratory, North University of China, Taiyuan 030051, China
Department of Mechanical and Biomedical Engineering, City University of Hong Kong, Kowloon 999077, Hong Kong

Qiang Zhang
Key Laboratory of Instrumentation Science and Dynamic Measurement of Ministry of Education, North University of China, Taiyuan 030051, China
Science and Technology on Electronic Test and Measurement Laboratory, North University of China, Taiyuan 030051, China

Zhibo Zhang
Key Laboratory of Instrumentation Science and Dynamic Measurement of Ministry of Education, North University of China, Taiyuan 030051, China
Science and Technology on Electronic Test and Measurement Laboratory, North University of China, Taiyuan 030051, China

Zhenyin Hai
Key Laboratory of Instrumentation Science and Dynamic Measurement of Ministry of Education, North University of China, Taiyuan 030051, China
Science and Technology on Electronic Test and Measurement Laboratory, North University of China, Taiyuan 030051, China

Libo Gao
Key Laboratory of Instrumentation Science and Dynamic Measurement of Ministry of Education, North University of China, Taiyuan 030051, China
Science and Technology on Electronic Test and Measurement Laboratory, North University of China, Taiyuan 030051, China

Ruiting Feng
Key Laboratory of Instrumentation Science and Dynamic Measurement of Ministry of Education, North University of China, Taiyuan 030051, China
Science and Technology on Electronic Test and Measurement Laboratory, North University of China, Taiyuan 030051, China

Jun Tang
Key Laboratory of Instrumentation Science and Dynamic Measurement of Ministry of Education, North University of China, Taiyuan 030051, China
Science and Technology on Electronic Test and Measurement Laboratory, North University of China, Taiyuan 030051, China

Jun Liu
Key Laboratory of Instrumentation Science and Dynamic Measurement of Ministry of Education, North University of China, Taiyuan 030051, China
Science and Technology on Electronic Test and Measurement Laboratory, North University of China, Taiyuan 030051, China

Wendong Zhang
Key Laboratory of Instrumentation Science and Dynamic Measurement of Ministry of Education, North University of China, Taiyuan 030051, China
Science and Technology on Electronic Test and Measurement Laboratory, North University of China, Taiyuan 030051, China

Dong Sun
Key Laboratory of Instrumentation Science and Dynamic Measurement of Ministry of Education, North University of China, Taiyuan 030051, China

Jingqiang Li
Department of Physics and Astronomy, Rice University, Houston, TX 77005, USA

Sithara S. Wijeratne
Department of Physics and Astronomy, Rice University, Houston, TX 77005, USA

Xiangyun Qiu
Department of Physics, George Washington University, Washington, DC 20052, USA

Ching-Hwa Kiang
Department of Physics and Astronomy, Rice University, Houston, TX 77005, USA
Department of Bioengineering, Rice University, Houston, TX 77005, USA

Tomomi Takeshima
Graduate School of Environmental Science, Hokkaido University, Hokkaido 060-0810, Japan
Nissei Bio Co. Ltd., Hokkaido 061-1374, Japan

Yuya Tada
Nissei Bio Co. Ltd., Hokkaido 061-1374, Japan

Norihito Sakaguchi
Division of Materials Science and Engineering, Faculty of Engineering, Hokkaido University, Hokkaido 060-8628, Japan

Fumio Watari
Graduate School of Dental Medicine, Hokkaido University, Hokkaido 060-8586, Japan

Bunshi Fugetsu
Graduate School of Environmental Science, Hokkaido University, Hokkaido 060-0810, Japan
Policy Alternative Research Institute, The University of Tokyo, Tokyo 113-0032, Japan

Richard M. A. Manara
School of Chemistry, University of Southampton, Highfield Campus, Southampton SO17 1BJ, UK

Susana Tomasio
School of Chemistry, University of Southampton, Highfield Campus, Southampton SO17 1BJ, UK

Syma Khalid
School of Chemistry, University of Southampton, Highfield Campus, Southampton SO17 1BJ, UK

Kazuo Umemura
Biophysics Section, Department of Physics, Tokyo University of Science, 1-3 Kagurazaka,Shinjuku,Tokyo 1628601, Japan

Dilru R. Ratnaweera
Department of Chemistry, Clemson University, Clemson, SC 29634-0973, USA

Chaitra Mahesha
Department of Mechanical Engineering, Clemson University, Clemson, SC 29634-0921,USA

David A. Zumbrunnen
Department of Mechanical Engineering, Clemson University, Clemson, SC 29634-0921,USA

Dvora Perahia
Department of Chemistry, Clemson University, Clemson, SC 29634-0973, USA

Magdalena Graczyk-Zajac
Institut für Materialwissenschaft, Technische Universität Darmstadt, Jovanka-Bontschits-Straße 2, 64287 Darmstadt, Germany

Lukas Mirko Reinold
Institut für Materialwissenschaft, Technische Universität Darmstadt, Jovanka-Bontschits-Straße 2, 64287 Darmstadt, Germany

Jan Kaspar
Institut für Materialwissenschaft, Technische Universität Darmstadt, Jovanka-Bontschits-Straße 2, 64287 Darmstadt, Germany

Pradeep Vallachira Warriam Sasikumar
Institut für Materialwissenschaft, Technische Universität Darmstadt, Jovanka-Bontschits-Straße 2, 64287 Darmstadt, Germany

Gian-Domenico Soraru
Dipartimento di Ingegneria Industriale, Università di Trento, Via Sommarive 9, 38123 Trento, Italy

Ralf Riedel
Institut für Materialwissenschaft, Technische Universität Darmstadt, Jovanka-Bontschits-Straße 2, 64287 Darmstadt, Germany

Stephen F. Ralph
School of Chemistry, University of Wollongong, Northfields Avenue, Wollongong 2522, NSW, Australia

Charu Vashisth Rohatgi
Ian Wark Institute, University of South Australia, Mawson Lakes 5095, Australia

Naba K. Dutta
Ian Wark Institute, University of South Australia, Mawson Lakes 5095, Australia

Namita Roy Choudhury
Ian Wark Institute, University of South Australia, Mawson Lakes 5095, Australia

Thirunavukkarasu Somanathan
Department of Nanoscience, School of Basic Sciences, Vels University, Pallavaram, Chennai, Tamil Nadu 600117, India
Center for Advanced Research and Development (CARD), Vels University, Pallavaram, Chennai, Tamil Nadu 600117, India

Karthika Prasad
Department of Nanoscience, School of Basic Sciences, Vels University, Pallavaram, Chennai, Tamil Nadu 600117, India
Institute of Future Environments, Nanotechnology and Molecular Science, School of Chemistry, Physics and Mechanical Engineering, Queensland University of Technology, Brisbane, QLD 4000, Australia

Kostya (Ken) Ostrikov
Institute of Future Environments, Nanotechnology and Molecular Science, School of Chemistry, Physics and Mechanical Engineering, Queensland University of Technology, Brisbane, QLD 4000, Australia

Arumugam Saravanan
Department of Nanoscience, School of Basic Sciences, Vels University, Pallavaram, Chennai, Tamil Nadu 600117, India

Vemula Mohana Krishna
Department of Nanoscience, School of Basic Sciences, Vels University, Pallavaram, Chennai, Tamil Nadu 600117, India

Hayriye Yilmaz
Department of Biomedical Devices and Technology, Kayseri Vocational School, Erciyes University, Kayseri 38039, Turkey

Interdisciplinary Center for Nanotoxicity, Department of Chemistry and Biochemistry, Jackson State University, Jackson, MS 39217, USA

Bakhtiyor Rasulev
Interdisciplinary Center for Nanotoxicity, Department of Chemistry and Biochemistry, Jackson State University, Jackson, MS 39217, USA
Center for Computationally Assisted Science and Technology (CCAST), North Dakota State University, Fargo, ND 58108, USA

Jerzy Leszczynski
Interdisciplinary Center for Nanotoxicity, Department of Chemistry and Biochemistry, Jackson State University, Jackson, MS 39217, USA

Daniel S. Weadick
Department of Chemistry, Waterloo Institute for Nanotechnology, Waterloo, ON N2L 3G1, Canada

Juewen Liu
Department of Chemistry, Waterloo Institute for Nanotechnology, Waterloo, ON N2L 3G1, Canada

Giuseppe Chichiriccò
Department of Life, Health and Environmental Sciences, University of L'Aquila, Via Vetoio, I-67010 Coppito, L'Aquila, Italy

Anna Poma
Department of Life, Health and Environmental Sciences, University of L'Aquila, Via Vetoio, I-67010 Coppito, L'Aquila, Italy

Enrique Maciá
Materials Physics Department, Faculty of Physics, University Complutense of Madrid, E-28040 Madrid, Spain

Fuchun Zhang
College of Physics and Electronic Information, Yanan University, Yanan 716000, China

Dandan Chao
College of Physics and Electronic Information, Yanan University, Yanan 716000, China

Hongwei Cui
College of Physics and Electronic Information, Yanan University, Yanan 716000, China

Weihu Zhang
College of Physics and Electronic Information, Yanan University, Yanan 716000, China

Weibin Zhang
Department of Physics, Dongguk University, Seoul 100-715, Korea

Moataz Mekawy
Department of Neurosurgery, Graduate School of Medicine, Tohoku University, 2-1 Seiryo-machi, Aoba-ku, Sendai, Miyagi 980-8575, Japan
Center of Materials Science, Zewail City of Science and Technology, Sheikh Zayed District, 12588, 6th of October City, Giza, Egypt

Atsushi Saito
Department of Neurosurgery, Aomori Prefecture Central Hospital, 2 1 1 Higashitsukurimichi, Aomori 0308553, Japan

Hiroaki Shimizu
Department of Neurosurgery, Graduate School of Medicine, Akita University, 1-1-1 Hondo, Akita 010-8543, Japan

Teiji Tominaga
Department of Neurosurgery, Graduate School of Medicine, Tohoku University, 2-1 Seiryo-machi, Aoba-ku, Sendai, Miyagi 980-8575, Japan